Titles in This Series

Volume

1 **Markov random fields and their applications,** Ross Kindermann and J. Laurie Snell

2 **Proceedings of the conference on integration, topology, and geometry in linear spaces,** William H. Graves, Editor

3 **The closed graph and P-closed graph properties in general topology,** T. R. Hamlett and L. L. Herrington

4 **Problems of elastic stability and vibrations,** Vadim Komkov, Editor

5 **Rational constructions of modules for simple Lie algebras,** George B. Seligman

6 **Umbral calculus and Hopf algebras,** Robert Morris, Editor

7 **Complex contour integral representation of cardinal spline functions,** Walter Schempp

8 **Ordered fields and real algebraic geometry,** D. W. Dubois and T. Recio, Editors

9 **Papers in algebra, analysis and statistics,** R. Lidl, Editor

10 **Operator algebras and K-theory,** Ronald G. Douglas and Claude Schochet, Editors

11 **Plane ellipticity and related problems,** Robert P. Gilbert, Editor

12 **Symposium on algebraic topology in honor of José Adem,** Samuel Gitler, Editor

13 **Algebraists' homage: Papers in ring theory and related topics,** S. A. Amitsur, D. J. Saltman, and G. B. Seligman, Editors

14 **Lectures on Nielsen fixed point theory,** Boju Jiang

15 **Advanced analytic number theory. Part I: Ramification theoretic methods,** Carlos J. Moreno

16 **Complex representations of GL(2, K) for finite fields K,** Ilya Piatetski-Shapiro

17 **Nonlinear partial differential equations,** Joel A. Smoller, Editor

18 **Fixed points and nonexpansive mappings,** Robert C. Sine, Editor

19 **Proceedings of the Northwestern homotopy theory conference,** Haynes R. Miller and Stewart B. Priddy, Editors

20 **Low dimensional topology,** Samuel J. Lomonaco, Jr., Editor

21 **Topological methods in nonlinear functional analysis,** S. P. Singh, S. Thomeier, and B. Watson, Editors

22 **Factorizations of $b^n \pm 1$, $b = 2, 3, 5, 6, 7, 10, 11, 12$ up to high powers,** John Brillhart, D. H. Lehmer, J. L. Selfridge, Bryant Tuckerman, and S. S. Wagstaff, Jr.

23 **Chapter 9 of Ramanujan's second notebook—Infinite series identities, transformations, and evaluations,** Bruce C. Berndt and Padmini T. Joshi

24 **Central extensions, Galois groups, and ideal class groups of number fields,** A. Fröhlich

25 **Value distribution theory and its applications,** Chung-Chun Yang, Editor

26 **Conference in modern analysis and probability,** Richard Beals, Anatole Beck, Alexandra Bellow, and Arshag Hajian, Editors

27 **Microlocal analysis,** M. Salah Baouendi, Richard Beals, and Linda Preiss Rothschild, Editors

28 **Fluids and plasmas: geometry and dynamics,** Jerrold E. Marsden, Editor

29 **Automated theorem proving,** W. W. Bledsoe and Donald Loveland, Editors

30 **Mathematical applications of category theory,** J. W. Gray, Editor

31 **Axiomatic set theory,** James E. Baumgartner, Donald A. Martin, and Saharon Shelah, Editors

32 **Proceedings of the conference on Banach algebras and several complex variables,** F. Greenleaf and D. Gulick, Editors

Titles in This Series

Volume

33 **Contributions to group theory,** Kenneth I. Appel, John G. Ratcliffe, and Paul E. Schupp, Editors

34 **Combinatorics and algebra,** Curtis Greene, Editor

35 **Four-manifold theory,** Cameron Gordon and Robion Kirby, Editors

36 **Group actions on manifolds,** Reinhard Schultz, Editor

37 **Conference on algebraic topology in honor of Peter Hilton,** Renzo Piccinini and Denis Sjerve, Editors

38 **Topics in complex analysis,** Dorothy Browne Shaffer, Editor

39 **Errett Bishop: Reflections on him and his research,** Murray Rosenblatt, Editor

40 **Integral bases for affine Lie algebras and their universal enveloping algebras,** David Mitzman

41 **Particle systems, random media and large deviations,** Richard Durrett, Editor

42 **Classical real analysis,** Daniel Waterman, Editor

43 **Group actions on rings,** Susan Montgomery, Editor

44 **Combinatorial methods in topology and algebraic geometry,** John R. Harper and Richard Mandelbaum, Editors

45 **Finite groups–coming of age,** John McKay, Editor

46 **Structure of the standard modules for the affine Lie algebra $A_1^{(1)}$,** James Lepowsky and Mirko Primc

47 **Linear algebra and its role in systems theory,** Richard A. Brualdi, David H. Carlson, Biswa Nath Datta, Charles R. Johnson, and Robert J. Plemmons, Editors

48 **Analytic functions of one complex variable,** Chung-chun Yang and Chi-tai Chuang, Editors

49 **Complex differential geometry and nonlinear differential equations,** Yum-Tong Siu, Editor

50 **Random matrices and their applications,** Joel E. Cohen, Harry Kesten, and Charles M. Newman, Editors

51 **Nonlinear problems in geometry,** Dennis M. DeTurck, Editor

52 **Geometry of normed linear spaces,** R. G. Bartle, N. T. Peck, A. L. Peressini, and J. J. Uhl, Editors

53 **The Selberg trace formula and related topics,** Dennis A. Hejhal, Peter Sarnak, and Audrey Anne Terras, Editors

54 **Differential analysis and infinite dimensional spaces,** Kondagunta Sundaresan and Srinivasa Swaminathan, Editors

55 **Applications of algebraic K-theory to algebraic geometry and number theory,** Spencer J. Bloch, R. Keith Dennis, Eric M. Friedlander, and Michael R. Stein, Editors

56 **Multiparameter bifurcation theory,** Martin Golubitsky and John Guckenheimer, Editors

57 **Combinatorics and ordered sets,** Ivan Rival, Editor

58.I **The Lefschetz centennial conference. Part I: Proceedings on algebraic geometry,** D. Sundararaman, Editor

58.II **The Lefschetz centennial conference. Part II: Proceedings on algebraic topology,** S. Gitler, Editor

58.III **The Lefschetz centennial conference. Part III: Proceedings on differential equations,** A. Verjovsky, Editor

59 **Function estimates,** J. S. Marron, Editor

60 **Nonstrictly hyperbolic conservation laws,** Barbara Lee Keyfitz and Herbert C. Kranzer, Editors

61 **Residues and traces of differential forms via Hochschild homology,** Joseph Lipman

62 **Operator algebras and mathematical physics,** Palle E. T. Jorgensen and Paul S. Muhly, Editors

Titles in This Series

Volume

63 **Integral geometry,** Robert L. Bryant, Victor Guillemin, Sigurdur Helgason, and R. O. Wells, Jr., Editors
64 **The legacy of Sonya Kovalevskaya,** Linda Keen, Editor
65 **Logic and combinatorics,** Stephen G. Simpson, Editor
66 **Free group rings,** Narian Gupta
67 **Current trends in arithmetical algebraic geometry,** Kenneth A. Ribet, Editor
68 **Differential geometry: The interface between pure and applied mathematics,** Mladen Luksic, Clyde Martin, and William Shadwick, Editors
69 **Methods and applications of mathematical logic,** Walter A. Carnielli and Luiz Paulo de Alcantara, Editors
70 **Index theory of elliptic operators, foliations, and operator algebras,** Jerome Kaminker, Kenneth C. Millett, and Claude Schochet, Editors
71 **Mathematics and general relativity,** James A. Isenberg, Editor
72 **Fixed point theory and its applications,** R. F. Brown, Editor
73 **Geometry of random motion,** Rick Durrett and Mark A. Pinsky, Editors
74 **Geometry of group representations,** William M. Goldman and Andy R. Magid, Editors
75 **The finite calculus associated with Bessel functions,** Frank M. Cholewinski
76 **The structure of finite algebras,** David C. Hobby and Ralph Mckenzie
77 **Number theory and its applications in China,** Wang Yuan, Yang Chung-chun, and Pan Chengbiao, Editors
78 **Braids,** Joan S. Birman and Anatoly Libgober, Editors
79 **Regular differential forms,** Ernst Kunz and Rolf Waldi
80 **Statistical inference from stochastic processes,** N. U. Prabhu, Editor
81 **Hamiltonian dynamical systems,** Kenneth R. Meyer and Donald G. Saari, Editors
82 **Classical groups and related topics,** Alexander J. Hahn, Donald G. James, and Zhe-xian Wan, Editors
83 **Algebraic K-theory and algebraic number theory,** Michael R. Stein and R. Keith Dennis, Editors
84 **Partition problems in topology,** Stevo Todorcevic
85 **Banach space theory,** Bor-Luh Lin, Editor
86 **Representation theory and number theory in connection with the local Langlands conjecture,** J. Ritter, Editor

Representation Theory and Number Theory in Connection with the Local Langlands Conjecture

CONTEMPORARY MATHEMATICS

Volume 86

Representation Theory and Number Theory in Connection with the Local Langlands Conjecture

Proceedings of a Conference held December 8–14, 1985 with support from Stiftung Volkswagenwerk, Bayerisches Staatsministerium für Unterricht und Kultus, and Gesellschaft der Freunde der Universität Augsburg

J. Ritter, Editor

AMERICAN MATHEMATICAL SOCIETY

Providence · Rhode Island

The Conference on Representation Theory and Number Theory in Connection with the Local Langlands Conjecture was held at the University of Augsburg, Augsburg, West Germany on December 8–14, 1985 with support from Stiftung Volkswagenwerk, Bayerisches Staatsministerium für Unterricht und Kultus, and Gesellschaft der Freunde der Universität Augsburg.

1980 *Mathematics Subject Classification* (1985 *Revision*). Primary 11S37, 11S40, 11S45, 20G25, 22E50.

Library of Congress Cataloging-in-Publication Data

Representation theory and number theory in connection with the local langlands conjecture: proceedings of a conference held December 8–14, 1985/with support from Stiftung Volkswagenwerk, Bayerisches Staatsministerium für Unterricht und Kultus, and Gesellschaft der Freunde der Universität/J. Ritter, editor.

p. cm. –(Contemporary mathematics, ISSN 0271-4132; v. 86)

Papers from the Conference on Representation Theory and Number Theory in Connection with the Local Langlands Conjecture, held at the University of Augsburg, Augsburg, West Germany.

Bibliography: p.

ISBN 0-8218-5093-8 (alk. paper)

1. Algebraic number theory–Congresses. 2. Representations of groups–Congresses. I. Ritter, J. (Jürgen), 1943–. II. Conference on Representation Theory and Number Theory in Connection with the Local Langlands Conjecture (1985: University of Augsburg). III. Series.

QA241.R44 1989 88-39030
512′.74–dc 19 CIP

CONTENTS

PREFACE

The present volume reflects the contents of the talks given at the conference "Representation Theory and Number Theory in connection with the Local Langlands Conjecture" held at the University of Augsburg in December 1985*). The Langlands programme sums up those parts of mathematical research that belong to the representation theory of reductive groups and to class field theory, the two topics being tied up by the vision that, roughly speaking, the irreducible representations of the general linear group may well serve as parameters for the description of all number fields. In the local situation, i.e. when the base field is a given p-adic field K and where we think of the extension theory of K being determined by the irreducible representations of the absolute Galois group G_K of K, great progress has been achieved in establishing an arithmetic correspondence between the objects in question, that is, the supercuspidal representations of $GL(n,K)$ or, equally well, the irreducible representations of the multiplicative group of a division algebra D that is central and of index n over K, and, on the other hand, those irreducible representations of G_K whose degrees divide n:

- Howe-Corwin, Koch-Zink, and, with respect to the root numbers, Bushnell-Fröhlich, have settled the so-called tame case, that is, when $p \nmid n$.
- Henniart, Kutzko, and Moy have solved the case $n = p$.
- Deligne and Kazhdan have proved a matching theorem providing a one-to-one correspondence between the representation theory of $GL(n,K)$ and $D^{\times}$.
- Henniart, on the occasion of this conference, gave the proof of the numerical Langlands conjecture.

Since no book or paper was available presenting the different methods used so far nor even collecting the results at our disposal, there seemed to be a need for a conference reflecting what has been done in this aerea. The programme of the conference was divided into two parts:

(i) the representation theory of local division algebras and local Galois groups; the Langlands conjecture in the tame case

(ii) new results - the case $n = p$; the matching theorem; principal orders; tame Deligne representations; classification of representations of $GL(n)$; the numerical Langlands conjecture.

---

*) The meeting was run by G. Michler (Essen) and the editor, it was made possible by the generous support of Stiftung Volkswagenwerk, of Bayerisches Staatsministerium für Unterricht und Kultus, and of Gesellschaft der Freunde der Universität Augsburg; it took place in the Schwabenakademie Irsee near Augsburg.

The collection of talks in this volume gives a good account of what the state of affairs in the local Langlands programme is; we have only left out those talks which either were meant to merely provide concrete examples or the subject of which has by now appeared in the literature in detail, as for example the matching theorem in [ ]. Some of the conjectures stated in the talks have meanwhile been proved - we give the reference where the proof is going to be published.

J. Ritter, Augsburg, April 1987

List of participants

E. Becker, Dortmund

J. Brinkhuis, Rotterdam

C.Bushnell, London

J.W.S. Cassels, Cambridge

Ph. Cassou-Noguès, Bordeaux/Harvard

L. Corwin, New Brunswick

G.-M. Cram, Augsburg

C. Deninger, Regensburg

G. Everest, Norwich

A. Fröhlich, London

P. Gérardin, Paris

W.-D. Geyer, Erlangen

G. Henniart, Paris

G. Hiß, Aachen

K. Hoechsmann, Vancouver/Augsburg

U. Jannsen, Regensburg

M. Jarden, Tel Aviv

W. Jehne, Köln

R. Knörr, Essen

W. Kohnen, Augsburg

B. Külshammer, Dortmund

Ph. Kutzko, Iowa

M. Lorenz, MPI Bonn

D. Manderscheid, Iowa

J. Martinet, Bordeaux

B.H. Matzat, Karlsruhe

L. McCulloh, Urbana

G. Michler, Essen

C. Moreno, New York

A. Moy, Seattle

J. Neukirch, Regensburg

H. Opolka, Göttingen

J. Queyrut, Bordeaux

C. Riehm, Hamilton

J. Ritter, Augsburg

J. Rohlfs, Eichstätt

P. Sally, Chicago

R. Schertz, Augsburg

M. Taylor, Cambridge

W. Willems, Mainz

S. Wilson, Durham

K. Wingberg, Regensburg

Wen-Ch'ing Winnie Li, Pennsylvania

Contemporary Mathematics
Volume 86, 1989

THE IRREDUCIBLE REPRESENTATIONS OF THE MULTIPLICATIVE GROUP OF A TAME DIVISION ALGEBRA OVER A LOCAL FIELD (Following H. Koch and E.-W. Zink)

by E.Becker and B. Külshammer, Dortmund

1. Introduction

Let F denote a local field of residue characteristic p and D/F a central division algebra of dimension $[D:F] = n^2$. D is called tame if $p \nmid n$. R. Howe [46] gave a parametrization of the irreducible representations of $D^\times$. In [64] H. Koch and E.-W. Zink published their own version of R. Howe's work together with an application to the local Langlands' theory. In this paper we present the approach by H. Koch and E.-W. Zink.

Let D be a tame division algebra over F, $\mathcal{O}$ the valuation ring of D, $\mathcal{P}$ its maximal ideal, $V = V_o$ the group of units and $V_i = 1 + \mathcal{P}^i$ the i-th group of 1-units, cf. [11 , §1] or [11]. A finite-dimensional complex representation Π of $D^\times$ is continuous if and only if Π is trivial on some V_k. The smallest non-negative integer with this property is called the conductor of Π. The set of all (equivalence classes of) irreducible finite-dimensional continuous complex representations of $D^\times$ is denoted by Irr $D^\times$.

Let $\Pi \in$ Irr $D^\times$. If Π is 1-dimensional then, as is well-known, $\Pi = \chi \circ \mathrm{Nrd}_D$ for some character $\chi : F^\times \to \mathbb{C}^\times$ where $\mathrm{Nrd}_D : D^\times \to F^\times$ denotes the reduced norm. If Π is not 1-dimensional then, since $D^\times/V \cong \mathbb{Z}$, the conductor of Π is positive. It is an essential fact that in this case either there is a 1-dimensional representation λ of $D^\times$ such that $\lambda \otimes \Pi$ has smaller conductor than Π, or Π is induced from a representation $\Pi' \in$ Irr $D'^\times$ for a proper central division subalgebra D'/F' of D via a certain induction map $I^D_{D'}$. Using these two reduction techniques, the representations $\Pi \in$ Irr $D^\times$ can be obtained from characters of subfields of D. More precisely, Irr $D^\times$ can be parametrized by so-called admissible pairs $(K/F,\chi)$ with $K \subseteq D$. By definition, an admissible pair $(K/F,\chi)$ consists of a tame field extension K/F and a continuous character $\chi : K^\times \to \mathbb{C}^\times$ satisfying the following conditions:

(i) χ does not admit a proper norm factorization, i.e. if $\chi = \chi' \circ N_{K/L}$ for an intermediate field $F \subseteq L \subseteq K$ and a character $\chi' : L^\times \to \mathbb{C}^\times$ then $L = K$;

(ii) if $F \subseteq L \subseteq K$ is an intermediate field and if the restriction $\chi/_{U_K^1}$ to the 1-unit group U_K^1 of K admits a factorization over $N_{K/L}$ then K/L is unramified.

The set of all admissible pairs $(K/F,\chi)$ with $K \subseteq D$ is denoted by AP_D. Following [64] one can associate with any $(K/F,\chi) \in AP_D$ a representation $\Pi_D(K/F,\chi) \in \mathrm{Irr}\, D^\times$. Obviously $D^\times$ acts on AP_D via conjugation. Let $AP_D/D^\times$ denote the set of orbits. Then the main theorem can be stated as follows:

(1.1) Theorem. *For any tame division algebra D/F, the map* $(K/F,\chi) \mapsto \Pi_D(K/F,\chi)$ *induces a bijection* $AP_D/D^\times \xrightarrow{\sim} \mathrm{Irr}\, D^\times$.

Reversing the reduction process for representations $\Pi \in \mathrm{Irr}\, D^\times$ described above, the construction of $\Pi_D(K/F,\chi)$ is carried out by induction on the dimension $[D:F]$ and on the conductor of χ, the main step - due to L.Corwin and R.Howe [20] - being the definition of the induction map $I_{D'}^D$.

We now briefly describe the contents of this paper. Sections 2, 3 and 4 review some basic facts on local fields, tame division algebras and representation theory, respectively, and contain proofs of some auxiliary results. Section 5 deals with a special case of theorem (1.1). In section 6 we construct the induction map $I_{D'}^D$. In section 7 we consider a basic invariant of representations $\Pi \in \mathrm{Irr}\, D^\times$, the fundamental pair. Finally, in section 8, we present the parametrization of Irr $D^\times$ by $AP_D/D^\times$.

2. Local fields

In this section we list some properties of local fields, tame extensions and characters which are needed in subsequent sections. For proofs we usually refer to [64] and standard books on algebraic number theory.

Let K be a local field of residue characteristic p. We denote the valuation ring of K by $\mathcal{O}_K$, its maximal ideal by $\mathcal{P}_K$, its unit group by U_K and its i-th group of 1-units by $U_K^i = 1 + \mathcal{P}_K^i$ ($i \in \mathbb{N}$). Let π_K be a prime element in K and ζ_K the group of roots of unity in K with orders prime to p. Then $C := \langle \pi_K, \zeta_K \rangle$ is a complement of U_K^1 in $K^\times$. Any complement of U_K^1 in $K^\times$ arises in this way. The group C is called a complementary subgroup of K. We are going to use the following property of complementary subgroups with respect to tame extensions:

(2.1) <u>If</u> K/F <u>is a tame extension then, given any complementary subgroup</u> C_F <u>of</u> F, <u>there is a unique complementary subgroup</u> C_K <u>of</u> K <u>containing</u> C_F. <u>Moreover,</u> $K = F(C_K)$.

The following vanishing result for cohomology will be important for us:

(2.2) <u>If</u> K/F <u>is a tame Galois extension with Galois group</u> $G = \mathrm{Gal}(K/F)$ <u>then</u> $\hat{H}^q(G, U_K^i) = 0$ <u>for</u> $q \in \mathbb{Z}$, $i \in \mathbb{N}$.

As a consequence, we obtain:

(2.3) <u>For any tame extension</u> K/F <u>and</u> $i \in \mathbb{N}$, $N_{K/F}(U_K^i) = U_K^i \cap F$.

By a character of $K^\times$ we mean a continuous homomorphism $\chi : K^\times \to \mathbb{C}^\times$. We do not require that $|\chi(x)| = 1$ for $x \in K^\times$. If $\chi : K^\times \to \mathbb{C}^\times$ is a character then $U_K^{k+1} \subseteq \mathrm{Ker}\ \chi$ for some $k \geq 0$. The smallest non-negative integer $k = j(\chi)$ with this property is called the index of χ.

(2.4) Lemma. <u>Let</u> K/F <u>be a tame extension and</u> $\chi : K^\times \to \mathbb{C}^\times$ <u>a character of index</u> $j = j(\chi) > 0$.
<u>Then there is a smallest subfield</u> F' <u>between</u> K <u>and</u> F <u>such that</u> $\chi|_{U_K^j}$ <u>factors through</u> $N_{K/F'}$, <u>and there is a unique character</u> $\alpha : U_K^j \cap F' = N_{K/F'}(U_K^j) \to \mathbb{C}^\times$ <u>such that</u> $\chi|_{U_K^j} = \alpha \circ N_{K/F'}|_{U_K^j}$. <u>Moreover,</u> $e(K/F') = \gcd(e(K/F), j)$.

We shall refer to (F',α) as the fundamental pair of $(K/F,\chi)$.

Proof. Denote by $\hat{K}$ the Galois hull of K/F in some separable closure of K, and set $G := \mathrm{Gal}(\hat{K}/F)$, $\hat{\chi} := \chi \circ N_{\hat{K}/K}$. Denote by H the stabilizer of $\hat{\chi}|_{U^j_K}$ in G and by F' the fixed field of H. Since

$$\hat{\chi}(\sigma(x)) = \chi(N_{\hat{K}/K}(\sigma(x))) = \chi(N_{\hat{K}/K}(x)) = \hat{\chi}(x)$$

for $x \in \hat{K}$, $\sigma \in \mathrm{Gal}(\hat{K}/K)$, we get $\mathrm{Gal}(\hat{K}/K) \subseteq H$ and $F' \subseteq K$.

Let $x \in U^j_K$ such that $N_{K/F'}(x) = 1$. Since $\hat{K}/K$ is unramified we have $U^j_K = U^j_{\hat{K}} \cap K = N_{\hat{K}/K}(U^j_{\hat{K}})$. Thus we can find $y \in U^j_{\hat{K}}$ such that $x = N_{\hat{K}/K}(y)$. Since $N_{\hat{K}/F'}(y) = N_{K/F'}(x) = 1$ and $\hat{H}^{-1}(H,U^j_{\hat{K}}) = 0$ we may write $y = \prod_{\sigma\in H} (\sigma(y_\sigma)y_\sigma^{-1})^{\alpha_\sigma}$ with $y_\sigma \in U^j_{\hat{K}}$, $\alpha_\sigma \in \mathbb{Z}$. Thus

$$\chi(x) = \hat{\chi}(y) = \prod_{\sigma\in H} \hat{\chi}(\sigma(y_\sigma))^{\alpha_\sigma}\, \hat{\chi}(y_\sigma^{-\alpha_\sigma}) = 1.$$

This shows that the map $\alpha : U^j_K \cap F' = N_{K/F'}(U^j_K) \to \mathbb{C}^\times$, defined by $\alpha(N_{K/F'}(z)) := \chi(z)$, is well-defined and has the required properties.

Suppose now that $F \subseteq L \subseteq K$ is any intermediate field and $\chi' : U^j_K \cap L = N_{K/L}(U^j_K) \to \mathbb{C}^\times$ is any character such that $\chi = \chi' \circ N_{K/L}$ on U^j_K. For $\sigma \in \mathrm{Gal}(\hat{K}/L)$ and $x \in U^j_{\hat{K}}$ we then have

$$\hat{\chi}(\sigma(x)) = \chi(N_{\hat{K}/K}(\sigma(x))) = \chi'(N_{\hat{K}/L}(\sigma(x))) = \chi'(N_{\hat{K}/L}(x)) = \hat{\chi}(x).$$

Thus $\mathrm{Gal}(\hat{K}/L) \subseteq H$ and $F' \subseteq L$. Since $N_{K/F'}(U^j_K) = U^j_K \cap F'$ the map α is unique.

It remains to show that $e(K/F') = \gcd(e(K/F),j)$. We show first that $e(K/F')|j$. If this is not true then

$$N_{K/F'}(U^j_K) = U^j_K \cap F' = U^{j+1}_K \cap F' = N_{K/F'}(U^{j+1}_K)$$

and

$$\chi(U^j_K) = \alpha(N_{K/F'}(U^j_K)) = \alpha(N_{K/F'}(U^{j+1}_K)) = \chi(U^{j+1}_K) = 1,$$

a contradiction to $j(\chi) = j$. Thus $e(K/F')|j$. Setting $i := j/e(K/F')$ it remains to show that $\gcd(e(F'/F),i) = 1$. We know that $U^j_K \cap F' = U^i_{F'}$,

$U_K^{j+1} \cap F' = U_{F'}^{i+1}$, and that α is trivial on $U_{F'}^{i+1}$. Thus $\alpha: U_{F'}^{i}/U_{F'}^{i+1} \to \mathbb{C}^\times$. Assume that $\gcd(e(F'/F),i) \neq 1$, and let q be a prime factor of $\gcd(e(F'/F),i)$. Then there is an intermediate field $F \subseteq L \subseteq F'$ such that F'/L is totally ramified and $[F' : L] = q$. Now $N_{F'/L}$ induces an isomorphism $U_{F'}^{i} / U_{F'}^{i+1} \to U_{F'}^{i} \cap L/U_{F'}^{i+1} \cap L$. This leads to a further norm factorization of $\chi|_{U_K^j}$, a contradiction. So the lemma is proved.

Recall the definition of an admissible pair. The following assertions are straightforward consequences of the definition.

(2.5) <u>If</u> $(K/F,\chi)$ <u>is an admissible pair and</u> $\chi_F : F^\times \to \mathbb{C}^\times$ <u>is a character then</u> $(K/F,(\chi_F \circ N_{K/F}) \cdot \chi)$ <u>is an admissible pair.</u>

(2.6) <u>If</u> $(K/F,\chi)$ <u>is an admissible pair and</u> $F \subseteq F' \subseteq K$ <u>is an intermediate field then</u> $(K/F',\chi)$ <u>is an admissible pair.</u>

We shall need the following property later.

(2.7) <u>Let</u> $(K/F',\chi)$ <u>be an admissible pair with</u> $j = j(\chi) > 0$, <u>let</u> $\chi|_{U_K^j} = \alpha \circ N_{K/F'}|_{U_K^j}$, <u>let</u> F'/F <u>be a tame field extension and suppose that</u> α <u>admits no factorization through</u> $N_{F'/L}$ <u>for</u> $F \subseteq L \subseteq F'$. <u>Then</u> $(K/F,\chi)$ <u>is an admissible pair with fundamental pair</u> (F',α).

<u>**Proof.**</u> By hypothesis, K/F is a tame field extension. Let (K_j,χ_j) be the fundamental pair of $(K/F,\chi)$. Since

$$\alpha \circ N_{K/F'}|_{U_K^j} = \chi|_{U_K^j} = \chi_j \circ N_{K/K_j}|_{U_K^j}$$

we conclude that $K_j \subseteq F'$ and $\alpha|_{U_K^j \cap F'} = \chi_j \circ N_{F'/K_j}|_{U_K^j \cap F'}$. Since α has no proper norm factorization we obtain $K_j = F'$ and $\chi_j = \alpha$. Thus (F',α) is the fundamental pair of $(K/F,\chi)$.

Suppose now that $F \subseteq L \subseteq K$ is an intermediate field and that $\chi' : L^\times \to \mathbb{C}^\times$ is a character such that $\chi|_{U_K^1} = \chi' \circ N_{K/L}|_{U_K^1}$. Then also $\chi|_{U_K^j} = \chi' \circ N_{K/L}|_{U_K^j}$. Since (F',α) is the fundamental pair of $(K/F,\chi)$ we obtain $F' \subseteq L$. Since $(K/F',\chi)$ is an admissible pair, K/L is unramified. In case $\chi = \chi' \circ N_{K/L}$ we similarly get $K = L$. Thus we have shown that $(K/F, \chi)$ is an admissible pair.

3. Tame division algebras

This section contains the basic facts on central division algebras over local fields we are going to use. Let F denote a local field of residue characteristic p, and let D/F be a central division algebra of dimension $[D : F] = n^2$. The valuation of F extends uniquely to a valuation of D. Let $\mathcal{O}$ denote the valuation ring of D, $\mathcal{P}$ its maximal ideal, V its group of units and $V_i = 1 + \mathcal{P}^i$ ($i \in \mathbb{N}$) its i-th group of 1-units, cf. [11 , § 1] or [11]. Let Nrd, Trd : D → F denote reduced norm and reduced trace, respectively. Then one has

$$\mathrm{Nrd}(D^\times) = F^\times, \quad \mathrm{Nrd}(V) = U_F,$$
$$\mathrm{Nrd}(V_i) = V_i \cap F \quad (i \in \mathbb{N}),$$
$$\mathrm{Trd}(\mathcal{P}^i) = \mathcal{P}^i \cap F \quad (i \in \mathbb{Z}).$$

Moreover, ramification index and residue degree are given by

$$e(D/F) = f(D/F) = n.$$

The division algebra D/F contains an unramified extension F_n/F of degree n and a prime element Π_D such that the map $F_n \to F_n$, $x \to \Pi_D x \Pi_D^{-1}$, generates $\mathrm{Gal}(F_n/F)$ and F_n, Π_D generate D over F. Let ζ_{F_n} denote the group of roots of unity in F_n of order prime to p. Then $C := \langle \zeta_{F_n}, \Pi_D \rangle$ is a complement to V_1 in $D^\times$, i.e. $D^\times = C \ltimes V_1$. Any complement of V_1 in $D^\times$ arises in this way. We call C a complementary subgroup of D. We now list several properties of complementary subgroups. The proofs follow from standard facts on division algebras, cf. [64].

(3.1) <u>If</u> C <u>is a complementary subgroup of</u> D <u>then</u> $C_F := C \cap F$ <u>is a complementary subgroup of</u> F. <u>The mapping</u> $x \to x^n$ <u>maps the set of prime elements in</u> C <u>onto the set of prime elements in</u> C_F. <u>Moreover,</u> C_F <u>is the center of</u> C.

(3.2) <u>Any complementary subgroup</u> C_F <u>of</u> F <u>is contained in a complementary subgroup</u> C_D <u>of</u> D, <u>and any two complementary subgroups of</u> D <u>containing</u> C_F <u>are conjugate in</u> $D^\times$.

(3.3) Let K/F be a tame field extension with $K \subseteq D$, and let C_D, C_K, C_F be complementary subgroups of D, K, F, respectively, such that $C_D \cap F = C_F = C_K \cap F$. Then $xC_Kx^{-1} \subseteq C_D$ for some $x \in D^\times$.

Using these properties of complementary subgroups one can now prove the following result.

(3.4) Lemma. Let $K_1/F, K_2/F$ be tame field extensions such that $K_1 \subseteq D$, $K_2 \subseteq D$ and $D_1^\times V_1 = D_2^\times V_1$ where D_i denotes the centralizer of K_i in D. Then $dK_1d^{-1} = K_2$ for some $d \in V_1$.

Proof. Let C_D, C_{D_i}, C_i, C_F be complementary subgroups of D, D_i, K_i, F, respectively, such that $C_D \cap F = C_F = C_i \cap F$, $C_{D_i} \cap K_i = C_i$. The isomorphism $C_{D_i} \cong D_i^\times V_1/V_1$ maps C_i onto $K_i^\times V_1/V_1$. Thus, by (3.1), $K_i^\times V_1/V_1$ is the center of $D_i^\times V_1/V_1$; in particular, $K_1^\times V_1 = K_2^\times V_1$. By (3.3) we can find $x_i \in D^\times$ such that $x_i C_i x_i^{-1} \subseteq C_D$. Since $D^\times = C_D V_1$ we may take $x_i \in V_1$. Replacing K_i by $x_i K_i x_i^{-1}$ we may assume that $C_F \subseteq C_i \subseteq C_D$. Let $a \in C_1 \subseteq K_1^\times V_1 = K_2^\times V_1 = C_2 V_1$, and write $a = cv$ with $c \in C_2$, $v \in V_1$. Since $a, c \in C_D$, $v = 1$. Thus $C_1 \subseteq C_2$, and by (2.1), $K_1 = F(C_1) \subseteq F(C_2) = K_2$. By symmetry, $K_1 = K_2$.

Now let $1 \leq i \leq j \leq 2i$. Then the map $x \to 1 + x$ induces an isomorphism $P^i/P^j \to V_i/V_j$; in particular, V_i/V_j is abelian. We choose an additive character $\psi_F : F^+ \to \mathbb{C}^\times$, fixed throughout this paper, with conductor $\mathcal{O}_F$, i.e. $\mathcal{O}_F$ is the largest module among the P_F^i ($i \in \mathbb{Z}$) on which ψ_F vanishes. Then $\psi_D := \psi_F \circ \mathrm{Trd} : D^+ \to \mathbb{C}^\times$ is an additive character with conductor P^{1-n}. Given $c \in P^{-j+1-n}$ we consider the character

$$\theta_c : V_i \to \mathbb{C}^\times, \quad 1 + x \to \psi_D(cx).$$

Then θ_c is trivial on V_j, and the assignment $c \to \theta_c$ induces an isomorphism

$$P^{-j+1-n}/P^{-i+1-n} \xrightarrow{\sim} \mathrm{Hom}(V_i/V_j, \mathbb{C}^\times).$$

The reduced trace $\mathrm{Trd} : D \to F$ induces a non-degenerate symmetric bilinear form on D. For a subset X of D we let $X^\perp := \{d \in D : \mathrm{Trd}(dX) = 0\}$. Let F'/F be a tame field extension with $F' \subseteq D$, and denote the centralizer of F' in D by D'. Then we have:

(3.5) $P^i = (D' \cap P^i) \oplus (D'^\perp \cap P^i)$ for $i \in \mathbb{Z}$.

Proof. Since $\mathrm{Trd}|_{D'} = \mathrm{Tr}_{F'/F} \circ \mathrm{Trd}_{D'}$ and since F'/F is separable, D' is a non-degenerate subspace of D. Thus $D = D' \oplus D'^{\perp}$. Now let $i \in \mathbb{Z}$ and $a \in P^i$. We write $a = a_1 + a_2$ with $a_1 \in D'$, $a_2 \in D'^{\perp}$. Assume that $a_1 \notin P^i$ or $a_2 \notin P^i$. Since $a = a_1 + a_2 \in P^i$ we have $a_1 \notin P^i$ and $a_2 \notin P^i$. Thus $a_1 \in P^j$ for some $j < i$ and $a_2 \in P^j$. Then $-a_1^{-1} a_2 = 1 - a_1^{-1} a \in D'^{\perp} \cap V_1$. This implies that for any element x in the valuation ring O' of D',

$$\mathrm{Trd}(x) \equiv \mathrm{Trd}((1 - a_1^{-1} a)x) = \mathrm{Trd}(- a_1^{-1} a_2 x) = 0 \pmod{P_F}.$$

Thus

$$\mathrm{Tr}_{F'/F}(O_{F'}) = \mathrm{Tr}_{F'/F}(\mathrm{Trd}_{D'}(O')) = \mathrm{Trd}(O') \subseteq P_F$$

which is impossible since F'/F is tame.

The following somewhat technical results on conjugacy classes in $D^{\times}$ will be of importance later. Throughout the rest of this section, let C be a complementary subgroup of D and $c \in C$ such that $F(c)/F$ is tame. Denote the centralizer of $F' := F(c)$ in D by D'.

(3.6) <u>The map</u> $\varphi_c : x \to x - cxc^{-1}$ <u>is an automorphism of</u> $D'^{\perp} \cap P^i$ <u>for</u> $i \in \mathbb{Z}$.

Proof. It is easy to check that φ_c defines an endomorphism of $D'^{\perp} \cap P^i$. Since $D' \cap D'^{\perp} = 0$, φ_c is injective. We now show that φ_c induces an automorphism of $D'^{\perp} \cap P^i / D'^{\perp} \cap P^{i+1}$. Since this group is finite it suffices to check injectivity. Let $x \in D'^{\perp} \cap P^i$ such that $x - cxc^{-1} \in D'^{\perp} \cap P^{i+1}$ and assume that $x \notin P^{i+1}$. Choose a prime element π of D in C, and let $d \in C$ such that $y := \pi^i \cdot d \equiv x \pmod{P^{i+1}}$. Then $cyc^{-1} y^{-1} \in C$ and $cyc^{-1} y^{-1} \equiv cxc^{-1} x^{-1} \equiv 1 \pmod{P}$. Since $C \cap V_1 = 1$ we obtain $cyc^{-1} y^{-1} = 1$ and therefore $y \in D' \cap P^i$. Hence $x \in (D' \cap P^i) + P^{i+1} = (D' \cap P^i) \oplus (D'^{\perp} \cap P^{i+1})$, and $x \in P^{i+1}$. Thus φ_c induces an automorphism of $D'^{\perp} \cap P^i / D'^{\perp} \cap P^{i+1}$. By successive approximation we obtain that ϕ_c is an automorphism.

(3.7) <u>Let</u> $n \in \mathbb{N}$ <u>and</u> $v \in V_1$ <u>such that</u> $cv \in D' + cP^n$. <u>Then there is</u> $x_n \in P^n \cap D'^{\perp}$ <u>such that</u> $(1 + x_n)\, cv\, (1 + x_n)^{-1} \in D' + cP^{n+1}$.

Proof. Since

$$(D' + P^n) \cap V_1 = 1 + ((D' + P^n) \cap P) = 1 + ((D' \cap P) + P^n) =$$

$= 1 + (P \cap D') + (P^n \cap D'^{\perp})$

we may write $v = 1 + y + z$ with $y \in P \cap D'$, $z \in P^n \cap D'^{\perp}$. By (3.6) there is $x_n \in P^n \cap D$ such that $z = x_n - c^{-1} x_n c$. Then

$$(1 + x_n)\, cv\, (1 + x_n)^{-1} = (1 + x_n)\, cv \sum_{k=0}^{\infty} (-x_n)^k \equiv$$

$$\equiv cv - cx_n + x_n c = c(v - z) = c(1 + y) \pmod{cP^{n+1}}.$$

(3.8) <u>For</u> $n \in \mathbb{N}_0$ <u>and</u> $v \in V_1$ <u>there is</u> $x \in P \cap D'^{\perp}$ <u>such that</u> $(1 + x)\, cv\, (1 + x)^{-1} \in D' + c\, P^{n+1}$;
<u>moreover</u> x <u>is unique</u> mod $(P^{n+1} \cap D'^{\perp})$.

Proof. We argue by induction on n, the case $n = 0$ being trivial. So assume $n > 0$. By induction we have some $x' \in P \cap D'^{\perp}$ such that $(1 + x')\, cv\, (1 + x')^{-1} \in D' + cP^n$. By (3.7) we find some $x_n \in P^n \cap D'^{\perp}$ such that

$$(1 + x_n)(1 + x')\, cv\, (1 + x')^{-1} (1 + x_n)^{-1} \in D' + c\, P^{n+1}.$$

Setting $x := x_n + x'$ we obtain $(1 + x)\, cv\, (1 + x)^{-1} \in D' + c\, P^{n+1}$.
Suppose now that also $y \in P \cap D'^{\perp}$ such that $(1 + y)\, cv\, (1 + y)^{-1} \in D' + c\, P^{n+1} \subseteq D' + c\, P^n$.
By induction we can write $y = x + z$ with $z \in P^n \cap D'^{\perp}$. Then

$$0 \equiv (1 + y)\, cv\, (1 + y)^{-1} \equiv (1 + x)\, cv\, (1 + x)^{-1} + (zc - cz) \equiv$$

$$\equiv zc - cz \pmod{D' + c\, P^{n+1}}.$$

Thus

$$z - czc^{-1} \in P^n \cap (D' + P^{n+1}) = (P^n \cap D') + P^{n+1} = (P^n \cap D') \oplus$$

$$\oplus (P^{n+1} \cap D'^{\perp}).$$

Since

$$z - czc^{-1} \in D'^{\perp} \text{ and } D = D' \oplus D'^{\perp} \text{ we obtain } z - czc^{-1} \in P^{n+1} \cap D'^{\perp}.$$

By (3.6), $z \in P^{n+1} \cap D'^{\perp}$.

(3.9) Proposition. For $v \in V_1$, there is a unique $x \in P \cap D'^{\perp}$ such that $(1 + x)\, cv\, (1 + x)^{-1} \in D'$.

Proof. By (3.8) there are elements $x_n \in P \cap D'^{\perp}$ $(n \in \mathbb{N}_0)$ such that

$$(1 + x_n)\; cv\; (1 + x_n)^{-1} \in D' + c\, P^{n+1};$$

moreover, $x_n \equiv x_{n+1} \pmod{P^{n+1}}$. This shows that there is $x \in D$ such that $x \equiv x_n \pmod{P^{n+1}}$ for $n \in \mathbb{N}_0$. Obviously $x \in P \cap D'^{\perp}$ and $(1 + x)\, cv\, (1 + x)^{-1} \in D'$. The uniqueness of x is proved similarly.

(3.10) Let $n \in \mathbb{N}$, $v \in V_1 \cap D'$ and $d \in D^\times$ such that $dcvd^{-1} \equiv cv \pmod{cP^n}$. Then $d \in D'^\times V_n$.

Proof. We argue by induction on n. In case $n = 1$ we may assume that $v = 1$ and $d \in C$. Then $dcd^{-1}c^{-1} \in C$ and $dcd^{-1}c^{-1} \equiv 1 \pmod{P}$, so $d \in D'$. Suppose that $n > 1$. By induction we may write $d = d'(1 + x)$ with $d' \in D'^\times$, $x \in P^{n-1}$. Then $(1 + x)\, cv\, (1 + x)^{-1} \in D' + c\, P^n$. Write $x = y + z$ with $y \in D' \cap P^{n-1}$, $z \in D'^{\perp} \cap P^{n-1}$. Since $1 + x = (1+y)(1+(1+y)^{-1}z)$ we obtain $(1 + (1 + y)^{-1}z)\, cv\, (1 + (1 + y)^{-1} z)^{-1} \in D' + c\, P^n$. By (3.8), $(1 + y)^{-1} z \in P^n$ and $d \in D'^\times V_n$.

4. Facts from group representation theory

We are dealing with representations of the topological groups $D^\times$, $F^\times$ etc.. Thus the natural category of representations is that of the continuous homomorphisms $\Pi : D^\times \to GL(m, \mathbb{C})$, $m \in \mathbb{N}$. Since $D^\times$ has a topology defined by the normal subgroups V_k, a homomorphism $\Pi : D^\times \to GL(m, \mathbb{C})$ is continuous if and only if $V_{k+1} \subseteq \operatorname{Ker} \Pi$ for some $k \geq 0$. The smallest non-negative integer $k = j(\Pi)$ with this property is called the index of Π. If $k = j(\Pi)$ then we can consider Π as a homomorphism $D^\times / V_{k+1} \to GL(m, \mathbb{C})$. Conversely any homomorphism $\Pi : D^\times / V_{k+1} \to GL(m, \mathbb{C})$ gives rise to a continuous homomorphism $D^\times \to GL(m, \mathbb{C})$. Thus essentially we have to deal with finite-dimensional complex representations of discrete groups. The set of all (equivalence classes of) irreducible representations of $D^\times$ will be denoted by Irr $D^\times$.

The appropriate framework is thus the following: G is a discrete group and Irr G denotes the set of equivalence classes of finite-dimensional irreducible complex representations of G. We often do not distinguish between a representation and its equivalence class. For proofs of the results in this section we refer to [64] or standard books on representation theory.

Let N be a normal subgroup of G. Then G acts on Irr N by conjugation:

$$\Gamma^g(x) := \Gamma(gxg^{-1}) \text{ for } g \in G,\ x \in N,\ \Gamma \in \text{Irr } N.$$

Given $\Gamma \in \text{Irr } N$ let $\text{St}_G(\Gamma) = \text{St}(\Gamma) := \{g \in G : \Gamma^g = \Gamma\}$ denote the stabilizer of Γ in G.

(4.1) For $\Pi \in \text{Irr } G$, the irreducible constituents of $\Pi|_N$ are conjugate under G. Thus, if $\Gamma \in \text{Irr } N$ is an irreducible constituent of $\Pi|_N$ then $|G : \text{St}_G(\Gamma)| < \infty$.

For $\Gamma \in \text{Irr } N$, let $\text{Irr}(G|\Gamma)$ denote the set of $\Pi \in \text{Irr } G$ such that Γ is a constituent of $\Pi|_N$.

(4.2) If $\Gamma \in \text{Irr } N$ and $|G : \text{St}(\Gamma)| < \infty$ then, for any subgroup U of G with $\text{St}(\Gamma) \subseteq U$ the induction map for representations induces a bijection $\text{Ind}_U^G : \text{Irr}(U/\Gamma) \rightarrow \text{Irr}(G/\Gamma)$.

We next deal with the case of Γ being stable under G, i.e. $\text{St}(\Gamma) = G$. This assumption is kept fixed throughout the rest of this section.

(4.3) If Γ extends to a representation $\hat{\Gamma}$ of G then the mapping

$$\text{Irr}(G/N) \to \text{Irr}(G|\Gamma),\ \Sigma \to \Sigma \otimes \hat{\Gamma},$$

is a bijection.

We further need criteria for the extendability of Γ to G. Here we assume $P = G/N$ finite what is enough for the applications. Given $\Gamma \in \text{Irr}(N)$ with $\text{St}(\Gamma) = G$ there is a Mackey obstruction $\bar{w}_\Gamma \in H^2(P, \mathbb{C}^\times)$ for the extendability.

(4.4) If $G = P \ltimes N$ is a semidirect product then the order of $\bar{w}_\Gamma$ in $H^2(P, \mathbb{C}^\times)$ divides $\dim \Gamma$.

Now assume that G is not necessarily a semidirect product but that P is abelian. Then the Universal Coefficient Theorem induces a bijection between $H^2(P, \mathbb{C}^\times)$ and the set of $\mathbb{C}$-valued symplectic bilinear forms on P, cf. also [64].

(4.5) <u>If the radical of the symplectic form associated with</u> $\bar{w}_\Gamma$ <u>is trivial then</u> $|G : N|$ <u>is a square and</u> $\mathrm{Ind}_N^G(\Gamma) = |G : N|^{1/2}\Pi$ <u>for some</u> $\Pi \in \mathrm{Irr}(G)$.

This fact will be important for the construction of the map $I_{D'}^{D}$.

5. Representations of index 0

Let F be a local field and D/F a central division algebra with $[D : F] = n^2$. In this section we are going to determine the irreducible representations Π of $D^\times$ with $j(\Pi) = 0$, i.e. $\Pi = 1$ on V_1. In the proof we will use the fact that $[D^\times, D^\times]$ is the kernel of the reduced norm map $\mathrm{Nrd} : D^\times \to F^\times$.

Recall that AP_D denotes the set of admissible pairs $(K/F,\chi)$ with $K \subseteq D$. For $(K/F,\chi) \in AP_D$ with $j(\chi) = 0$ we define a representation $\Pi_D(K/F,\chi)$ of $D^\times$ in the following way: Denote by D' the centralizer of K in D and set $\tilde{\chi} := \chi \circ \mathrm{Nrd}_{D'}$. Then extend $\tilde{\chi}$ to $\tilde{\chi}_\perp \in \mathrm{Irr}\, D'^\times V_1$ by setting $\tilde{\chi}_\perp(d'v) := \tilde{\chi}(d')$ for $d' \in D'^\times$, $v \in V_1$; this is well-defined since $\tilde{\chi}$ vanishes on $D'^\times \cap V_1$. Finally set

$$\Pi_D(K/F,\chi) := \mathrm{Ind}_{D'^\times V_1}^{D^\times}(\tilde{\chi}_\perp).$$

(5.1) Theorem.

(i) <u>For any</u> $(K/F,\chi) \in AP_D$ <u>with</u> $j(\chi) = 0$, $\Pi_D(K/F,\chi) \in \mathrm{Irr}\, D^\times$ <u>and</u> $j(\Pi_D(K/F,\chi)) = 0$.

(ii) <u>Any</u> $\Pi \in \mathrm{Irr}\, D^\times$ <u>with</u> $j(\Pi) = 0$ <u>can be obtained in this way.</u>

(iii) <u>The map</u> $(K/F,\chi) \to \Pi_D(K/F,\chi)$ <u>induces a bijection</u>

$$\{(K/F,\chi) \in AP_D : j(\chi) = 0\}/D^\times \leftrightarrows \{\Pi \in \mathrm{Irr}\, D^\times : j(\Pi) = 0\}.$$

Proof.

(i) Since $(K/F,\chi)$ is admissible and $\chi|_{U_K^1}$ is trivial, K/F must be unramified. Choose an unramified extension F_n/F of degree n with $K \subseteq F_n \subseteq D$. Then D contains a prime element π_D such that the map $\sigma : F_n \to F_n$, $x \to \pi_D x \pi_D^{-1}$, generates $G(F_n/F)$. Since $F_n \subseteq D'$ and since F_n and D have the same residue field we get $V \subseteq D'^{\times} V_1$. We next apply (4.2) to the situation $G = D^{\times}$, $N = V$, $\Gamma = \tilde{\chi}_{\perp}|_V$. Obviously $D'^{\times} V_1 \subseteq St(\Gamma)$. By (4.2) it suffices to show that $D'^{\times} V_1 = St(\Gamma)$ since $j(\Pi_D(K/F,\chi)) = 0$.

Let $x \in St(\Gamma)$. Since $V \subseteq St(\Gamma)$ we may assume that $x = \pi_D^k$ for some $k \in \mathbb{Z}$. We show first that σ^k stabilizes $\chi|_{U_k}$. So let $u \in U_K$. Since F_n/K is unramified we have $\hat{H}^0(G(F_n/K), U_{F_n}) = 0$. Thus $u = N_{F_n/K}(v)$ for some $v \in U_{F_n}$. Since $Nrd_{D'}|_{F_n} = N_{F_n/K}$ we get

$$\chi(\sigma^k u) = \chi(\sigma^k N_{F_n/K}(v)) = \chi(N_{F_n/K}(\sigma^k v)) =$$

$$\chi(Nrd_{D'}(\pi_D^k v \pi_D^{-k})) = \Gamma(\pi_D^k v \pi_D^{-k}) = \Gamma(v) = \chi(u).$$

We show next that $\chi : K^{\times} \to \mathbb{C}^{\times}$ factors through $N_{K/L}$ where L denotes the fixed field of $\sigma^k|_K$. Let $y \in K$ with $N_{K/L}(y) = 1$; in particular, $y \in U_K$. Since K/L is unramified we have $\hat{H}^{-1}(\langle\sigma^k\rangle, U_K) = 0$. Thus $y = z^{-1}\sigma^k(z)$ for some $z \in U_K$. This implies that $\chi(y) = \chi(z^{-1})\chi(\sigma^k z) = 1$. Therefore χ factors through $N_{K/L}$. Since $(K/F,\chi)$ is admissible we conclude that $K = L$. Thus $\pi_D^k w \pi_D^{-k} = \sigma^k w = w$ for $w \in K$, and $\pi_D^k \in D'^{\times}$ which remained to be proved.

(ii) Let $\Pi \in Irr\, D^{\times}$ with $j(\Pi) = 0$. Since Π is trivial on V_1 and V/V_1 is abelian, the restriction $\Pi|_V$ is a sum of 1-dimensional representations of V. Let Γ be one of these. We first determine $St(\Gamma)$. Clearly $F^{\times} V = \langle\pi_D^n, V\rangle \subseteq St(\Gamma)$. Thus $St(\Gamma) = \langle\pi_D^f, V\rangle$ for some $f|n$. Let $K \subseteq D$ be an unramified extension of F with $[K : F] = f$, and denote the centralizer of K in D by D'. Then $D'^{\times} V = \langle\pi_D^f, V\rangle = St(\Gamma)$ since $e(D/D') = f(K/F) = f$. By (4.2), $\Pi = Ind_{D'^{\times}V}^{D^{\times}}(\Pi_0)$ for some $\Pi_0 \in Irr(D'^{\times} V|\Gamma)$. Since Π_0 is trivial on V_1 and $D'^{\times}V = D'^{\times}V_1$, $\Pi' := \Pi_0|_{D'^{\times}} \in Irr(D'^{\times}|\Gamma')$ where $\Gamma' := \Gamma|_{D'\cap V}$; moreover, $j(\Pi') = 0$. Since Γ' is stable in $D'^{\times}$ and since $D'^{\times} \cap V$ has a complement in $D'^{\times}$, Γ' extends to a character $\hat{\Gamma} : D'^{\times} \to \mathbb{C}^{\times}$. Thus, by (4.3), $\Pi' = \Sigma \otimes \hat{\Gamma}$ with an irreducible representation Σ of $D'^{\times}/D' \cap V \cong \mathbb{Z}$. Hence $\dim \Pi' = \dim \Sigma = 1$; in particular, Π' is trivial on $[D'^{\times}, D'^{\times}]$. Since $[D'^{\times}, D'^{\times}]$ is the kernel of $Nrd_{D'}$, $\Pi' = \chi \circ Nrd_{D'}$ for some character $\chi : K^{\times} \to \mathbb{C}^{\times}$ with $j(\chi) = 0$. Now obviously $\Pi_0 = \tilde{\chi}_{\perp}$ and

$\Pi = \mathrm{Ind}_{D'^{\times}V_1}^{D^{\times}}(\tilde{\chi}_{\perp})$.

It remains to show that the pair $(K/F,\chi)$ is admissible. Let $K \supseteq L \supseteq F$ be an intermediate field and $\chi' : L^{\times} \to \mathbb{C}^{\times}$ a character such that $\chi = \chi' \circ N_{K/L}$. Denote the centralizer of L in D by D". Then $\tilde{\chi} = \chi' \circ N_{K/L} \circ \mathrm{Nrd}_{D'} = \chi' \circ \mathrm{Nrd}_{D''}|_{D'^{\times}}$. This implies that $D''^{\times} \subseteq \mathrm{St}(\Gamma) = D'^{\times}V_1$. By (3.4) we obtain $K = L$.

(iii) Obviously, conjugate admissible pairs lead to equivalent representations. Let conversely $(K/F,\chi)$, $(K'/F,\chi') \in AP_D$ with $j(\chi) = 0 = j(\chi')$ and $\Pi_D(K/F,\chi) = \Pi_D(K'/F,\chi')$ Π. We had seen in the proof of (i) that $K/F, K'/F$ are unramified and that $D'^{\times}V_1 = \mathrm{St}(\tilde{\chi}_{\perp}|_V)$, $D''^{\times}V_1 = \mathrm{St}(\tilde{\chi}'_{\perp}|_V)$ where D" denotes the centralizer of K' in D. Since $\tilde{\chi}_{\perp}|_V$, $\tilde{\chi}'_{\perp}|_V$ are irreducible constituents of $\Pi|_V$ we may assume by (4.1) that $\tilde{\chi}_{\perp}|_V = \tilde{\chi}'_{\perp}|_V$. Then $D'^{\times}V_1 = \mathrm{St}(\tilde{\chi}_{\perp}|_V) = D''^{\times}V_1$. By (3.4), K and K' are conjugate by an element in V_1. So we may assume that $K = K'$ and $D' = D''$. Now (4.2) implies that $\tilde{\chi}_{\perp} = \tilde{\chi}'_{\perp}$, and $\chi = \chi'$ follows.

It is now easy to check that the map $(K/F,\chi) \to \Pi_D(K/F,\chi)$ has the following properties:

(5.2) $\Pi_D(F/F,\chi) = \chi \circ \mathrm{Nrd}_D$ <u>for any character</u> $\chi : F^{\times} \to \mathbb{C}^{\times}$ <u>with</u> $j(\chi) = 0$.

(5.3) $\Pi_D(K/F, (\chi_F \circ N_{K/F})\chi) = \tilde{\chi}_F \otimes \Pi_D(K/F,\chi)$ <u>for any</u> $(K/F,\chi) \in AP_D$ with $j(\chi) = 0$ <u>and any character</u> $\chi_F : F^{\times} \to \mathbb{C}^{\times}$ <u>with</u> $j(\chi_F) = 0$.

(5.4) $\omega_{\Pi_D(K/F,\chi)} = \chi^{n/[K:F]}|_{F^{\times}}$ <u>for any</u> $(K/F,\chi) \in AP_D$ <u>with</u> $j(\chi) = 0$. <u>Here</u> ω_{Π} <u>denotes the unique irreducible constituent of</u> $\Pi|_{F^{\times}}$ <u>for</u> $\Pi \in \mathrm{Irr}\, D^{\times}$.

6. <u>The induction map</u> $I_{D'}^{D}$.

In this section we study the following situation: F is a local field of residue characteristic p, D/F is a tame central division algebra, F'/F is a field extension with $F' \subseteq D$, and we are given a non-trivial character $\alpha : U_{F'}^{j'}/U_{F'}^{j'+1} \to \mathbb{C}^{\times}$ for some $j' > 0$ such that α does not factor through $N_{F'/L}$ for $F' \supsetneq L \supseteq F$. Arguing as in the proof of (2.4) we see that this implies $\gcd(j', e(F'/F)) = 1$.

We denote the centralizer of F' in D by D' and write $[D:F] = n^2$, $[D':F'] = n'^2$ so that $n = n'[F':F]$. Setting $k := j'e(D/F') = j'n'f(F'/F)$ we have a character

$$\tilde{\alpha} := \alpha \circ \mathrm{Nrd}_{D'} : V_k \cap D'/V_{k+1} \cap D' \to \mathbb{C}^\times.$$

Our aim in this section is to construct a map $I_{D'}^{D} : \mathrm{Irr}(D'^\times|\tilde{\alpha}) \to \mathrm{Irr} D^\times$. The map $I_{D'}^{D}$ will be used in order to extend the map $(K/F,\chi) \to \Pi_D(K/F,\chi)$ to all of AP_D. The construction of $I_{D'}^{D}$ is carried out in several steps.

By (3.5), $\mathcal{P}^i = (D' \cap \mathcal{P}^i) \oplus (D'^\perp \cap \mathcal{P}^i)$ for $i \in \mathbb{Z}$. Now let $1 \leq i \leq j \leq 2i$ and apply the isomorphism $\mathcal{P}^i/\mathcal{P}^j \to V_i/V_j$ induced by $x \to 1 + x$. Then the following facts are easily verified:

(6.1) $V_i/V_j = (D' \cap V_i)V_j/V_j \times W_iV_j/V_j$ _where_ $W_i = 1 + (D'^\perp \cap \mathcal{P}^i)$. _Moreover,_ W_iV_j _is a normal subgroup of_ $D'^\times V_i$ _such that_ $D'^\times V_i = D'^\times W_iV_j$ _and_ $D'^\times V_j \cap W_iV_j = V_j$; _in particular,_

$$D'^\times V_i/V_j = D'^\times V_j/V_j \ltimes W_iV_j/V_j.$$

Suppose now that $1 \leq i \leq k + 1 \leq 2i$. For any character $\chi : V_i \cap D'/V_{k+1} \cap D' \to \mathbb{C}^\times$ define a character $\chi_\perp : V_i/V_{k+1} \to \mathbb{C}^\times$ by

$$\chi_\perp(vw) := \chi(v) \text{ for } v \in V_i \cap D',\ w \in W_iV_{k+1};$$

this is well-defined by (6.1). In particular, we get the character $\tilde{\alpha}_\perp : V_k/V_{k+1} \to \mathbb{C}^\times$. We shall construct $I_{D'}^{D}$ in such a way that it yields a bijection $\mathrm{Irr}(D'^\times|\tilde{\alpha}) \xrightarrow{\sim} \mathrm{Irr}(D^\times|\tilde{\alpha}_\perp)$. For $\chi \in \mathrm{Irr}(V_i \cap D|\tilde{\alpha})$ obviously $\chi_\perp \in \mathrm{Irr}(V_i|\tilde{\alpha}_\perp)$. We collect the basic properties of the map

$$\mathrm{Irr}(V_i \cap D'|\tilde{\alpha}) \to \mathrm{Irr}(V_i|\tilde{\alpha}_\perp),\ \chi \mapsto \chi_\perp.$$

(6.2) Proposition.

(i) $St_{D^\times}(\chi_\perp) = St_{D'^\times}(\chi)V_{k+1-i}$ _for_ $\chi \in \mathrm{Irr}(V_i \cap D'|\tilde{\alpha})$.

(ii) _The map_ $\chi \to \chi_\perp$ _induces a bijection_

$$\mathrm{Irr}(V_i \cap D'|\tilde{\alpha})/D'^\times \xrightarrow{\sim} \mathrm{Irr}(V_i|\tilde{\alpha}_\perp)/D^\times.$$

Proof. We start with some preliminary considerations. Fix complementary subgroups $C_F, C_{F'}$ of F, F', respectively, such that $C_F \subseteq C_{F'}$. By (3.3) there is a complementary subgroup C_D of D containing $C_{F'}$. It is easy to see that there is $c \in F'^{\times}$ such that $\alpha(1+x) = \psi_F(Tr_{F'/F}(cx))$ for $x \in U_{F'}^{j'}$; moreover, we may assume that $c \in C_{F'}$. Since $\psi_F \circ Tr_{F'/F}$ has conductor $P_{F'}^{1-e(F'/F)}$, $c\mathcal{O}_{F'} = P_{F'}^{-j'-e(F'/F)}$ and $c\mathcal{O} = P^{-k-n}$. We have

$$\tilde{\alpha}(1+x) = \alpha(Nrd_{D'}(1+x)) = \alpha(1+Trd_{D'}(x)) = \psi_F(Tr_{F'/F}(Trd_{D'}(cx))) \\ = \psi_D(cx)$$

for $x \in P^k \cap D'$ since $Nrd_{D'}(1+x) \equiv 1+Trd_{D'}(x) \pmod{P^{k+1} \cap D'}$ for $x \in P^k \cap D'$. Since $\psi_D(cx) = 1 = \tilde{\alpha}_{\perp}(1+x)$ for $x \in P^k \cap D'^{\perp}$ we obtain $\tilde{\alpha}_{\perp}(1+x) = \psi_D(cx)$ for $x \in P^k$.

Denoting the centralizer of $F'' := F(c)$ in D by D'' we obviously have $D''^{\times} \subseteq St_{D^{\times}}(\tilde{\alpha}_{\perp})$. Since $[V_1, V_k] \subseteq V_{k+1}$ we also have $V_1 \subseteq St_{D^{\times}}(\tilde{\alpha}_{\perp})$. We prove that in fact $St_{D^{\times}}(\tilde{\alpha}_{\perp}) = D''^{\times} V_1$. Let $s \in St_{D^{\times}}(\tilde{\alpha}_{\perp})$. Since $V_1 \subseteq St_{D^{\times}}(\tilde{\alpha}_{\perp})$ we may assume that $s \in C_D$. Then $\psi_D(cx) = \psi_D(csxs^{-1}) = \psi_D(s^{-1}csx)$ for $x \in P^k$, so $s^{-1}cs \equiv c \pmod{P^{-k-n+1}}$ and $c^{-1}s^{-1}cs \in C_D \cap V_1 = 1$. Thus $s \in D''^{\times}$, and $St_{D^{\times}}(\tilde{\alpha}_{\perp}) = D''^{\times} V_1$ is proved.

We show next that $F'' = F'$. Define $\beta : V_k \cap F'' \to \mathbb{C}^{\times}$ by $\beta(1+x) := \psi_F(Tr_{F''/F}(cx))$ for $x \in P^k \cap F''$. Then

$$\alpha(Nrd_{D'}(1+x)) = \psi_D(cx) = \psi_F(Tr_{F''/F}(Trd_{D''}(cx))) =$$

$$\beta(1+Trd_{D''}(x)) = \beta(Nrd_{D''}(x)) = \beta(N_{F'/F''}(Nrd_{D'}(x)))$$

for $x \in P^k \cap D'$, as above. Therefore $\alpha = \beta \circ N_{F'/F''}|_{V_k \cap F'}$. Since α has no proper norm factorization this implies $F' = F'' = F(c)$; in particular, $D' = D''$ and $St_{D^{\times}}(\tilde{\alpha}_{\perp}) = D'^{\times} V_1$.

(i) Now let $\chi \in Irr(V_i \cap D' | \tilde{\alpha})$ with stabilizer S in $D'^{\times}$. It is easy to see that there is $d \in D'^{\times}$ such that $\chi(1+x) = \psi_F(Tr_{F'/F}(Trd_{D'}(dx))) = \psi_D(dx)$ for $x \in P^i \cap D'$. By (6.1), $\chi_{\perp}(1+x) = \psi_D(dx)$ for $x \in P^i$. Hence

$$\psi_D(cx) = \tilde{\alpha}_{\perp}(1+x) = \chi_{\perp}(1+x) = \psi_D(dx)$$

for $x \in P^k$. Therefore $d \equiv c \pmod{P^{-k-n+1}}$ and $c^{-1}d \in V_1 \cap D'$. Thus we may write $d = cv$ with $v \in V_1 \cap D'$.

Obviously $S \subseteq St_{D^\times}(\chi_\perp)$. Since $[V_i, V_{k+1-i}] \subseteq V_{k+1}$ we also have $V_{k+1-i} \subseteq St_{D^\times}(\chi_\perp)$. Suppose conversely that $t \in St_{D^\times}(\chi_\perp)$. Then $\psi_D(cvx) = \psi_D(t^{-1}cvtx)$ for $x \in P^i$, so $t^{-1}cvt \equiv cv(\mathrm{mod}\ P^{i+1-n})$. Since $c\mathcal{O} = P^{-k-n}$ we have $P^{-i+1-n} = c\,P^{k+1-i}$. Now (3.10) implies that $t \in D'^\times V_{k+1-i}$. Since $V_{k+1-i} \subseteq St_{D^\times}(\chi_\perp)$ we may thus assume that $t \in D'^\times$. It is then clear that $t \in S$.

(ii) For $d' \in D'^\times$ we clearly have $(\chi^{d'})_\perp = (\chi_\perp)^{d'}$. Hence $\chi \to \chi_\perp$ does indeed induce a map $\mathrm{Irr}(V_i \cap D'|\tilde{\alpha})/D'^\times \to \mathrm{Irr}(V_i|\tilde{\alpha}_\perp)/D^\times$. We prove that this map is injective. So let $\chi' \in \mathrm{Irr}(V_i \cap D'|\tilde{\alpha})$ such that $\chi'_\perp$ and $\chi_\perp$ are conjugate in $D^\times$, say $\chi'_\perp = (\chi_\perp)^z$ with $z \in D^\times$. We have to show that χ and χ' are conjugate under $D'^\times$. Restricting to V_k we have $(\tilde{\alpha}_\perp)^z = \tilde{\alpha}_\perp$. We have seen above that this implies $z \in D'^\times V_1$. Replacing χ by a conjugate under $D'^\times$ we may thus assume $z \in V_1$. Write $z = 1 + a + b$ with $a \in D' \cap P$, $b \in D'^\perp \cap P$. Then $z = (1+a)(1+(1+a)^{-1}b)$ with $(1+a)^{-1}b \in D'^\perp \cap P$. Replacing χ by a conjugate again we may even assume $z \in 1 + (D'^\perp \cap P)$.

Arguing as in the proof of (i) we see that there is $v' \in V_1 \cap D'$ such that $\chi'_\perp(1+x) = \psi_D(cv'x)$ for $x \in P^i$. Thus

$$\psi_D(z^{-1}cvzx) = \chi_\perp(z(1+x)z^{-1}) = \chi'_\perp(1+x) = \psi_D(cv'x)$$

for $x \in P^i$ which shows that $z^{-1}cvz \equiv cv'\ (\mathrm{mod}\ c\,P^{k+1-i})$. By (3.8), $z \in 1 + (P^{k+1-i} \cap D'^\perp) \subseteq V_{k+1-i} \subseteq St_{D^\times}(\chi_\perp)$. Hence $\chi'_\perp = (\chi_\perp)^z = \chi_\perp$ and $\chi = \chi'$. Thus injectivity is proved.

To prove surjectivity let $\gamma \in \mathrm{Irr}(V_i|\tilde{\alpha}_\perp)$. We have to show that γ is conjugate to $\eta_\perp$ for some $\eta \in \mathrm{Irr}(V_i \cap D'|\tilde{\alpha})$. There is $g \in D^\times$ such that $\gamma(1+x) = \psi_D(gx)$ for $x \in P^i$; in particular, $\psi_D(gx) = \psi_D(cx)$ for $x \in P^k$. Thus $g \equiv c\ (\mathrm{mod}\ P^{-k+1-n})$ and $c^{-1}g \equiv 1\ (\mathrm{mod}\ P)$. Hence $g = cu$ for some $u \in V_1$. Replacing γ by a conjugate we may assume by (3.9) that $u \in V_1 \cap D'$. Then obviously $\eta : V_i \cap D'/V_{k+1} \cap D' \to \mathbb{C}^\times$, defined by $\eta(1+x) := \psi_D(cux)$ for $x \in P^i \cap D'$, is a character in $\mathrm{Irr}(V_i \cap D'|\tilde{\alpha})$ such that $\eta_\perp = \gamma$. This yields surjectivity.

Now we set $i := (k+1)/2$ if k is odd, $i := k/2+1$ if k is even. Then i is as small as possible such that $1 \leq i \leq k+1 \leq 2i$. Moreover, we set $k' := j' \cdot n'$. It is clear that any $\Pi \in \mathrm{Irr}(D'^\times|\tilde{\alpha})$ has index $j(\Pi) = k'$. We extend any $\Pi \in \mathrm{Irr}\ D'^\times$ with $j(\Pi) = k'$ to $\Pi_\perp \in \mathrm{Irr}\ D'^\times V_i$ by setting

$$\Pi_\perp(dw) := \Pi(d) \text{ for } d \in D'^\times, \ w \in W_i V_{k+1};$$

this is well-defined by (6.1). Similarly, we extend any $\Pi' \in \operatorname{Irr} D'^\times$ with $j(\Pi') < k'$ to $\Pi'_\perp \in \operatorname{Irr} D'^\times V_{i-1}$ by setting

$$\Pi'_\perp(dw) := \Pi'(d) \text{ for } d \in D'^\times, \ w \in W_{i-1} V_k$$

if k is even. The map $I_{D'}^D$ will be defined by

$$I_{D'}^D(\Pi) := \operatorname{Ind}_{D'^\times V_{k+1-i}}^{D^\times} (\Omega(\Pi))$$

with some $\Omega(\Pi) \in \operatorname{Irr} D'^\times V_{k+1-i}$. Note that $k + 1 - i = i$ if k is odd, and $k + 1 - i = i - 1$ if k is even. Thus in case k is odd we simply define $\Omega(\Pi) = \Pi'_\perp$. The definition of $\Omega(\Pi)$ in the case where k is even is slightly more complicated and will make use of (4.5). For simplicity we set $V_1' := V_1 \cap D$.

(6.3) <u>Suppose that</u> k <u>is even, and let</u> $\beta : F'^\times \to \mathbb{C}^\times$ <u>be an extension of</u> α. <u>Set</u> $\tilde{\beta} := \beta \circ \operatorname{Nrd}_{D'} : D'^\times \to \mathbb{C}^\times$ <u>and extend</u> $\tilde{\beta}$ <u>to</u> $\tilde{\beta}_\perp : D'^\times V_i \to \mathbb{C}^\times$ <u>as above. Then</u> $V_1'V_i \leq V_1'V_{i-1}$, $V_1'V_{i-1}/V_1'V_i$ <u>is abelian,</u> $\tilde{\beta}_\perp/_{V_1'V_i}$ <u>is stable under</u> $V_1'V_{i-1}$, <u>and the radical of the associated symplectic bilinear form</u> $V_1'V_{i-1} \times V_1'V_{i-1} \to \mathbb{C}^\times$ <u>is</u> $V_1'V_i$.

Proof. Since $[V_1'V_{i-1}, V_1'V_{i-1}] \subseteq V_1'V_i$, $V_1'V_i$ is a normal subgroup of $V_1'V_{i-1}$ with abelian factor group. Let C_F, $C_{F'}$ be complementary subgroups of F, F', respectively, such that $C_F \subseteq C_{F'}$. By (3.3) there is a complementary subgroup C_D of D containing $C_{F'}$. As in the proof of (6.2) there is $c \in C_{F'}$ such that $F' = F(c)$, $c\mathcal{O} = P^{-k-n}$ and $\tilde{\alpha}_\perp(1+x) = \psi_D(cx)$ for $x \in P^k$; moreover, there is $v \in U_{F'}^1$ such that $\tilde{\beta}_\perp(1+x) = \psi_D(cvx)$ for $x \in P^i$.

In order to check that $\tilde{\beta}_\perp|_{V_1'V_i}$ is stable under $V_1'V_{i-1}$ we have to show that $\tilde{\beta}_\perp$ is trivial on $[V_1'V_i, V_1'V_{i-1}]$. Since $[V_i, V_{i-1}] \subseteq V_{k+1}$ it is enough to show that $\tilde{\beta}_\perp$ is trivial on $[V_1', V_{i-1}]$. For $y \in P$, $z \in P^{i-1}$ we have

$$[1 + y, 1 + z] \equiv 1 + (yz - zy) \pmod{P^{k+1}}.$$

Thus

$$\tilde{\beta}_\perp([1+y, 1+z]) = \psi_D(cv(yz-zy)) = \psi_D((cvy-ycv)z) = 1$$

for $y \in D' \cap P$, $z \in P^{i-1}$. This shows that $\tilde{\beta}_\perp|_{V_1'V_i}$ is stable in $V_1'V_{i-1}$. It remains to compute the radical of the symplectic bilinear form

$$V_1'V_{i-1} \times V_1'V_{i-1} \to \mathbb{C}^\times, \ (a,b) \to \tilde{\beta}_\perp([a,b]).$$

Suppose that $y \in P^{i-1}$ satisfies

$$1 = \tilde{\beta}_\perp([1+y, 1+z]) = \tilde{\beta}_\perp(1+(yz-zy)) = \psi_D(c(yz-zy)) = \psi_D((cy-yc)z)$$

for $z \in P^{i-1}$. Then $yc - cy \in P^{2-n-i} = cP^i$. By (3.10), $1 + y \in D'^\times V_i \cap V_{i-1} = (D'^\times \cap V_{i-1})V_i$. This shows that the symplectic form above has radical $V_1'V_i$.

Similarly as above, we can extend any $\Pi_1 \in \operatorname{Irr} V_1'$ with $j(\Pi_1) = k'$ to $\Pi_{1_\perp} \in \operatorname{Irr} V_1'V_i$ by setting $\Pi_{1_\perp}(vw) := \Pi_1(v)$ for $v \in V_1'$, $w \in W_iV_{k+1}$. Also we can extend any $\Pi_1' \in \operatorname{Irr} V_1'$ with $j(\Pi_1') < k'$ to $\Pi_{1_\perp}' \in \operatorname{Irr} V_1'V_{i-1}$ by setting $\Pi_{1_\perp}'(vw) := \Pi_1'(v)$ for $v \in V_1'$, $w \in W_{i-1}V_k$, if k is even. Now we apply (4.5).

(6.4) <u>Suppose that</u> k <u>is even, and let</u> $\Pi_1 \in \operatorname{Irr}(V_1'|\tilde{\alpha})$. <u>Then there is</u> $\Omega_1(\Pi_1) \in \operatorname{Irr}(V_1'V_{i-1}|\tilde{\alpha}_\perp)$ <u>such that</u>

$$\operatorname{Ind}_{V_1'V_i}^{V_1'V_{i-1}}(\Pi_{1_\perp}) = |V_1'V_{i-1} : V_1'V_i|^{1/2}\,\Omega_1(\Pi_1).$$

<u>Moreover, the following properties hold:</u>

(i) $\operatorname{St}_{D'^\times V_{i-1}}(\Omega_1(\Pi_1)) = \operatorname{St}_{D'^\times}(\Pi_1)\,V_{i-1}$.

(ii) <u>If</u> $\Pi_1 = \lambda_1 \otimes \Pi_1'$ <u>where</u> $\lambda_1 \in \operatorname{Irr}(V_1'|\tilde{\alpha})$ <u>with</u> $\dim \lambda_1 = 1$ <u>and where</u> $\Pi_1' \in \operatorname{Irr} V_1'$ <u>has index</u> $j(\Pi_1') < k'$ <u>then</u> $\Omega_1(\Pi_1) = \Omega_1(\lambda_1) \otimes \Pi_{1_\perp}'$.

Proof. Let $\beta : F'^\times \to \mathbb{C}^\times$ be an extension of α. Then $\tilde{\beta} : D'^\times \to \mathbb{C}^\times$ is an extension of $\tilde{\alpha}$, so $\Pi_1' := \tilde{\beta}^{-1}|_{V_1'} \otimes \Pi_1 \in \operatorname{Irr} V_1'$ has index $j(\Pi_1') < k'$. Hence $\Pi_{1_\perp}' \in \operatorname{Irr} V_1'V_{i-1}$ extends $\tilde{\beta}_\perp^{-1}|_{V_1'V_i} \otimes \Pi_{1_\perp}$. Since $\Pi_{1_\perp} = \tilde{\beta}_\perp|_{V_1'V_i} \otimes \Pi_{1_\perp}'|_{V_1'V_i}$, $\Pi_{1_\perp}$ and $\tilde{\beta}_\perp|_{V_1'V_i}$ have the same stabilizer and the same Mackey obstruction in $V_1'V_{i-1}$, and the radicals of their associated symplectic bilinear forms coincide. By (4.5) and (6.3), there is $\Omega_1(\Pi_1) \in \operatorname{Irr} V_1'V_{i-1}$ such that

$$\mathrm{Ind}_{V_1'V_i}^{V_1'V_{i-1}}(\Pi_{1_\perp}) = |V_1'V_{i-1} : V_1'V_i|^{1/2}\,\Omega_1(\Pi_1).$$

It is clear that $\Omega_1(\Pi_1) \in \mathrm{Irr}(V_1'V_{i-1}|\tilde{\alpha}_\perp)$.

(i) Obviously $S := \mathrm{St}_{D'^\times}(\Pi_1)$ fixes $\Pi_{1_\perp}$ and $\mathrm{Ind}_{V_1'V_i}^{V_1'V_{i-1}}(\Pi_{1_\perp})$. Thus S also fixes $\Omega_1(\Pi_1)$, and $SV_{i-1} \subseteq \mathrm{St}_{D'^\times V_{i-1}}(\Omega_1(\Pi_1)) =: T$. Conversely, let $t \in T$. Since $V_{i-1} \subseteq T$ we may assume $t \in D'^\times$. Then t fixes the unique irreducible constituent $\Pi_{1_\perp}$ of $\Omega_1(\Pi_1)|_{V_1'V_i}$. Thus $t \in S$.

(ii) Suppose now that $\Pi_1 = \lambda_1 \otimes \Pi_1'$ as above. Then $\Pi_{1_\perp}' \in \mathrm{Irr}\, V_1'V_{i-1}$ extends $\Pi_{1_\perp}'|_{V_1'V_i} = \lambda_{1_\perp}^{-1} \otimes \Pi_{1_\perp}$. Thus

$$|V_1'V_{i-1} : V_1'V_i|^{1/2}\,\Omega_1(\Pi_1) = \mathrm{Ind}_{V_1'V_i}^{V_1'V_{i-1}}(\lambda_{1_\perp} \otimes \Pi'_{1_\perp}|_{V_1'V_i}) =$$

$$= \mathrm{Ind}_{V_1'V_i}^{V_1'V_{i-1}}(\lambda_{1_\perp}) \otimes \Pi_{1_\perp}' = |V_1'V_{i-1} : V_1'V_i|^{1/2}\,\Omega_1(\Pi_1) \otimes \Pi_{1_\perp}',$$

and the result follows.

We construct $\Omega(\Pi)$ first for $\Pi \in \mathrm{Irr}(D'^\times|\tilde{\alpha})$ with $\dim \Pi = 1$.

(6.5) <u>If k is even then, for any extension $\beta : F'^\times \to \mathbb{C}^\times$ of α, there is a unique extension Ω_β of $\Omega_1(\tilde{\beta}|_{V_1'})$ to $D'^\times V_{i-1}$ satisfying $\omega_{\Omega_\beta} = \tilde{\beta}|_{F^\times}$ and</u> $\det \Omega_\beta|_{D'^\times} = \tilde{\beta}^{\dim\Omega_\beta}$.

Proof. Let C_F, $C_{F'}$, $C_{D'}$ be complementary subgroups of F, F', D', respectively, such that $C_F \subseteq C_{F'} \subseteq C_{D'}$. Let $\Pi_{D'}$ be a prime element of D' in $C_{D'}$. It is well-known that $\mathrm{Nrd}_{D'}(\Pi_{D'}) = (-1)^{n'-1}\,\Pi_{D'}^{n'}$. Since D/F is tame it follows easily that $\mathrm{Nrd}_{D'}(C_{D'}) \subseteq C_{F'}$.

Obviously $\tilde{\beta}|_{V_1'} \in \mathrm{Irr}(V_1'|\tilde{\alpha})$. Suppose first that β is trivial on $C_{F'}$. By (6.4), $\Omega_1(\tilde{\beta}|_{V_1'}) \in \mathrm{Irr}\, V_1'V_{i-1}$ is stable under $D'^\times V_{i-1}$, and $\dim\Omega_1(\tilde{\beta}|V_1')$ is a power of p. We denote by $\hat{\Omega}_1(\tilde{\beta}|_{V_1'})$ the unique extension of $\Omega_1(\tilde{\beta}|_{V_1'})$ to $C_F \times V_1'V_{i-1}$ which is trivial on C_F, and compute its Mackey obstruction in $D'^\times V_{i-1}$. Since $D'^\times V_{i-1}/C_F \times V_1'V_{i-1} \cong C_{D'}/C_F$ is a p'-group, its order is prime to p. On the other hand, $D'^\times V_{i-1}/C_F \cong$

$C_{D'}/C_F \ltimes V_1'V_{i-1}$ is a semidirect product. By (4.4), the order of the obstruction divides $\dim \hat{\Omega}_1(\tilde{\beta}|_{V_1'})$. Hence the Mackey obstruction is trivial, and $\hat{\Omega}_1(\tilde{\beta}|_{V_1'})$ extends to $\Omega_\beta \in \operatorname{Irr} D'^{\times}V_{i-1}$.

Choose $m \in \mathbb{Z}$ such that $m \cdot \dim \Omega_\beta \equiv -1 (\operatorname{mod} |C_{D'}/C_F|)$ and set $\lambda := (\det\Omega_\beta)|_{C_{D'}}$. Then $\lambda \otimes \Omega_\beta$ is another extension of $\hat{\Omega}_1(\tilde{\beta}|_{V_1'})$, and $\det(\lambda \otimes \Omega_\beta)|_{C_{D'}} = \lambda^{\dim \Omega_\beta}(\det\Omega_\beta)|_{C_{D'}}$ is trivial. Replacing Ω_β by $\lambda \otimes \Omega_\beta$ we may thus assume that $\det \Omega_\beta|_{C_{D'}}$ is trivial. Using $\operatorname{Nrd}_{D'}(C_{D'}) \subseteq C_{F'}$, it is now easy to verify that Ω_β satisfies $\omega_{\Omega_\beta} = \tilde{\beta}|_{F^\times}$ and $\det \Omega_\beta|_{D'^\times} = \tilde{\beta}^{\dim \Omega_\beta}$.

Now let $\beta : F'^{\times} \to \mathbb{C}^{\times}$ be an arbitrary extension of α, and decompose $\beta = \beta_0 \otimes \beta_1$ according to $F'^{\times} = C_{F'} \times U^1_{F'}$. We construct Ω_{β_1} as above and consider $\tilde{\beta}_0$ as a character of $C_{D'}$ and of $D'^{\times}V_{i-1} = C_{D'} \ltimes V_1'V_{i-1}$. Then we set $\Omega_\beta := \tilde{\beta}_0 \otimes \Omega_{\beta_1}$ and check again that $\omega_{\Omega_\beta} = \tilde{\beta}|_{F^\times}$ and $(\det\Omega_\beta)|_{D'^\times} = \tilde{\beta}^{\dim\Omega_\beta}$.

It remains to prove uniqueness of Ω_β. Let Ω'_β be another extension of $\Omega_1(\tilde{\beta}|_{V_1'})$ such that $\omega_{\Omega'_\beta} = \tilde{\beta}|_{F^\times}$ and $(\det\Omega'_\beta)|_{D'^\times} = \tilde{\beta}^{\dim\Omega_\beta}$. Then Ω_β and Ω'_β coincide on $C_F \times V_1'V_{i-1}$. By (4.3), $\Omega'_\beta = \mu \otimes \Omega_\beta$ for some character μ of $D'^{\times}V_{i-1}/C_F \times V_1'V_{i-1} \cong C_{D'}/C_F$. Since

$$(\det \Omega_\beta')|_{C_{D'}} = (\det \mu \otimes \Omega_\beta)|_{C_{D'}} = \mu^{\dim\Omega_\beta}(\det \Omega_\beta)|_{C_{D'}},$$

$\mu^{\dim\Omega_\beta}$ is trivial on $C_{D'}/C_F$. Since $\dim \Omega_\beta$ is a power of p and $C_{D'}/C_F$ is a p'-group, μ is trivial. Hence $\Omega'_\beta = \Omega_\beta$.

Now we define $\Omega(\Pi)$ for arbitrary $\Pi \in \operatorname{Irr}(D'^{\times}|\tilde{\alpha})$ in case k is even. Let $\beta : F'^{\times} \to \mathbb{C}^{\times}$ be an extension of α. Then $\Pi' := \tilde{\beta}^{-1} \otimes \Pi \in \operatorname{Irr} D'^{\times}$ has index $j(\Pi') < k'$, and $\Pi = \tilde{\beta} \otimes \Pi'$. We know that Π' extends to $\Pi'_\perp \in \operatorname{Irr} D'^{\times}V_{i-1}$. We set

$$\Omega(\Pi) := \Omega_\beta \otimes \Pi'_\perp.$$

(6.6) The definition of $\Omega(\Pi)$ is independent of the choice of β.

Proof. Let $\beta' : F'^{\times} \to \mathbb{C}^{\times}$ be another extension of α. Then $\beta'\beta^{-1} : F'^{\times} \to \mathbb{C}^{\times}$ has index $j(\beta'\beta^{-1}) < j'$, and $(\beta'\beta^{-1})^{\sim} : D'^{\times} \to \mathbb{C}^{\times}$ has index $j((\beta'\beta^{-1})^{\sim}) < k'$. Thus $(\beta'\beta^{-1})^{\sim}$ extends to $(\beta'\beta^{-1})^{\sim}_{\perp} : D'^{\times} V_{i-1} \to \mathbb{C}^{\times}$. We show that $\Omega_{\beta} \otimes (\beta'\beta^{-1})^{\sim}_{\perp} \in \operatorname{Irr} D'^{\times}V_{i-1}$ satisfies the defining properties of $\Omega_{\beta'}$. By (6.5) and (6.4),

$$\Omega_{\beta} \otimes (\beta'\beta^{-1})^{\sim}_{\perp}|_{V_1'V_{i-1}} = \Omega_1(\tilde{\beta}|_{V_1'}) \otimes (\beta'\beta^{-1})^{\sim}_{\perp}|_{V_1'V_{i-1}} =$$

$$\Omega_1(\tilde{\beta} \otimes (\beta'\beta^{-1})^{\sim}|_{V_1'})^{\sim} = \Omega_1(\tilde{\beta}'|_{V_1'}).$$

Moreover,

$$\omega_{\Omega_{\beta} \otimes (\beta'\beta^{-1})^{\sim}_{\perp}} = \omega_{\Omega_{\beta}} \otimes (\beta'\beta^{-1})^{\sim}_{\perp}|_{F^{\times}} = \tilde{\beta} \otimes (\beta'\beta^{-1})^{\sim}|_{F^{\times}} = \tilde{\beta}'|_{F^{\times}}$$

and

$$\det(\Omega_{\beta} \otimes (\beta'\beta^{-1})^{\sim}_{\perp}|_{D'^{\times}}) = (\det\Omega_{\beta}|_{D'^{\times}})(\beta'\beta^{-1})^{\sim\,\dim\Omega_{\beta}} =$$

$$(\tilde{\beta} \otimes (\beta'\beta^{-1})^{\sim})^{\dim\Omega_{\beta}} = \tilde{\beta}'^{\dim\Omega_{\beta}}.$$

Now (6.5) implies $\Omega_{\beta'} = \Omega_{\beta} \otimes (\beta'\beta^{-1})^{\sim}_{\perp}$. Thus

$$\Omega_{\beta'} \otimes (\tilde{\beta}'^{-1} \otimes \Pi)_{\perp} = \Omega_{\beta} \otimes (\beta'\beta^{-1})^{\sim}_{\perp} \otimes (\tilde{\beta}'^{-1} \otimes \Pi)_{\perp} = \Omega_{\beta} \otimes (\tilde{\beta}^{-1} \otimes \Pi)_{\perp}$$

which was to be proved.

We see in particular that $\Omega(\tilde{\beta}) = \Omega_{\beta}$. The following is the essential property of Ω.

(6.7) Let k be even, and let $\Pi_1 \in \operatorname{Irr}(V_1'|\tilde{\alpha})$. Then Ω induces a bijection $\operatorname{Irr}(D'^{\times}|\Pi_1) \xrightarrow{\sim} \operatorname{Irr}(D'^{\times}V_{i-1}|\Omega_1(\Pi_1))$.

Proof. We begin with some preliminary considerations. Let $\beta : F'^{\times} \to \mathbb{C}^{\times}$ be a fixed extension of α. Then $\Pi_1' := \tilde{\beta}^{-1}|_{V_1'} \otimes \Pi_1 \in \operatorname{Irr} V_1'$ with $j(\Pi_1') < k'$. By (6.4)

$$\Omega_1(\Pi_1) = \Omega_1(\tilde{\beta}|_{V_1'} \otimes \Pi_1') = \Omega_1(\tilde{\beta}|_{V_1'}) \otimes \Pi_{1\perp}'.$$

Obviously $St_{D'^\times}(\Pi_1') = St_{D'^\times}(\Pi_1) =: S$. By (6.4), $St_{D'^\times V_{i-1}}(\Omega_1(\Pi_1)) = S V_{i-1}$. Let C_F be a complementary subgroup of F. Then $C_F \times V_1' \subseteq S$. Extend Π_1' to $\hat{\Pi}_1' \in \operatorname{Irr} C_F \times V_1'$ such that $\hat{\Pi}_1'$ is trivial on C_F. We compute the Mackey obstruction of $\hat{\Pi}_1'$ in S. Since $S/C_F \times V_1'$ is a p'-group its order is prime to p. On the other hand the extension

$$1 \to V_1' \to S/C_F \to S/C_F \times V_1' \to 1$$

splits. By (4.4), the order of the obstruction divides $\dim\hat{\Pi}_1'$. Since V_1' is a pro-p group, $\dim\hat{\Pi}_1' = \dim\Pi_1'$ is a power of p. Thus the obstruction is trivial and $\hat{\Pi}_1'$ extends to $\tilde{\Pi}_1' \in \operatorname{Irr} S$. We extend $\tilde{\Pi}_1'$ further to $\tilde{\Pi}_{1\perp}' \in \operatorname{Irr} SV_{i-1}$ by setting $\tilde{\Pi}_{1\perp}'(sw) := \tilde{\Pi}_1'(s)$ for $s \in S$, $w \in W_{i-1} V_k$. By (6.5) and (6.4), $\Omega_\beta|_{SV_{i-1}} \otimes \tilde{\Pi}_{1\perp}'$ is an extension of $\Omega_1(\beta|_{V_1'}) \otimes \tilde{\Pi}_{1\perp}' = \Omega_1(\Pi_1) \in \operatorname{Irr} V_1' V_{i-1}$.

Now let $\Pi \in \operatorname{Irr}(D'^\times|\Pi_1) \subseteq \operatorname{Irr}(D'^\times|\tilde{\alpha})$. We show first that $\Omega(\Pi) \in \operatorname{Irr}(D'^\times V_{i-1}|\Omega_1(\Pi_1))$. Obviously $\Pi' := \tilde{\beta}^{-1} \otimes \Pi \in \operatorname{Irr}(D'^\times|\Pi_1')$. By (4.2), $\Pi' = \operatorname{Ind}_S^{D'^\times}(\Pi_0)$ for some $\Pi_0 \in \operatorname{Irr}(S|\Pi_1')$. By (4.3), $\Pi_0 = \tilde{\Pi}_1' \otimes \Sigma$ for some $\Sigma \in \operatorname{Irr} S/V_1'$. Hence $\Pi' = \operatorname{Ind}_S^{D'^\times}(\tilde{\Pi}_1' \otimes \Sigma)$ and $\Pi = \tilde{\beta} \otimes \operatorname{Ind}_S^{D'^\times}(\tilde{\Pi}_1' \otimes \Sigma)$.

We can regard Σ as a representation of $SV_{i-1}/V_1'V_{i-1} \cong S/V_1'$. Then obviously $\Pi_\perp' = \operatorname{Ind}_{SV_{i-1}}^{D'^\times V_{i-1}}(\tilde{\Pi}_1' \otimes \Sigma)$ and

$$\Omega(\Pi) = \Omega_\beta \otimes \operatorname{Ind}_{SV_{i-1}}^{D'^\times V_{i-1}}(\tilde{\Pi}_{1\perp}' \otimes \Sigma) = \operatorname{Ind}_{SV_{i-1}}^{D'^\times V_{i-1}}(\Omega_\beta|_{SV_{i-1\perp}} \otimes \tilde{\Pi}_{1\perp}' \otimes \Sigma);$$

here $\Omega_\beta|_{SV_{i-1}} \otimes \tilde{\Pi}_{1\perp}' \otimes \Sigma \in \operatorname{Irr}(SV_{i-1}|\Omega_1(\Pi_1))$ by (4.3) and $\Omega(\Pi) \in \operatorname{Irr}(D'^\times V_{i-1}|\Omega_1(\Pi_1))$ by (4.2). Thus Ω is well-defined.

We show next that Ω is injective. Let $\Gamma \in \operatorname{Irr}(D'^\times|\Pi_1)$ such that $\Omega(\Pi) = \Omega(\Gamma)$. Then $\Gamma' := \tilde{\beta}^{-1} \otimes \Gamma \in \operatorname{Irr} D'^\times$ with $j(\Gamma') < k'$. As above, $\Gamma' = \operatorname{Ind}_S^{D'^\times}(\tilde{\Pi}_1' \otimes \Lambda)$ for some $\Lambda \in \operatorname{Irr} S/V_1'$, and $\Omega(\Gamma) = \operatorname{Ind}_{SV_{i-1}}^{D'^\times V_{i-1}}(\Omega_\beta|_{SV_{i-1}} \otimes \tilde{\Pi}_{1\perp}' \otimes \Lambda)$ where $\Omega_\beta|_{SV_{i-1}} \otimes \tilde{\Pi}_{1\perp}' \otimes \Lambda \in \operatorname{Irr}(SV_{i-1}|\Omega_1(\Pi_1))$. Since $\Omega(\Pi) = \Omega(\Gamma)$, (4.2) and (4.3) imply that $\Sigma = \Lambda$. Hence $\Pi' = \Gamma'$ and $\Pi = \Gamma$.

To prove surjectivity of Ω let $\Phi \in \operatorname{Irr}(D'^\times V_{i-1}|\Omega_1(\Pi_1))$. By (4.2), $\Phi =$

$\mathrm{Ind}_{SV_{i-1}}^{D'^{\times}V_{i-1}}(\Phi_0)$ for some $\Phi_0 \in \mathrm{Irr}(SV_{i-1}|\Omega_1(\Pi_1))$. By (4.3), $\Phi_0 =$ $\Omega_\beta|_{SV_{i-1}} \otimes \tilde{\Pi}'_{1\perp} \otimes \Lambda$ for some $\Lambda \in \mathrm{Irr}\, SV_{i-1}/V'_1V_{i-1}$. We can regard Λ as a representation of $S/V'_1 \cong SV_{i-1}/V'_1V_{i-1}$. Then $\tilde{\Pi}'_1 \otimes \Lambda \in \mathrm{Irr}(S|\Pi'_1)$ by (4.3), and $\mathrm{Ind}_S^{D'^{\times}}(\tilde{\Pi}'_1 \otimes \Lambda) \in \mathrm{Irr}(D'^{\times}|\Pi'_1)$ by (4.2). Thus $\Pi :=$ $\tilde{\beta} \otimes \mathrm{Ind}_S^{D'}(\tilde{\Pi}'_1 \otimes \Lambda) \in \mathrm{Irr}(D'^{\times}|\Pi_1)$, and $\Omega(\Pi) = \Phi$. This finishes the proof of (6.7).

Let k be arbitrary again. We are now in a position to define the map $I_{D'}^{D}$. For $\Pi \in \mathrm{Irr}(D'^{\times}|\tilde{\alpha})$ we set

$$I_{D'}^{D}(\Pi) := \mathrm{Ind}_{D'^{\times}V_{k+1-i}}^{D^{\times}}(\Omega(\Pi)).$$

The following is the main result on $I_{D'}^{D}$.

(6.8) Proposition. <u>The map</u> $I_{D'}^{D}$ <u>defines a bijection</u> $\mathrm{Irr}(D'^{\times}|\tilde{\alpha}) \to \mathrm{Irr}(D^{\times}|\tilde{\alpha}_\perp)$.

Proof. Let $\Pi \in \mathrm{Irr}(D'^{\times}|\tilde{\alpha})$. We start by proving that $I_{D'}^{D}(\Pi) \in \mathrm{Irr}(D^{\times}|\tilde{\alpha}_\perp)$. Suppose first that k is odd, and choose a component $\chi \in \mathrm{Irr}(V_i \cap D'|\tilde{\alpha})$ of $\Pi|_{V_i \cap D'}$. Then $\chi_\perp \in \mathrm{Irr}(V_i|\alpha_\perp)$ is a component of $\Pi_\perp|_{V_i}$. By (6.2), $\mathrm{St}_{D^{\times}}(\chi_\perp) \subseteq D'^{\times}V_i$. Since $\Pi_\perp \in \mathrm{Irr}(D'^{\times}V_i|\chi_\perp)$, (4.2) implies

$$I_{D'}^{D}(\Pi) = \mathrm{Ind}_{D'^{\times}V_i}^{D^{\times}}(\Pi_\perp) \in \mathrm{Irr}(D^{\times}|\chi_\perp) \subseteq \mathrm{Irr}(D^{\times}|\tilde{\alpha}_\perp).$$

Suppose now that k is even. Choose a component $\Pi_1 \in \mathrm{Irr}(V'_1|\tilde{\alpha})$ of $\Pi|_{V'_1}$ and a component $\chi \in \mathrm{Irr}(V_i \cap D'|\tilde{\alpha})$ of $\Pi_1|_{V_i \cap D'}$. Then $\chi_\perp \in \mathrm{Irr}(V_i|\tilde{\alpha}_\perp)$ is a component of $\Pi_{1_\perp}|_{V_i}$, $\Pi_{1_\perp} \in \mathrm{Irr}(V'_1V_i|\tilde{\alpha}_\perp)$ is a component of $\Omega_1(\Pi_1)|_{V'_1V_i}$, and $\Omega_1(\Pi_1) \in \mathrm{Irr}(V'_1V_{i-1}|\tilde{\alpha}_\perp)$ is a component of $\Omega(\Pi)|_{V'_1V_{i-1}}$. Hence $\Omega(\Pi) \in \mathrm{Irr}(D'^{\times}V_{i-1}|\chi_\perp)$, and $\mathrm{St}_{D^{\times}}(\chi_\perp) \subseteq D'^{\times}V_{i-1}$ by (6.2). By (4.2),

$$I_{D'}^{D}(\Pi) = \mathrm{Ind}_{D'^{\times}V_{i-1}}^{D^{\times}}(\Omega(\Pi)) \in \mathrm{Irr}(D^{\times}|\chi_\perp) \subseteq \mathrm{Irr}(D^{\times}|\tilde{\alpha}_\perp).$$

To prove injectivity of $I_{D'}^{D}$ we suppose that we have $\Pi' \in \mathrm{Irr}(D'^{\times}|\tilde{\alpha})$

with $I^D_{D'}(\Pi) = I^D_{D'}(\Pi')$. Suppose first that k is odd, and choose a component $\chi' \in \mathrm{Irr}(V_i \cap D'|\tilde{\alpha})$ of $\Pi'|V_i$. Then $\chi_\perp$, $\chi'_\perp \in \mathrm{Irr}(V_i|\tilde{\alpha}_\perp)$ are components of $I^D_{D'}(\Pi)|_{V_i}$. By (4.1), $\chi_\perp$ and $\chi'_\perp$ are conjugate in $D^\times$. By (6.2), χ and χ' are conjugate in $D'^\times$. Hence we may assume $\chi = \chi'$ by (4.1). Now

$$\mathrm{Ind}^{D^\times}_{D'^\times V_i}(\Pi_\perp) = I^D_{D'}(\Pi) = I^D_{D'}(\Pi') = \mathrm{Ind}^{D^\times}_{D'^\times V_i}(\Pi'_\perp) \in \mathrm{Irr}(D^\times|\chi_\perp)$$

and $\mathrm{St}_{D^\times}(\chi_\perp) \subseteq D'^\times V_i$. Now (4.2) implies $\Pi_\perp = \Pi'_\perp$, and $\Pi = \Pi'$ follows.

Now consider the case where k is even. Choose a component $\Pi'_1 \in \mathrm{Irr}(V'_1|\tilde{\alpha})$ of $\Pi'|_{V'_1}$ and a component $\chi' \in \mathrm{Irr}(V_i \cap D'|\tilde{\alpha})$ of $\Pi'_1|_{V_i \cap D'}$. Then $\chi_\perp$, $\chi'_\perp \in \mathrm{Irr}(V_i|\tilde{\alpha}_\perp)$ are components of $I^D_{D'}(\Pi)|_{V_i}$. By (4.1), $\chi_\perp$ and $\chi'_\perp$ are conjugate in $D^\times$. By (6.2), χ and χ' are conjugate in $D'^\times$. Replacing Π_1 and χ by conjugates we may thus assume $\chi = \chi'$. Then $\mathrm{St}_{D^\times}(\chi_\perp) \subseteq D'^\times V_{i-1}$ and

$$\mathrm{Ind}^{D^\times}_{D'^\times V_{i-1}}(\Omega(\Pi)) = I^D_{D'}(\Pi) = I^D_{D'}(\Pi') = \mathrm{Ind}^{D^\times}_{D'^\times V_{i-1}}(\Omega(\Pi')) \in \mathrm{Irr}(D^\times|\chi_\perp).$$

By (4.2), $\Omega(\Pi) = \Omega(\Pi')$. By (6.7), $\Omega_1(\Pi_1)$, $\Omega_1(\Pi'_1) \in \mathrm{Irr}(V'_1V_{i-1}|\tilde{\alpha}_\perp)$ are components of $\Omega(\Pi)|_{V'_1V_{i-1}}$. By (4.1), $\Omega_1(\Pi_1)$ and $\Omega_1(\Pi'_1)$ are conjugate in $D'^\times V_{i-1}$. Hence $\Omega_1(\Pi'_1) = \Omega_1(\Pi_1)^d$ for some $d \in D'^\times$. Now $\chi_\perp$, $(\chi_\perp)^d \in \mathrm{Irr}\, V_i$ are components of $\Omega_1(\Pi'_1)|_{V_i}$. By (4.1), $\chi_\perp$ and $(\chi_\perp)^d$ are conjugate in V'_1V_{i-1}. By (6.2), $\chi_\perp = (\chi_\perp)^{dv}$ for some $v \in V'_1$. Hence $\chi_\perp$ is a component of $\Omega_1(\Pi'_1) = \Omega_1(\Pi_1)^{dv} = \Omega_1(\Pi_1^{dv})$. Replacing Π_1 by Π_1^{dv} we may thus assume $\Omega_1(\Pi'_1) = \Omega_1(\Pi_1)$. By (6.4), $\Pi_{1\perp}$ and $\Pi'_{1\perp}$ are the unique irreducible constituents of $\Omega_1(\Pi_1)|_{V'_1V_i}$ and $\Omega_1(\Pi'_1)|_{V'_1V_i}$, respectively. Therefore $\Pi_{1\perp} = \Pi'_{1\perp}$ and $\Pi_1 = \Pi'_1$. Now $\Omega(\Pi) = \Omega(\Pi') \in \mathrm{Irr}(D'^\times V_{i-1}|\Omega_1(\Pi_1))$, and (6.7) implies $\Pi = \Pi'$.

It remains to prove surjectivity of $I^D_{D'}$. Let $\Phi \in \mathrm{Irr}(D^\times|\tilde{\alpha}_\perp)$ and choose a component $\gamma \in \mathrm{Irr}(V_i|\tilde{\alpha}_\perp)$ of $\Phi|_{V_i}$. By (6.2) and (4.1) we may assume that $\gamma = \chi_\perp$ for some $\chi \in \mathrm{Irr}(V_i \cap D'|\tilde{\alpha})$. By (6.2), $\mathrm{St}_{D^\times}(\chi_\perp) \subseteq D'^\times V_{k+1-i}$. By (4.2), there is $\Phi_0 \in \mathrm{Irr}(D'^\times V_{k+1-i}|\chi_\perp)$ such that $\Phi = \mathrm{Ind}^{D^\times}_{D'^\times V_{k+1-i}}(\Phi_o)$.

Suppose first that k is odd. Since $\Phi_0 \in \mathrm{Irr}(D'^{\times}V_i|\chi_{\perp})$, Φ_0 is trivial on W_iV_{k+1}. By (6.1), $\Pi := \Phi_0|D'^{\times} \in \mathrm{Irr}(D'^{\times}|\chi) \subseteq \mathrm{Irr}(D'^{\times}|\tilde{\alpha})$ and $\Phi_0 = \Pi_{\perp}$. Hence $\Phi = \mathrm{Ind}_{D'^{\times}V_i}^{D^{\times}}(\Pi_{\perp}) = I_{D'}^{D}(\Pi)$.

Thus we may assume that k is even. In this case we choose a component $\Phi_{i-1} \in \mathrm{Irr}(V_1'V_{i-1}|\chi_{\perp})$ of $\Phi_0|_{V_1'V_{i-1}}$ and a component $\Phi_i \in \mathrm{Irr}(V_1'V_i|\chi_{\perp})$ of $\Phi_{i-1}|_{V_1'V_i}$. Then Φ_i is trivial on W_iV_{k+1}. By (6.1), $\Pi_1 := \Phi_i|_{V_1'} \in \mathrm{Irr}(V_1'|\chi) \subseteq \mathrm{Irr}(V_1'|\tilde{\alpha})$ and $\Phi_i = \Pi_{1\perp}$. Then Φ_{i-1} is a component of

$$\mathrm{Ind}_{V_1'V_i}^{V_1'V_{i-1}}(\Pi_{1\perp}) = |V_1'V_{i-1} : V_1'V_i|^{1/2}\, \Omega_1(\Pi_1)$$

which implies $\Phi_{i-1} = \Omega_1(\Pi_1)$. Now $\Phi_0 \in \mathrm{Irr}(D'^{\times}V_{i-1}|\Omega_1(\Pi_1))$. By (6.7), $\Phi_0 = \Omega(\Pi)$ for some $\Pi \in \mathrm{Irr}(D'^{\times}|\tilde{\alpha})$. Hence $\Phi = \mathrm{Ind}_{D'^{\times}V_{i-1}}^{D^{\times}}(\Omega(\Pi)) = I_{D'}^{D}(\Pi)$.

The following property of $I_{D'}^{D}$ will be used later.

(6.9) <u>If $F' \neq F$ then $I_{D'}^{D}(\tilde{\lambda}|_{D'^{\times}} \otimes \Pi) = \tilde{\lambda} \otimes I_{D'}^{D}(\Pi)$ for any $\Pi \in \mathrm{Irr}(D'^{\times}|\tilde{\alpha})$ and any character $\lambda : F^{\times} \to \mathbb{C}^{\times}$ satisfying $j(\lambda \circ N_{F'/F}) \leq j'$.</u>

Proof. Since α has no proper norm factorization, $\alpha' := (\lambda \circ N_{F'/F}|_{U_{F'}^{j'}})\cdot\alpha :$

$U_{F'}^{j'}/U_{F'}^{j'+1} \to \mathbb{C}^{\times}$ is a non-trivial character admitting no factorization through $N_{F'/L}$ for $F' \supsetneq L \supseteq F$. Since $\tilde{\lambda}|_{D'^{\times}} = \lambda \circ N_{F'/F} \circ \mathrm{Nrd}_{D'}$, $\tilde{\lambda}|_{D'^{\times}} \otimes \Pi \in \mathrm{Irr}(D'^{\times}|\tilde{\alpha}')$; in particular, $I_{D'}^{D}(\tilde{\lambda}|_{D'^{\times}} \otimes \Pi)$ is defined.

Let C_F, $C_{F'}$ be complementary subgroups of F, F', respectively, such that $C_F \subseteq C_{F'}$. By (3.3), there is a complementary subgroup C_D of D containing $C_{F'}$. In the proof of (6.1) we had found $c \in C_{F'}$ such that $F' = F(c)$. Now (3.6) implies that for $y \in D'^{\perp} \cap P^i$ there is $x \in D'^{\perp} \cap P^i$ such that $y = x - cxc^{-1}$. Then

$$(1 + x)\, c\, (1 + x)^{-1}\, c^{-1} \equiv 1 + x - cxc^{-1} = 1 + y \pmod{P^{k+1}}.$$

Hence $\tilde{\lambda}|_{W_i}$ is trivial, and $(\tilde{\lambda}|D'^{\times} \otimes \Pi)_{\perp} = \tilde{\lambda}|_{D'^{\times}V_i} \otimes \Pi_{\perp}$. If k is odd then

$$I_{D'}^{D}(\tilde{\lambda}|_{D'^{\times}} \otimes \Pi) = \mathrm{Ind}_{D'^{\times}V_i}^{D^{\times}}(\tilde{\lambda}|_{D'^{\times}V_i} \otimes \Pi_{\perp}) = \tilde{\lambda} \otimes \mathrm{Ind}_{D'^{\times}V_i}^{D^{\times}}(\Pi_{\perp}) =$$

$$= \tilde{\lambda} \otimes I_{D'}^{D}(\Pi).$$

Therefore we may assume that k is even. Let $\beta : F'^{\times} \to \mathbb{C}^{\times}$ be an extension of α. Then $\beta' := (\lambda \circ N_{F'/F})\beta : F'^{\times} \to \mathbb{C}^{\times}$ is an extension of α'; moreover, $\tilde{\beta}' = \tilde{\lambda}|_{D'^{\times}} \otimes \tilde{\beta}$. By (6.4),

$$|V_1'V_{i-1} : V_1'V_i|^{1/2}\, \Omega_1(\tilde{\beta}'|_{V_1'}) = \mathrm{Ind}_{V_1'V_i}^{V_1'V_{i-1}}(\tilde{\lambda}|_{V_1'V_i} \otimes \tilde{\beta}_{\perp}|_{V_1'V_i}) =$$

$$= \tilde{\lambda}|_{V_1'V_{i-1}} \otimes \mathrm{Ind}_{V_1'V_i}^{V_1'V_{i-1}}(\tilde{\beta}_{\perp}|_{V_1'}) =$$

$$= |V_1'V_{i-1} : V_1'V_i|^{1/2}\, \tilde{\lambda}|_{V_1'V_{i-1}} \otimes \Omega_1(\tilde{\beta}|_{V_1'}).$$

Hence $\Omega_1(\tilde{\beta}'|_{V_1'}) = \tilde{\lambda}|_{V_1'V_{i-1}} \otimes \Omega_1(\tilde{\beta}|_{V_1'})$. We now prove that $\tilde{\lambda}|_{D'^{\times}V_{i-1}} \otimes \Omega_{\beta}$ satisfies the defining properties of $\Omega_{\beta'}$; indeed,

$$(\tilde{\lambda}|_{D'^{\times}V_{i-1}} \otimes \Omega_{\beta})|_{V_1'V_{i-1}} = \tilde{\lambda}|_{V_1'V_{i-1}} \otimes \Omega_1(\tilde{\beta}|_{V_1'}) = \Omega_1(\tilde{\beta}'|_{V_1'}).$$

Similarly,

$$\omega_{\tilde{\lambda}|_{D'^{\times}V_{i-1}} \otimes \Omega_{\beta}} = \tilde{\lambda}|_{F^{\times}} \otimes \omega_{\Omega_{\beta}} = \tilde{\lambda}|_{F^{\times}} \otimes \tilde{\beta}|_{F^{\times}} = \tilde{\beta}'|_{F^{\times}}.$$

and

$$\det(\tilde{\lambda}|_{D'^{\times}V_{i-1}} \otimes \Omega_{\beta})|_{D'^{\times}} = \tilde{\lambda}^{\dim\Omega_{\beta}} \det \Omega_{\beta}|_{D'^{\times}} = (\tilde{\lambda}|_{D'^{\times}} \otimes \tilde{\beta})^{\dim\Omega_{\beta}} =$$

$$= \tilde{\beta}'^{\dim\Omega_{\beta}}.$$

By (6.5), $\Omega_{\beta'} = \tilde{\lambda}|_{D'^{\times}V_{i-1}} \otimes \Omega_{\beta}$. Now write $\Pi = \tilde{\beta} \otimes \Pi'$ where $\Pi' \in \mathrm{Irr}\, D'^{\times}$ with $j(\Pi') < k'$. Since $\tilde{\lambda}|_{D'^{\times}} \otimes \Pi = \tilde{\beta}' \otimes \Pi'$ we obtain

$$\Omega(\tilde{\lambda}|_{D'^{\times}} \otimes \Pi) = \Omega_{\beta'} \otimes \Pi'_{\perp} = \tilde{\lambda}|_{D'^{\times}V_{i-1}} \otimes \Omega_{\beta} \otimes \Pi'_{\perp} = \tilde{\lambda}|_{D'^{\times}V_{i-1}} \otimes \Omega(\Pi).$$

Hence

$$I_{D'}^{D}(\tilde{\lambda}|_{D'^{\times}} \otimes \Pi) = \mathrm{Ind}_{D'^{\times}V_{i-1}}^{D^{\times}}(\tilde{\lambda}|_{D'^{\times}V_{i-1}} \otimes \Omega(\Pi)) =$$

$$= \tilde{\lambda} \otimes \mathrm{Ind}_{D'^{\times}V_{i-1}}^{D^{\times}}(\Omega(\Pi)) = \tilde{\lambda} \otimes I_{D'}^{D}(\Pi).$$

7. Representations of positive index.

Let F be a local field of residue characteristic p, and let D/F be a tame central division algebra of dimension $[D : F] = n^2$. In this section we prove that any $\Pi \in \mathrm{Irr}D^\times$ with $j(\Pi) > 0$ can be written in the form $\Pi = I_{D'}^{D}(\Pi')$ with $\Pi' \in \mathrm{Irr}D'^\times$ for a suitable division subalgebra D' of D.

(7.1) Proposition. *Let $\gamma : V_k/V_{k+1} \to \mathbb{C}^\times$ be a non-trivial character for some $k > 0$.*
Then there are a field extension F'/F with $F' \subseteq D$ and a non-trivial character $\alpha : V_k \cap F' \to \mathbb{C}^\times$ admitting no factorization through $N_{F'/L}$ for $F' \supsetneq L \supseteq F$ such that $\gamma = \tilde{\alpha}_{\perp}$. Moreover, the pair (F',α) is unique up to conjugation with elements in V_1, and $e(F'/F) = n/\gcd(k,n)$.

Proof. Let C_F, C_D be complementary subgroups of F,D, respectively, such that $C_F \subseteq C_D$. There is $c \in P^{-k-n} \setminus P^{-k-n+1}$ such that $\gamma(1+x) = \psi_D(cx)$ for $x \in P^k$; we may assume that $c \in C_D$. Set $F' := F(c)$, $e' := e(F'/F)$ and $e := n/\gcd(k,n)$. Since $cO = P^{-k-n}$, $F(c^e)/F$ is unramified. Hence

$$e' = e(F(c)/F(c^e)) \leq [F(c) : F(c^e)] \leq e.$$

On the other hand, writing $cO_{F'} = P_{F'}^{\ell}$, we have

$$P^{(-k-n)e'} = (cO)^{e'} = (P_{F'}^{\ell}O)^{e'} = P^{e(D/F')\ell e'} = P^{n\ell}.$$

Thus $n \mid e'(-k-n)$ and $e \mid e'$. We conclude that $e(F'/F) = n/\gcd(k,n) = [F(c) : F(c^e)]$, $\ell = (-k-n)/\gcd(k,n)$, $V_k \cap F' = U_{F'}^{j'}$ and $V_{k+1} \cap F' = U_{F'}^{j'+1}$ where $j' = k/\gcd(k,n)$. Now define $\alpha : V_k \cap F' \to \mathbb{C}^\times$ by $\alpha(1+x) := \psi_F(\mathrm{Tr}_{F'/F}(cx))$ for $x \in P^k \cap F'$. Since $c(P^{k+1} \cap F') = P_{F'}^{1-e}$, α is trivial on $V_{k+1} \cap F'$.

Let $F' \supsetneq L \supseteq F$ be an intermediate field, and let $\beta : V_k \cap L \to \mathbb{C}^\times$ be a character such that $\alpha = \beta \circ N_{F'/L}|_{V_k \cap F'}$. By (2.1), there is a unique complementary subgroup C_L of L containing C_F, and there is a unique complementary subgroup $C_{F'}$ of F' containing C_L. It is not difficult to see that $c \in C_{F'}$. There is $d \in L^\times$ such that $\beta(1+x) = \psi_F(\mathrm{Tr}_{L/F}(dx))$ for $x \in P^k \cap L$; we may assume that $d \in C_L$. Since β is trivial on $V_{k+1} \cap L$,

$$\psi_F(\mathrm{Tr}_{F'/F}(cx)) = \alpha(1+x) = \beta(N_{F'/L}(1+x)) = \beta(1+\mathrm{Tr}_{F'/L}(x)) =$$

$$\psi_F(\mathrm{Tr}_{F'/F}(dx))$$

for $x \in P^k \cap F'$. Hence $c \equiv d(\mathrm{mod}\ P_{F'}^{1-j'-e})$ and $c^{-1}d \in C_{F'} \cap U_{F'}^1 = 1$. We conclude that $c = d \in L$ and $F' = F(c) \subseteq L$. Thus α has no proper norm factorization.

We have $\gamma(1+x) = \psi_D(cx) = 1 = \tilde{\alpha}_\perp(x)$ for $x \in P^k \cap D'^\perp$ and

$$\gamma(1+x) = \psi_F(\mathrm{Tr}_{F'/F}(\mathrm{Trd}_{D'}(cx))) = \alpha(1+\mathrm{Trd}_{D'}(x)) = \tilde{\alpha}(1+x)$$

for $x \in P^k \cap D'$. Hence $\tilde{\alpha}$ and α are non-trivial, and $\gamma = \tilde{\alpha}_\perp$.

It remains to show that the pair (F',α) is unique up to conjugation. Let (K,α') be another such pair, and denote the centralizer of K in D by D''. By (6.2),

$$D'^\times V_1 = St_{D^\times}(\tilde{\alpha}_\perp) = St_{D^\times}(\gamma) = St_{D^\times}(\tilde{\alpha}'_\perp) = D''^\times V_1.$$

By (3.4), F' and K are conjugate by an element in V_1. Hence we may assume that $F' = K$. Then $\alpha \circ \mathrm{Nrd}_{D'}|_{V_k \cap D'} = \gamma|_{V_k \cap D'} = \alpha' \circ \mathrm{Nrd}_{D'}|_{V_k \cap D'}$, and $\alpha = \alpha'$ follows.

We call (F',α) a fundamental pair of γ. By (7.1) it is unique up to conjugation with elements in V_1. Now let $\Pi \in \mathrm{Irr}\ D^\times$ with $j(\Pi) = k > 0$. Then we call any fundamental pair of any irreducible constituent $\gamma : V_k/V_{k+1} \to \mathbb{C}^\times$ of $\Pi|_{V_k}$ a fundamental pair of Π. By (4.1), the fundamental pairs of Π are unique up to conjugacy in $D^\times$. The following fact is now immediate from (7.1) and (6.8).

(7.2) <u>Let</u> $\Pi \in \mathrm{Irr}\ D^\times$ <u>with</u> $j(\Pi) > 0$, <u>let</u> (F',α) <u>be a fundamental pair of</u> Π, <u>and denote the centralizer of</u> F' <u>in</u> D <u>by</u> D'. <u>Then</u> $\Pi = I_{D'}^{D}(\Pi')$ <u>for some</u> $\Pi' \in \mathrm{Irr}(D'^\times|\tilde{\alpha})$.

8. The parametrization of Irr $D^\times$

Let F be a local field of resdidue characteristic p, and let D/F be a tame central division algebra, $[D : F] = n^2$. In this section we are going to define a map

$$AP_D \to \mathrm{Irr}\ D^\times,\ (K/F,\chi) \mapsto \Pi_D(K/F,\chi),$$

satisfying the following properties:

(8.1.1) $j(\Pi_D(K/F,\chi)) = j(\chi)e(D/K)$ _for_ $(K/F,\chi) \in AP_D$;

(8.1.2) $\Pi_D(K/F, (\chi_F \circ N_{K/F})\chi) = \tilde{\chi}_F \otimes \Pi_D(K/F,\chi)$
for any $(K/F,\chi) \in AP_D$ _and any character_ $\chi_F : F^\times \to \mathbb{C}^\times$;

(8.1.3) _The fundamental pair of_ $(K/F,\chi)$ _is a fundamental pair of_ $\Pi_D(K/F,\chi)$, _for any_ $(K/F,\chi) \in AP_D$ _with_ $j(\chi) > 0$.

In section 5 we had defined $\Pi_D(K/F,\chi) \in \mathrm{Irr}\, D^\times$ for $(K/F,\chi) \in AP_D$ with $j(\chi) = 0$ and verified (8.1.1) and (8.1.2) in this case; note that (8.1.3) is vacuous if $j(\chi) = 0$. The general definition of $\Pi_D(K/F,\chi)$ proceeds by induction on $[D : F] + j(\chi)$. In the induction process we shall make use of the fact that, for any tame central division algebra D' over a local field F' and any $(K'/F',\chi') \in AP_{D'}$, with $[D' : F'] + j(\chi') < [D : F] + j(\chi)$, the representation $\Pi_{D'}(K'/F',\chi') \in \mathrm{Irr}\, D'^\times$ is already defined and that the following properties hold:

(8.2.1) $j(\Pi_{D'}(K'/F',\chi')) = j(\chi')e(D'/K')$;

(8.2.2) $\Pi_{D'}(K'/F',(\chi'_{F'} \circ N_{K'/F'})\chi') = \tilde{\chi}'_{F'} \otimes \Pi_{D'}(K'/F',\chi')$
for any character $\chi'_{F'} : F'^\times \to \mathbb{C}^\times$ with $e(K'/F')j(\chi'_{F'}) \leq j(\chi')$;

(8.2.3) if $j(\chi') > 0$ then the fundamental pair of $(K'/F',\chi')$ is a fundamental pair of $\Pi_{D'}(K'/F',\chi')$.

In order to define $\Pi_D(K/F,\chi)$ in case $j = j(\chi) > 0$ we denote by (F',α) the fundamental pair of $(K/F,\chi)$ and distinguish between the cases $F' = F$ and $F' \neq F$. Suppose first that $F' = F$ so that $\chi|_{U_K^j} = \alpha \circ N_{K/F}|_{U_K^j}$, and choose an extension $\beta : F^\times \to \mathbb{C}^\times$ of α. By (2.5), $(K/F,(\beta^{-1} \circ N_{K/F})\chi) \in AP_D$, and $j((\beta^{-1} \circ N_{K/F})\chi) < j(\chi)$. So by induction $\Pi_D(K/F, (\beta^{-1} \circ N_{K/F})\chi) \in \mathrm{Irr}\, D^\times$ is already defined. We set

$$\Pi_D(K/F,\chi) := \tilde{\beta} \otimes \Pi_D(K/F, (\beta^{-1} \circ N_{K/F})\chi) \in \mathrm{Irr}\, D^\times.$$

(8.3) _The definition of_ $\Pi_D(K/F,\chi)$ _does not depend on the choice of_ β.

Proof. Let $\beta' : F^\times \to \mathbb{C}^\times$ be another extension of α. We have to show that

$$\tilde{\beta} \otimes \Pi_D(K/F, (\beta^{-1} \circ N_{K/F})\chi) = \tilde{\beta}' \otimes \Pi_D(K/F, (\beta'^{-1} \circ N_{K/F})\chi).$$

By symmetry we may assume that $j((\beta'^{-1} \circ N_{K/F})\chi) \leq j((\beta^{-1} \circ N_{K/F})\chi)$. Then

$$e(K/F)j(\beta'\beta^{-1}) = j((\beta'\beta^{-1}) \circ N_{K/F}) \leq j((\beta^{-1} \circ N_{K/F})\chi) < j(\chi).$$

By (8.2.2),

$$\Pi_D(K/F, (\beta'^{-1} \circ N_{K/F})\chi) = (\beta'^{-1}\beta)^{\sim} \otimes \Pi_D(K/F, (\beta^{-1} \circ N_{K/F})\chi),$$

from which our claim follows.

Now we check that $\Pi_D(K/F,\chi)$ satisfies properties analogous to (8.2).

(8.4.1) $j(\Pi_D(K/F,\chi)) = j(\chi)e(D/K)$.

__Proof.__ By (8.2.1)

$$j(\Pi_D(K/F,(\beta^{-1} \circ N_{K/F})\chi) = j((\beta^{-1} \circ N_{K/F})\chi)e(D/K) < j(\chi)e(D/K)$$

$$= nj(\alpha) = nj(\beta) = j(\tilde{\beta}).$$

Thus $j(\Pi_D(K/F,\chi)) = j(\tilde{\beta}) = j(\chi)e(D/K)$.

(8.4.2) $\Pi_D(K/F,(\chi_F \circ N_{K/F})\chi) = \tilde{\chi}_F \otimes \Pi_D(K/F,\chi)$
__for any character__ $\chi_F : F^\times \to \mathbb{C}^\times$ __with__ $e(K/F)j(\chi_F) \leq j(\chi)$.

__Proof.__ Since $j(\chi_F \circ N_{K/F}) = e(K/F)j(\chi_F) \leq j(\chi)$ we conclude that $j((\chi_F \circ N_{K/F})\chi \leq j(\chi)$. Suppose first that $j((\chi_F \circ N_{K/F})\chi) < j(\chi)$. Then $\chi_F^{-1} : F^\times \to \mathbb{C}^\times$ is an extension of α. By (8.3), $\Pi_D(K/F,\chi) = \tilde{\chi}_F^{-1} \otimes \Pi_D(K/F,(\chi_F \circ N_{K/F})\chi)$, and the result is true in this case. Suppose now that $j((\chi_F \circ N_{K/F})\chi) = j(\chi)$. Then $(K/F,(\chi_F \circ N_{K/F})\chi) \in AP_D$ has fundamental pair $(F,\alpha\chi_F|_{U_K^j \cap F})$. Since $\beta\chi_F$ is an extension of $\alpha\chi_F|_{U_K^j \cap F}$, (8.3) implies that

$$\Pi_D(K/F,(\chi_F \circ N_{K/F})\chi) = (\beta\chi_F)^{\sim} \otimes \Pi_D(K/F, (\beta^{-1} \circ N_{K/F})\chi) =$$

$$\tilde{\chi}_F \otimes \Pi_D(K/F,\chi).$$

(8.4.3) If $j(\chi) > 0$ then (F,α) is a fundamental pair of $\Pi_D(K/F,\chi)$.

Proof. We have seen in the proof of (8.4.1) that

$$j(\Pi_D(K/F,\chi)) = j(\tilde{\beta}) > j(\Pi_D(K/F,(\beta^{-1} \circ N_{K/F})\chi)).$$

Therefore $\Pi_D(K/F,\chi)$ and $\tilde{\beta}$ have the same fundamental pairs, and obviously (F,α) is the unique fundamental pair of $\tilde{\beta}$.

Now we define $\Pi_D(K/F,\chi)$ in the remaining case $F' \neq F$. By (2.4), $e(K/F') = \gcd(e(K/F),j)$, and, setting $j' := j/e(K/F')$, $\alpha : U_K^j \cap F' = U_{F'}^{j'} \to \mathbb{C}^\times$ is a non-trivial character admitting no factorization through $N_{F'/L}$ for $F' \supsetneq L \supseteq F$. Denote the centralizer of F' in D by D'. Then obviously $(K/F',\chi) \in AP_{D'}$ with fundamental pair (F',α). By induction, $\Pi_{D'}(K/F',\chi) \in \mathrm{Irr}\, D'^\times$ is already defined. By (8.2.3), (F',α) is a fundamental pair of $\Pi_{D'}(K/F',\chi)$. By (7.1) it is in fact the only one, and $\Pi_{D'}(K/F',\chi) \in \mathrm{Irr}(D'^\times/\tilde{\alpha})$. Thus by (6.8) we can define

$$\Pi_D(K/F,\chi) := I_{D'}^{D}(\Pi_{D'}(K|F',\chi)) \in \mathrm{Irr}\,(D^\times|\tilde{\alpha}_\perp).$$

Then $j(\Pi_D(K/F,\chi)) = j'e(D/F') = je(D/K)$, and (F',α) is a fundamental pair of $\Pi_D(K/F,\chi)$ by (7.1). We now check that $\Pi_D(K/F,\chi)$ satisfies the remaining property (8.2.2).

(8.5) $\Pi_D(K/F, (\chi_F \circ N_{K/F})\chi) = \tilde{\chi}_F \otimes \Pi_D(K/F,\chi)$
<u>for any character</u> $\chi_F : F^\times \to \mathbb{C}^\times$ <u>with</u> $e(K/F)j(\chi_F) \leq j(\chi)$.

Proof. Since $j(\chi_F \circ N_{K/F}) = e(K/F)j(\chi_F) \leq j(\chi)$ we know that $j((\chi_F \circ N_{K/F})\chi) \leq j(\chi) = j$. Assume that $j((\chi_F \circ N_{K/F})\chi) < j(\chi)$. Then $\chi|_{U_K^j} = \chi_F^{-1} \circ N_{K/F}$, and $(F,\chi_F^{-1}|_{U_K^j \cap F})$ would be the fundamental pair of $(K/F,\chi)$, a contradiction since $F' \neq F$. Thus $j((\chi_F \circ N_{K/F})\chi) = j$. Let (L,β) be the fundamental pair of $(K/F, (\chi_F \circ N_{K/F})\chi) \in AP_D$. Since

$$\beta \circ N_{K/L}|_{U_K^j} = (\chi_F \circ N_{K/F})\chi|_{U_K^j} = ((\chi_F \circ N_{F'/F})\alpha) \circ N_{K/F'}|_{U_K^j}$$

we obtain $L \subseteq F'$ and $(\chi_F \circ N_{F'/F})\alpha|_{U_K^j \cap F'} = \beta \circ N_{F'/L}|_{U_K^j \cap F'}$. Since α has no proper norm factorization we conclude that $F' = L$. Therefore $(F',(\chi_F \circ N_{F'/F})\alpha)$ is the fundamental pair of $(K/F,(\chi_F \circ N_{K/F})\chi) \in AP_D$. By definition,

$$\Pi_D(K/F,(\chi_F \circ N_{K/F})\chi) = I_{D'}^{D}(\Pi_{D'}(K/F',(\chi_F \circ N_{K/F})\chi)),$$

and by (8.2.2),

$$\Pi_{D'}(K/F',(\chi_F \circ N_{K/F})\chi) = (\chi_F \circ N_{F'/F} \circ \mathrm{Nrd}_{D'}) \otimes \Pi_{D'}(K/F',\chi) =$$

$$= \tilde{\chi}_F|_{D'^\times} \otimes \Pi_{D'}(K/F',\chi).$$

Now by (6.9),

$$I_{D'}^{D}(\tilde{\chi}_F|_{D'^\times} \otimes \Pi_{D'}(K/F',\chi)) = \tilde{\chi}_F \otimes I_{D'}^{D}(\Pi_{D'}(K/F',\chi)) =$$

$$= \tilde{\chi}_F \otimes \Pi_D(K/F,\chi),$$

and the result is proved.

Thus we have now extended the map $(K/F,\chi) \mapsto \Pi_D(K/F,\chi)$ to all of AP_D. Moreover, it follows immediately from (5.1), (5.3), (8.4) and (8.5) that

(8.6) <u>The assignment</u>

$$AP_D \to \mathrm{Irr}\ D^\times,\ (K/F,\chi) \mapsto \Pi_D(K/F,\chi),$$

<u>satisfies the properties listed in (8.1).</u>

We now prove the main result of this paper.

(8.7) Theorem. <u>The map</u>

$$AP_D \to \mathrm{Irr}\ D^\times,\ (K/F,\chi) \mapsto \Pi_D(K/F,\chi),$$

<u>induces a bijection</u> $AP_D/D^\times \xrightarrow{\sim} \mathrm{Irr}\ D^\times$.

Proof. It is routine to verify that conjugate admissible pairs lead to equivalent representations. Thus we have indeed a map $AP_D/D^\times \to \mathrm{Irr}\ D^\times$. We now prove injectivity. Let $(K/F,\chi)$, $(K'/F,\chi') \in AP_D$ with $\Pi_D(K/F,\chi) = \Pi_D(K'/F,\chi')$. We argue by induction on $[D : F] + j(\chi)$. If $j(\chi) = 0$ then

$$0 = j(\Pi_D(K/F,\chi)) = j(\Pi_D(K'/F,\chi')) = j(\chi')e(D/K')$$

by (8.1.1); hence $j(\chi') = 0$. By (5.1), $(K/F,\chi)$ and $(K'/F,\chi')$ are conjugate in $D^\times$. Thus we may assume that $j(\chi) > 0$ and $j(\chi') > 0$. Denote the fundamental pairs of $(K/F,\chi)$ and $(K'/F,\chi')$ by (F',α) and (F'',α'), respectively. By (8.1.3), (F',α) and (F'',α') are fundamental pairs of $\Pi_D(K/F,\chi) = \Pi_D(K'/F,\chi')$. By (7.1), (F',α) and (F'',α') are conjugate in $D^\times$. Replacing $(K/F,\chi)$ by a conjugate we may thus assume that $(F',\alpha) = (F'',\alpha')$. We distinguish between two cases.

Suppose first that $F' = F$. In this case extend α to a character $\beta : F^\times \to \mathbb{C}^\times$. Then $(K/F,(\beta^{-1} \circ N_{K/F})\chi)$, $(K'/F,(\beta^{-1} \circ N_{K'/F})\chi') \in AP_D$ with $j((\beta^{-1} \circ N_{K/F})\chi) < j(\chi)$, and

$$\Pi_D(K/F,(\beta^{-1} \circ N_{K/F})\chi) = \tilde{\beta}^{-1} \otimes \Pi_D(K/F,\chi) = \tilde{\beta}^{-1} \otimes \Pi_D(K'/F,\chi')$$

$$= \Pi_D(K'/F,(\beta^{-1} \circ N_{K'/F})\chi').$$

By induction, $(K/F,(\beta^{-1} \circ N_{K/F})\chi)$ and $(K'/F,(\beta^{-1} \circ N_{K'/F})\chi')$ are conjugate in $D^\times$; hence $(K/F,\chi)$ and $(K'/F,\chi')$ are conjugate by the same element.

We are left with the case $F' \neq F$. Denoting the centralizer of F' in D by D' we have

$$I_{D'}^{D}(\Pi_{D'}(K/F',\chi)) = \Pi_D(K/F,\chi) = \Pi_D(K'/F,\chi') = I_{D'}^{D}(\Pi_{D'}(K'/F',\chi')).$$

Now (6.8) implies that $\Pi_{D'}(K/F',\chi) = \Pi_{D'}(K'/F',\chi')$. By induction $(K/F',\chi)$, $(K'/F',\chi') \in AP_{D'}$ are conjugate in $D'^\times$. Then $(K/F,\chi)$ and $(K'/F,\chi')$ are conjugate by the same element. This completes the proof of injectivity.

To prove surjectivity we let $\Pi \in \mathrm{Irr}\, D^\times$ be arbitrary. We argue by induction on $[D : F] + j(\Pi)$. By (5.1) we may assume that $j(\Pi) > 0$. Let (F',α) be a fundamental pair of Π. Suppose first that $F' = F$. In this case let $\beta : F^\times \to \mathbb{C}^\times$ be an extension of α. Then $\tilde{\beta}^{-1} \otimes \Pi \in \mathrm{Irr}\, D^\times$ with $j(\tilde{\beta}^{-1} \otimes \Pi) < j(\Pi)$. By induction there is $(K/F,\chi) \in AP_D$ with $\tilde{\beta}^{-1} \otimes \Pi = \Pi_D(K/F,\chi)$. Then $(K/F,(\beta \circ N_{K/F})\chi) \in AP_D$ with

$$\Pi = \tilde{\beta} \otimes \Pi_D(K/F,\chi) = \Pi_D(K/F,(\beta \circ N_{K/F})\chi),$$

so we are done in this case.

Suppose now that $F' \neq F$, and denote the centralizer of F' in D by D'. By (7.2) there is $\Pi' \in \mathrm{Irr}(D'^{\times} | \tilde{\alpha})$ such that $\Pi = I_{D'}^{D}(\Pi')$. It is clear that (F', α) is the unique fundamental pair of Π'. Since $j(\Pi') \leq j(\Pi)$ and $[D' : F'] < [D : F]$ we obtain $(K/F', \chi) \in AP_{D'}$ with $\Pi' = \Pi_{D'}(K/F', \chi)$ by induction. By (8.1), (F', α) is the fundamental pair of $(K/F', \chi)$. Since α admits no factorization through $N_{F'/L}$ for $F \subseteq L \subsetneq F'$, (2.7) implies that $(K/F, \chi) \in AP_D$ with fundamental pair (F', α). By definition,

$$\Pi_D(K/F, \chi) = I_{D'}^{D}(\Pi_{D'}(K/F', \chi)) = I_{D'}^{D}(\Pi') = \Pi,$$

and the result is proved.

For completeness we list some further properties of the map

$$AP_D \to \mathrm{Irr}\, D^{\times}, \ (K/F, \chi) \mapsto \Pi_D(K/F, \chi).$$

(8.8.1) $\Pi_D(F/F, \chi) = \tilde{\chi} = \chi \circ \mathrm{Nrd}_D$ for $(F/F, \chi) \in AP_D$;

(8.8.2) $\Pi_D(K/F, \chi^{-1}) = \Pi_D(K/F, \chi)^{\vee}$ <u>where</u> $\check{\Pi}$ <u>denotes the contragredient (dual) representation of</u> Π;

(8.8.3) $\omega_{\Pi_D(K/F, \chi)} = \chi^{n/[K:F]} |_{F^{\times}}$;

(8.8.4) $\mathrm{Im}\, \Pi_D(K/F, \chi)$ <u>is finite if and only if</u> $\mathrm{Im}\, \chi$ <u>is finite.</u>

References

[11] C.J. Bushnell and A. Fröhlich, Gauss Sums and p-adic Division Algebras, Springer LNM 987, Berlin 1983.

[20] L. Corwin and R. Howe, Computing characters of tamely ramified p-adic division algebras, Pacific J. Math. 73 (1977), 461 - 477.

[46] R. Howe, Representation theory for division algebras over local fields (tamely ramified case), Bull. AMS 17 (1971), 1063 - 1066.

[64] H. Koch und E.-W. Zink, Zur Korrespondenz von Darstellungen der Galoisgruppen und der zentralen Divisionsalgebren über lokalen Körpern, ZIMM-Report R-03/79.

[111] A. Weil, Basic Number Theory, Springer-Verlag, Berlin 1974.

Contemporary Mathematics
Volume 86, 1989

SEQUENCES OF EISENSTEIN POLYNOMIALS AND ARITHMETIC IN LOCAL DIVISION ALGEBRAS

by J. Rohlfs, Eichstätt

Introduction

Let F be a local field and n an integer. Then the local Langlands conjecture predicts a bijection between the classes of irreducible representations of the absolute Galois group of G_F of F of a dimension dividing n and the classes of irreducible representations of the multiplicative group D^* of any central division algebra D over F with $[D : F] = n^2$. The numerical Langlands conjecture which is proved in Henniart's talk [these proceedings] states that at least the numbers of such irreducible representations of G_F and D^* with corresponding conductors coincide.

In this paper we explain the algebraic preparations which are necessary to count numbers of irreducible representations of D^*. We follow Koch's papers [61], [62].

If Π is a finite dimensional (continuous) representation of D^* then there is a minimal $k \in \mathbb{N}$ such that Π is trivial on $1 + P^k$, where P is the maximal ideal in D. It is classical that the number of irreducible representations of $D^*/1 + P^k$ is equal to the number of conjugacy classes of $D^*/1 + P^k$. If a conjugacy class is represented by $y \in D^*$, then $y(1 + P^k) = y + P^j$ for some $j \in \mathbb{Z}$ and obviously it suffices to understand conjugacy classes mod P^j in D. We choose a $y_j \in y + P^j$ such that $[F(y_j) : F] \leq [F(z) : F]$ for all $z \in y + P^j$ and call such a y_j minimal. Let $\phi_j(X)$ be the irreducible polynomial of y_j over a fixed unramified closure F_∞ of F. It is explained in Geyer's talk [these proceedings] that $\phi_j(X)$ essentially determines the conjugacy class of y mod P^j in D. In this talk we explain another construction of the ϕ_j's due to Koch.

The approximating polynomials ϕ_j where first used in a special case by Corwin [19] and have been introduced in full generality by Koch in [62]. In [61] Koch defines the polynomials ϕ_j without reference to D and the minimal elements y_j. We will follow this approach. As an application we give Koch's [62] description of a filtration of the infinitesimal centralizer of y mod P^j which is important for the construction of representations, see [19].

In general we will give complete definitions and statements. For the details of proofs we often refer to [61].

§ 1. Approximating Polynomials

1.1. Let F be a local field with fixed algebraic closure $\bar{F}$ and denote by $F_\infty \subset \bar{F}$ the unramified closure of F in $\bar{F}$. Let π_F be a prime of F and v a valuation of $\bar{F}$ such that $v(\pi_F) = 1$.

We fix $y \in \bar{F}$ and denote by $\varphi(X) \in F_\infty[X]$ the monic irreducible polynomial of y over F_∞. One can view φ as the monic polynomial of minimal degree such that $v(\varphi(y)) = \infty$ is maximal. In the next proposition a sequence of polynomials is constructed which approximate φ with respect to this maximality property.

1.2. Proposition. *Suppose $y \in \bar{F}$. Then there is a sequence $\varphi_1(X),\dots,\varphi_s(X) \in F_\infty[X]$ of monic polynomials with coefficients in F_∞ of degree $\deg(\varphi_i) = \varepsilon_i \le \varepsilon_s$ where $\varepsilon_1 = 1$ such that*

(i) $\varphi_s(x)$ *is the irreducible polynomial of* y *over* F_∞.

(ii) *for all monic* $f(X) \in F_\infty[X]$ *of degree* ε_i *we have* $v(\varphi_i(y)) \ge v(f(y))$.

(iii) *If* $i < s$ *and* $\nu_i := v(\varphi_i(y)) = \dfrac{n_i}{\varepsilon'_{i+1}}$ *is the reduced quotient where*

$0 < \varepsilon'_{i+1} \in \mathbb{N}$, *then* $\varepsilon_{i+1} = \mathrm{scm}\,(\varepsilon_i, \varepsilon'_{i+1})$ *is the smallest common multiple of* ε_i *and* ε'_{i+1}.

Moreover we have for all $i < s$:

α) $\nu_i\, \varepsilon_i \notin \mathbb{Z}$

β) $\nu_{i+1} > \nu_i \dfrac{\varepsilon_{i+1}}{\varepsilon_i}$ where $\nu_s = \infty$

γ) $\varepsilon_i < \varepsilon_{i+1}$.

For the details of the proof see [61]. We recall some observations which will be needed later.

Since ε_s is the smallest degree of a monic polynomial with $v(\varphi_s(y)) = \infty$ we have $v(\varphi_i(y)) < \infty$ for $i < s$ and (iii) makes sense.

The sequence of the φ_i's is constructed inductively. Suppose that $\varphi_1,\dots,\varphi_t$ have been constructed. Then ε_{t+1} is given by (iii). The following elementary observation is crucial, see [61, 1.2.1]

Let $\Lambda_t := \{\lambda = (\lambda_1,\dots,\lambda_t) \mid \lambda_i \in \mathbb{N}\ ,\ 0 \le \lambda_i < k_i\}$ where we abbreviate $k_i = \varepsilon_{i+1}/\varepsilon_i$. Then $\{\sum_{i=1}^{t} \lambda_i\, \nu_i \mid \lambda \in \Lambda_t\}$ is a full system of representatives for $\frac{1}{\varepsilon_{t+1}}\mathbb{Z}/\mathbb{Z}$ and each $i \in \{0,1,2,\dots,\varepsilon_{t+1}-1\}$ can be written $i = \sum_{j=1}^{t} \lambda_j\, \varepsilon_j$ for a unique $\lambda \in \Lambda_t$.

Therefore every monic polynomial f in $F_\infty[X]$ of degree ε_{t+1} can be written as $f(X) = \varphi_t(X)^{k_t} + \sum_{\lambda\in\Lambda_t} a_\lambda \varphi_\lambda(X)$, where the $a_\lambda \in F_\infty$ are uniquely determined by f and where $\varphi_\lambda(X) = \prod_{i=1}^{t} \varphi_i(X)^{\lambda_i}$. If $t+1 < s$ and $v(f(y))$ is maximal then necessarily

$$k_t \nu_t - \sum_{i=1}^{t} \lambda_i \nu_i \le v(a_\lambda) .$$

This inequality implies the existence of φ_{t+1} .

1.3. We now fix a certain choice of the φ_i's . Denote by π_F the fixed prime in F and let ζ_∞ be the roots of unity in F_∞ of order prime to the characteristic of the residue field of F_∞ . Then we can write uniquely

$$a_\lambda = \sum_{\varkappa \ge v(a_\lambda)} \zeta_{\varkappa\lambda}\, \pi_F^{\varkappa} \quad \text{with} \quad \zeta_{\varkappa\lambda} \in \zeta_\infty \cup \{0\} .$$

If $j_\lambda := [\nu_{t+1} - \sum_{i=1}^{t} \nu_i\, \lambda_i]$ and if $t+1 < s$ then it is easy to see that a_λ is uniquely determined mod $\pi_F^{j_\lambda+1}$. We introduce $\mu_{t+1} := \nu_{t+1} - \sum_{i=1}^{t} \nu_i(k_i-1)$, $i_\lambda := \sum_{i=1}^{t} \nu_i(k_i - 1 - \lambda_i)$ and choose $a_\lambda = \sum_{\varkappa < \mu_{t+1}+i_\lambda} \zeta_{\varkappa\lambda}\, \pi_F^{\varkappa}$ for all $\lambda \in \Lambda_t$ and all $t < s$. Then the uniquely determined sequence $\varphi_1,\dots,\varphi_s$ is called the sequence of Eisenstein polynomials given by y .

1.4. We assume now that $v(y) \in \frac{1}{n}\mathbb{Z}$ and refine our sequence φ_i by truncating the coefficients a_λ . Observe that $v(y) \in \frac{1}{n}\mathbb{Z}$ for any $y \in D^*$ if $[D : F] = n^2$.

Since $\mu_{t+1} - \mu_t = \nu_{t+1} - k_t \nu_t > 0$ for all $t < s$ the sequence $\mu_1, \mu_2, \ldots, \mu_{s-1}$ is strictly increasing. We introduce $\mu_0 = -\infty$ and $\mu_s = \infty$. For $j \in \mathbb{Z}$ and $\frac{j}{n} \leq \mu_1 = \nu_1$ we define

$$\phi_j(X) = X + \sum_{\varkappa < \frac{j}{n}} \zeta_\varkappa \pi_F^\varkappa \quad \text{where}$$

$$\varphi_1(X) = X + \sum_{\varkappa < \mu_1} \zeta_\varkappa \pi_F^\varkappa, \; \zeta_\varkappa \in \zeta_\infty \cup \{0\}$$

and we put

$$F_{f_j} := F(\zeta_\varkappa, \varkappa < \tfrac{j}{n}), \quad f_j := [F_{f_j} : F] \text{ and } \varepsilon_j = 1 .$$

For $\mu_{t-1} < \frac{j}{n} \leq \mu_t$ and $t \geq 2$ we define

$$\phi_j(X) = \varphi_{t-1}(X)^{k_{t-1}} + \sum_{\lambda \in \Lambda_{t-1}} \; \sum_{\varkappa < \frac{j}{n} + i_\lambda} \zeta_{\varkappa\lambda} \pi_F^\varkappa \varphi_\lambda(X)$$

where

$$\varphi_t(X) = \varphi_{t-1}(X)^{k_{t-1}} + \sum_{\lambda \in \Lambda_{t-1}} \; \sum_{\varkappa < \mu_t + i_\lambda} \zeta_{\varkappa\lambda} \pi_F^\varkappa \varphi_\lambda(X) .$$

We define

$$e_j := \deg(\phi_j) = \varepsilon_t \quad \text{and}$$

$$f_j := [F_{f_j} : F] \text{ where } F_{f_j} := F_{f_{j-1}}(\zeta_{\varkappa\lambda}, \lambda \in \Lambda_{t-1}, \varkappa < \tfrac{j}{n} + i_\lambda) .$$

<u>The sequence of polynomials</u> $\phi_j(X) \in F_\infty[X]$, $j \in \mathbb{Z}$ <u>is called the sequence of approximating polynomials.</u>

By construction we have $\lim_{j\to\infty} \phi_j(X) = \varphi_s(X)$.

<u>We call</u> j <u>a jump if</u> $e_{j+1} f_{j+1} > e_j f_j$ <u>and say that</u> j <u>is a ramified jump, if</u> $e_{j+1} > e_j$. Denote by E the set of jumps and by $S \subset E$ the set of ramified jumps.

We now collect some easy observations:

Proposition 1.5. Assume $\mu_{t-1} < \frac{j}{n} \leq \mu_t$.

(i) We have $v(\phi_j(y)) \geq \frac{j}{n} + \sum_{i=1}^{t-1} (k_i-1)\, \nu_i$. If $j \in E$ we have an equality.

(ii) If $\frac{j}{n} < \mu_t$ then

$\phi_{j+1}(X) = \phi_j(X) + \zeta_{\varkappa\lambda}\, \pi_F^{\varkappa}\, \varphi_\lambda(X)$ if $\varkappa = \frac{j}{n} + i_\lambda$ is solvable for $\lambda \in \Lambda_{t-1}$

and $\phi_{j+1} = \phi_j$ if $\varkappa = \frac{j}{n} + i_\lambda$ is not solvable.

(iii) If $\frac{j}{n} = \mu_t$ we abbreviate $d_j = e_{j+1}/e_j$ and have

$\phi_{j+1}(X) = \phi_j(X)^{d_j} + \zeta_{\varkappa\lambda}\, \pi_F^{\varkappa}\, \varphi_\lambda(X)$ where $\lambda = (\lambda_1 \ldots \lambda_{t-1}, 0) \in \Lambda_t$

is the unique solution of $\frac{j}{n} = \varkappa = \mu_t + i_\lambda$.

(iv) $F_{f_{j+1}} = F_{f_j}(\zeta_{\varkappa\lambda})$ in the notation of (ii) resp. (iii).

Next we show that the embedding of $F(y)$ into $\bar{F}$ is irrelevant for the determination of the ϕ_j's .

Proposition 1.6. For all j the coefficients of ϕ_j lie in $F(y)$.

Proof. Assume inductively that $F_{f_j} \subset F(y)$. Suppose that $F_{f_{j+1}} \supsetneq F_{f_j}$.
Using 1.5. (ii) and (iii) we have

$$\frac{\phi_{j+1}(y)}{\pi_F^{\varkappa}\varphi_\lambda(y)} = \frac{\phi_j(y)^{d_j}}{\pi_F^{\varkappa}\varphi_\lambda(y)} + \zeta_{\varkappa\lambda} .$$

We have $v(\phi_{j+1}(y)/\pi_F^{\varkappa}\varphi_\lambda(y)) > 0$ by 1.5. (i) and $\phi_j(y)^{d_j}/\pi_F^{\varkappa}\varphi_\lambda(y) \in F(y)$ by induction hypothesis. Hence $\zeta_{\varkappa\lambda} \in F(y)$ by Hensel's lemma and the claim follows by 1.5. (iv).

§ 2. Generalities on Division Algebras

Suppose that D is a central division algebra over a local field F of degree n, where $n^2 = [D : F]$. We recall some classical results, see [11], and fix our notation.

2.1. Let $y \in D$ be a fixed element. For $z \in D$ let D_z be the centralizer of z in D with center $F(z)$. We write $m(z) = [F(z) : F]$ and $m(z) = e(z)\, f(z)$ where $e(z)$ is the ramification index of $F(z)$ over F and $f(z)$ the inertia degree. We abbreviate $m = m(y)$, $e = e(y)$, $f = f(y)$. Denote by v the valuation of D with $v(\pi_F) = 1$, where π_F is a prime of F.

Since $[D_y : F(y)] = (\frac{n}{m})^2$ we have an over F unramified extension $L \subset D_y$ with $[L : F] = \frac{n}{e}$. Choose a maximal over F unramified subfield F_n of D such that $F_n \cap D_y = L$. If $h|n$ we denote the unique subfield E of F_n with $[F_n : E] = h$ by $F_{n/h}$. Then $F_n \cap D_y = F_{n/e}$. Let $\zeta_{n/h}$ be the roots of unity in $F_{n/h}$ of order prime to the residue characteristic of F_n.

It is classical that there is a prime π contained in the maximal ideal P of D and a generator σ of the Galois group of F_n over F such that $\pi^n = \pi_F$ and $\pi x \pi^{-1} = {}^{\sigma}x$ for all $x \in F_n$.

We have $F_{f_j} \subset F_f$, where f_j is as introduced in 1.4, and we denote the centralizer of F_{f_j} resp. F_f by D_j resp. D_f. We observe that D_f resp. D_j has inertia degree n over F and ramification index $\frac{n}{f}$ resp. $\frac{n}{f_j}$.

Since $e_{j+1}|e_j|e|n$ we have trace $tr_{e_j} : F_n \longrightarrow F_{n/e_j}$ and norm $N_{e_j|e_{j+1}} : F_{n/e_j} \longrightarrow F_{n/e_{j+1}}$.

2.2. We use the notation given up to now. Then we have the following easy observations.

(i) if $j \in S$ then $e_{j+1} = \mathrm{scm}(e_j, \frac{n}{(n,j)})$, where (n,j) is the greatest common divisor of n and j.

(ii) if $j \in S$ then σ^j generates the Galois group of F_{n/e_j} over $F_{n/e_{j+1}}$

(iii) if $j, s \in E$ then $f_s | f | \frac{n}{e} | j$.

We recall $\frac{j}{n} = \mu_t = \nu_t + \sum_{i=1}^{t-1} (k_i - 1)\nu_i$, i.e. $\frac{j}{n} = \frac{n_t}{\varepsilon'_{t+1}} + \frac{r}{\varepsilon_t}$, $\mathbb{N} \ni r \neq 0$. If we write $\frac{j}{n} = \frac{j'}{n/(n,j)}$ the first claim follows. The second claim then follows from the equation $j = n_t \frac{n}{\varepsilon'_{t+1}} + r \frac{n}{\varepsilon_t}$. Since $f_j | f$ by Prop. 1.6. and $f | \frac{n}{e}$ the last claim follows from (i) using $\frac{n}{(n,j)} | e_{j+1} | e$.

§ 3. Special approximations of y and useful primes

3.1. In this chapter we show that $\phi_j(X)$ has a root $v_j \in D_j$ such that $y \equiv v_j \bmod P^j$. A root v_j is called a special approximation of y . Moreover we show that we can find primes $\pi_j \in D_j$ which allow to connect the arithmetic in D_{v_j} , D_j and D_k for $k > j$.

The element v_j is constructed by successive approximations. To carry them out, we need a Taylor formula for ϕ_j .

3.2. Definition. Assume that $\mu_{t-1} < \frac{j}{n} \leq \mu_t$ and define $\Delta_j(X) = 1$ if $t \leq 1$ and

$$\Delta_j(X) = \prod_{i=1}^{t-1} \varphi_i(X)^{k_i - 1} \quad \text{if} \quad t > 1 .$$

We observe that $\deg(\Delta_j(X)) = e_j - 1$ and abbreviate $v(\Delta_j(y)) := h_j/n$.

3.3. Proposition. Assume that $v \in (y + P^j) \cap D_j$, $j' \geq j$, $z \equiv v + \beta\omega \bmod P^{j'+1}$ where $\omega \in D_v \cap P^{j'}$ and $\beta \in F_n$. Then

$$\phi_j(z) \equiv \phi_j(v) + tr_{e_j}(\beta)\, \omega\, \Delta_j(v) \bmod P^{h_j + j' + 1} \quad \text{where}$$

$$\phi_j(v) \in P^{h_j + j} \quad \text{and} \quad tr_{e_j}(\beta)\, \omega\, \Delta(v) \in P^{h_j + j'} .$$

Up to the justification of several congruences the idea of the proof is as follows: Suppose that the above statement is true for j and $j + 1 \leq j'$. Then by 1.5. (ii), (iii)

$$\phi_{j+1}(z) = \phi_j(z)^{d_j} + \zeta_{\varkappa\lambda}\, \pi_F^{\varkappa}\, \varphi_\lambda(z)$$

and by assumption

$$\phi_j(z)^{d_j} \equiv (\phi_j(v) + tr_{e_j}(\beta)\,\omega\,\Delta_j(v))^{d_j}$$

$$\equiv \phi_j(v)^{d_j} + \sum_{i=0}^{d_j-1} tr_{e_j}(\beta)^{\sigma^{iu}}\,\omega\,\Delta_j(v)\,\phi_j(v)^{d_j-1} .$$

where $\frac{u}{n} = v(\phi_j(v))$. It can be shown that we may replace $\varphi_\lambda(z)$ by $\varphi_\lambda(v)$ in our congruences and we get:

$$\phi_{j+1}(z) \equiv \phi_{j+1}(v) + \sum_{i=0}^{d_j-1} tr_{e_j}(\beta)^{\sigma^{iu}}\,\omega\,\Delta_j(v)\phi^{d_j-1} .$$

If $d_j = 1$ we are done. If $d_j > 1$ then $\phi_j = \varphi_t$ and σ^u generates the Galois group of F_{n/e_j} over $F_{n/e_{j+1}}$, see 2.2. (ii), and the claim follows.

3.4. Remark. Using 3.3 for $\beta = 0$ and $v = y$ we get for $j \in E$: $v(\phi_j(z)) = v(\phi_j(y)) = (h_j+j)/n$. We abbreviate $g_j := h_j + j$.

As an application of 3.3 we prove the existence of special approximations.

3.5. Proposition. For every $j \in \mathbb{Z}$ there exists a special approximation $v_j \in D_j$, i.e. $v_j \equiv y \bmod P^j$ and $\phi_j(v_j) = 0$. The extension $F(v_j)|F$ has inertia degree f_j and ramification index e_j .

The claim on $F(v_j)/F$ is an easy consequence of the first claim. To prove the first one we successively approximate v_j by starting with y . We apply 3.3. for $v = y$, $\omega = \phi_j(y)/\Delta_j(y)$, $\nu(\omega) = j'/n$ with $j' \geq j$ and $\beta \in \zeta_n$ where $tr_{e_j}(\beta) = -1$. For $z = y + \beta\omega \in y + P^{j'}$ we have by 3.3

$$\phi_j(z) \equiv \phi_j(y) + tr_{e_j}(\beta)\,\omega\,\Delta_j(v) \equiv 0 \bmod P^{h_j+j'+1} .$$

Replacing y by z in this argument we construct a new z and so on. The first claim follows.

Next, we state the desired connection between arithmetic in D_{v_j}, D_j and D_k for $k > j$.

3.6. Proposition - Definition.

(i) There exists a prime π_j of D_j such that for all ramified jumps $s < j$ we have

$$\psi_s(y) \equiv \rho_s \pi_j^{g_s/f_j} \bmod P^{g_s+1} \quad \text{and} \quad \rho_s \in \zeta_{f_j} .$$

Such a π_j is called a pleasant prime.

(ii) If $\pi_j \in D_j$ is a pleasant prime then there is a prime $\eta \in D_{v_j}$ such that $\eta \equiv \pi_j \bmod P^{f_j+1}$.

(iii) If j is a jump there exists a pleasant prime $\pi_j \in D_j$ such that for $y \equiv v_j + \beta_j \pi_j^{j/f_j} \bmod P^{j+1}$, $\beta_j \in \zeta_n$, there exists a $\gamma_j \in \zeta_{f_{j+1}}$ such that $tr_{e_j}(\beta_j) \equiv \gamma_j \bmod P$.

Such a π_j is called a useful prime.

(iv) If $j < k$ are jumps and if π_j and π_k are useful primes, then

$$\pi_k \pi_j^{-f_k/f_j} \in \zeta_{n/e_{j+1}} .$$

For the details of the proof see [61, § 9]. We point out that according to 3.4.: $v(\psi_j(y)) = g_j/m$. So 3.6 (i) is true with $\rho_s \in \zeta_n$ and some prime for D_j. To prove 3.6. (i) one has to make a clever choice of a $\xi \in \zeta_n$ such that $\pi_j \equiv \xi \pi^{f_j}$ has the desired properties. Similarily for (iii). The statements (ii) and (iv) then are relatively simple consequences.

3.7. Remark. Since D_{v_s} has an over F unramified subfield isomorphic to F_{n/e_s} we see using 3.6. (ii) that an element $x_s \in D_{v_s}$ can be written $x_s \equiv \zeta_s \pi_s^{a/f_s} \bmod P^{a+1}$ where $\zeta_s \in F_{n/e_s}$ and vice versa: i.e. if $\zeta_s \in F_{n/e_s}$ and $a \in \mathbb{Z}$, $f_s | a$, is given there is an element $x_s \in D_{v_s}$ such that

$$x_s \equiv \zeta_s \pi_s^{a/f_s} \mod P^{a+1} .$$

Next we explain a congruence of central importance, which has been observed first by Corwin [19] if $n = p^2$, p a prime.

3.8. Proposition. _Suppose_ $j \in \mathbb{Z}$, $f_j | j$ _and write_ $y = v_j + \beta \pi_j^{j/f_j} \mod P^{j+1}$, _where_ $\pi_j \in D_j$ _is a pleasant prime and_ $\beta \in \zeta_n \cup \{o\}$. _Let_ $\gamma \in \zeta_{n/e_j} \cup \{o\}$ _and_ $tr_{e_j}(\beta) \equiv \gamma \mod P$. _Then_

$$\psi_j(y) \equiv \gamma \pi_j^{j/f_j} \Delta_j(y) \mod P^{g_j+1} \quad \text{and}$$

$$F_{f_{j+1}} = F_{f_j}(N_{e_j|e_{j+1}}(\gamma)) .$$

The congruence results from the Taylor formula 3.3. where we have to put $j' = j$, $v = v_j$, $\omega = \eta^{j/f_j}$ where η is as in 3.6. (ii) and $y = z$. To prove the second claim one uses the congruence

$$\psi_{j+1}(y) = \psi_j(y)^{d_j} + \zeta_{\varkappa\lambda} \varphi_\lambda(y) \equiv 0 \mod P^{g_j+1}$$

and substitutes the expression for $\psi_j(y)$ given above. Then a congruence mod P for $N_{e_j|e_{j+1}}(\gamma)$ arises which shows that up to a factor out of F_{f_j} the roots of unity $\zeta_{\varkappa\lambda}$ and $N_{e_j|e_{j+1}}(\gamma)$ coincide. Hence $F_{f_{j+1}} = F_{f_j}(\zeta_{\varkappa\lambda}) = F_{f_j}(N_{e_j|e_{j+1}}(\gamma))$.

We now reproduce a main result contained in [62] which is a description of a certain filtration of the infinitesimal centralizer mod P^ℓ of y . The proof of this result depends heavily on 3.6. and 3.8.

3.9. Proposition. _Let_ $\ell \in \mathbb{Z}$ _and abbreviate_ $A_\ell := \{x \in D / xy - yx \in P^\ell\}$. _Suppose that_ $t \in E$ _is a jump. Then_

$$A_\ell \cap P^{\ell-t} = \sum_{\substack{s \in E \\ s \leq t}} D_{v_s} \cap P^{\ell-s} ,$$

where $v_s \in y + P^s$, $s \leq t$, _are special approximating elements._

Proof. We have $P^{\ell-t} \subset P^{\ell-s}$ and if $x_s \in D_{v_s} \cap P^{\ell-s}$ we can write $y = v_s + a$ where $a \in P^s$. Then

$$x_s y - yx_s = x_s(v_s + a) - (v_s + a)x_s = x_s a - ax_s \in P^{\ell}$$

and the right side of the claimed identity is contained in it's left side.

We prove the opposite inclusion by induction with respect to t . If $t \in E$ is minimal then $D_{v_t} = D$ and the inclusion is obvious. Assume now that $u \in E$ is the maximal jump smaller than t . Then we have by induction assumption

$$P^{\ell-t} \cap A_{\ell} \subset P^{\ell-(t-u)-u} \cap A_{\ell-(t-u)} \subset \sum_{\substack{s \le u \\ s \in E}} D_{v_s} \cap P^{\ell-(t-u)-s} \subset D_{v_u} \cap P^{h-u} + P^{h-u+1}$$

where we abbreviate $h = \ell - (t - u)$ and $h - u = \ell - t$. We first show:

$$P^{\ell-t} \cap A_{\ell} \subset D_{v_t} \cap P^{\ell-t} + P^{\ell-t+1} . \tag{1}$$

We write $x \in P^{\ell-t} \cap A_{\ell}$ as $x = \sum_{s \le u} x_s$ where $x_s \in D_{v_s} \cap P^{h-s}$. If f_u does not divide $h - u$ then (1) holds. So let us assume that $f_u | h - u$. Then $f_s | f_n | h - s$ for all $s \le u$ and using 3.7. we can write $x_s \equiv \zeta_s \pi_s^{h-s/f_s}$ where $\zeta_s \in F_{n/e_s}$ and π_s is a useful prime. According to 3.6. (iii) we can write $y \equiv v_s + \beta_s \pi_s^{s/f_s} \bmod P^{s+1}$, $\beta_s \in \zeta_n$, $\gamma_s \equiv tr_{e_s} \beta_s \bmod P$ and $\gamma_s \in \zeta_{f_{s+1}}$.

Then we have

$$xy - yx = \sum_{s \le u} x_s y - yx_s = \sum_{s \le u} x_s(y - v_s) - (y - v_s) x_s$$

$$\equiv \sum_{s \le u} (\zeta_s \beta_s^{\sigma^{h-s}} - \beta_s \zeta_s^{\sigma^s}) \pi_s^{h/f_s} \bmod P^{h+1} .$$

Since $\pi_u^{-h/f_u} \pi_s^{h/f_s} =: \delta_s \in \zeta_{n/e_{s+1}}$ by 3.6. (iv) we get

$$\sum_{s \le u} (\zeta_s \beta_s^{\sigma^{h-s}} - \beta_s \zeta_s^{\sigma^s}) \delta_s \equiv 0 \bmod P .$$

If $s < u$ then $\gamma_s \in F_{n/e_u}$ and $\delta_s \in F_{n/e_u}$. Hence if we apply $tr_{e_u} = tr_{e_u|e_s} \circ tr_{e_s}$ in obvious notation to the last equation we get $\sum_{s<u} tr_{e_u|e_s}(\zeta_s - \zeta_s^{\sigma^s})\gamma_s\delta_s = 0$ and obtain $\zeta_u \gamma_u^{\sigma^{h-u}} - \gamma_u \zeta_u^{\sigma^u} \equiv 0 \bmod P$ or equivalently

$$\zeta_u^{\sigma^u - 1} \equiv \gamma_u^{\sigma^{h-u} - 1} \bmod P . \tag{2}$$

We apply $N_{e_u|e_{u+1}}$ to (2), write $N(\gamma) := N_{e_u|e_{u+1}}(\gamma_u)$ and recall from 3.8. that $N(\gamma)$ generates $F_{f_{u+1}} = F_{f_t}$ over F_{f_u}. Hence we deduce from (2) that $f_t | h-u$. If $e_t = e_u$ then $\zeta_u \in F_{n/e_t}$. If $e_t > e_u$ then σ^u generates the Galois group of F_{n/e_u} over F_{n/e_t}, see 2.2. (ii), and from (2) we deduce $\zeta_u^{\sigma^u - 1} = 1$. Therefore $\zeta_u \in F_{n/e_t}$ and there is an $x_t \in D_{v_t} \cap P^{\ell-t}$ such that $x_u \equiv x_t \bmod P^{\ell-t+1}$. Hence (1) holds. Now we can complete our induction with respect to t since

$$x - x_t \in A_\ell \cap P^{\ell-t+1} = P^{\ell-(t-1)} \cap A_\ell \subset P^{\ell-u} \cap A_\ell = \sum_{\substack{s \le u \\ s \in E}} D_{v_s} \cap P^{\ell-s} .$$

References

[19] L. Corwin: Representations of division algebras over local fields II; Pacific Journ. Math. 101, (1982), 49 - 70.

[61] H. Koch: Eisensteinsche Polynomfolgen und Arithmetik in Divisionsalgebren über lokalen Körpern; Math. Nachr. 104

[62] H. Koch: Zur Arithmetik von Divisionsalgebren über lokalen Körpern; Math. Nachr. 100 (1981), 9 - 19.

Contemporary Mathematics
Volume 86, 1989

KOCH'S CLASSIFICATION OF THE PRIMITIVE REPRESENTATIONS OF A GALOIS GROUP OF A LOCAL FIELD

by Moshe Jarden, Tel Aviv University

Let F be a local field, i. e., a finite extension of $\mathbb{Q}_p$ or of $\mathbb{F}_p((t))$. Let F_{tr} be the maximal tamely ramified extension of F, and let $T_F = \mathrm{Gal}(F_{tr}/F)$. Consider an $\mathbb{F}_p[T_F]$-module M, finite dimensional over $\mathbb{F}_p$. M is <u>symplectic</u> if it is equipped with a non-degenerate bilinear form $(,): M \times M \to \mathbb{F}_p$ such that $(x,x) = 0$, and $(\tau x, \tau y) = (x,y)$ for $x,y \in M$ and $\tau \in T_F$.

A subset M_o of M is a <u>submodule</u> if it is closed under addition and if it is invariant under the action of T_F. If in addition the restriction of the bilinear form $(,)$ to M_o is non-degenerate, then M_o is a <u>symplectic submodule</u>.

M is <u>simple</u> (as a symplectic module) if it contains no proper symplectic submodule. M is <u>isotropic</u> if $(x,y) = 0$ for all $x,y \in M$. M is completely <u>anisotropic</u> if the only isotropic submodule of M is 0.

If M contains a proper symplectic submodule M_o, then $M = M_o \perp M_o^{\perp}$. So M is the direct sum of simple symplectic submodules.

M is <u>irreducible</u> (as a T_F-module) if it does not contain a proper submodule.

Lemma. <u>If M is a simple completely anisotropic symplectic module, then M is irreducible.</u>

Proof. If M_1 is a proper submodule of M and $M_o = \{a \in M_1 | a \perp M_1\}$, then M_o is an isotropic submodule of M. Thus $M_o = 0$. It follows that M_1 is non-degenerated. Thus $M_1 = 0$.

M is <u>primary</u> if it is the direct sum of isomorphic irreducible modules.

Let $\Pi: G_F \to GL(n,\mathbb{C})$ be a continuous representation. Since $GL(n,\mathbb{C})$ contains an open neighbourhood of 1 which contains no non-trivial subgroups, $\mathrm{Ker}(\Pi)$ is an open subgroup of G_F . So Π may be considered as a faithful representation $\Pi: G \to GL(n,\mathbb{C})$, with $G = Gal(L/F)$ and L/F finite Galois.

The representation Π is <u>primitive</u> if it is irreducible and not induced from a proper subgroup of G .

The center of $GL(n,\mathbb{C})$ is the group C of scalar matrices

$$\begin{pmatrix} z & & \\ & \ddots & \\ & & z \end{pmatrix}, \qquad z \in \mathbb{C}^\times .$$

The representation Π defines a <u>projective representation</u> $\bar{\Pi}: G \to PGL(n,\mathbb{C})$. Two representations $\Pi_1, \Pi_2: G_F \to GL(n,\mathbb{C})$ are <u>projectively equivalent</u> if there exists $g \in GL(n,\mathbb{C})$ and a character $\chi \in Hom(G_F, \mathbb{C}^\times)$ such that $\Pi_2(x) = \chi(x)g^{-1}\Pi_1(x)g$. Each projective representation $\bar{\Pi}: G \to PGL(n,\mathbb{C})$ can be lifted to a representation $\Pi: G_F \to GL(n,\mathbb{C})$.

<u>**Theorem.**</u> **a)** <u>The degree of each primitive representation</u> $\Pi: G_F \to GL(n,\mathbb{C})$ <u>is a power of</u> p .

b) <u>To each primitive representation of</u> Π <u>of degree</u> n <u>we associate a completely anisotropic symplectic</u> $\mathbb{F}_p(T_F)$<u>-module</u> $M(\Pi)$ <u>of order</u> n^2 .

c) <u>If</u> Π_1 <u>and</u> Π_2 <u>are projectively equivalent, then</u> $M(\Pi_1) \simeq M(\Pi_2)$.

d) <u>To each symplectic completely anisotropic module</u> M <u>there exists</u> Π <u>such that</u> M <u>is isomorphic to</u> $M(\Pi)$.

<u>Construction of $M(\Pi)$</u>

Let $\Pi: G_F \to GL(n,\mathbb{C})$ be a primitive representation. Then the fixed field, L , of $Ker(\Pi)$ is a finite Galois extension of F . For $G = Gal(L/F)$ we may consider Π as a faithful primitive representation $\Pi: G \to GL(n,\mathbb{C})$. Identify G with its image. If F_0 (resp., F_1) is the maximal unramified (resp., tamely ramified) extension of L , then F_0/F and F_1/F_0 are cyclic extensions. Hence $Gal(F_1/F)$ is metacyclic and $H = Gal(L/F_1)$ is a p-group. In particular, G/H is supersolvable. By a theorem of Clifford, $res_H\Pi$ is irreducible. By a theorem of Rigby $n = p^d$; $H/Z(H) \simeq (\mathbb{Z}/p\mathbb{Z})^{2d}$; in particular $[H,H] \le Z(H)$. Use the identity $[ab,c] = [a,c][[a,c],b][b,c]$ to conclude that $[ab,c] = [a,c][b,c]$. Similarly $[a,bc] = [a,b][a,c]$ for $a, b, c, \in H$. The theorem of Rigby also says that $Z(H) = \langle v \rangle$ is cyclic and v is a scalar matrix. Choose $c \in Z(H)$ of order p . There exist $x_1, y_1, \dots, x_d, y_d \in H$ such that $H = \langle v, x_1, y_1, \dots, x_d, y_d \rangle$ and

$$[x_i, y_j] = c^{\delta_{ij}} \qquad [x_i,x_j] = [y_i y_j] = 1 . \tag{1}$$

Let $\varepsilon\colon \langle c \rangle \to \mathbb{F}_p^+$ be the isomorphism defined by $\varepsilon(c) = 1$. Denote the reduction of $GL(n,\mathbb{C})$ modulo its center by a bar:

$$\begin{array}{ccc}
GL(n,C) & \longrightarrow & PGL(n,C) \\
| & & | \\
G & \longrightarrow & \bar{G} \\
| & & | \\
H & \longrightarrow & \bar{H} \\
| & & | \\
Z(H) & \longrightarrow & 1 \\
| & & \\
\langle c \rangle & &
\end{array}$$

Let $M(\Pi)$ be an additive copy of $\bar{H} = H/Z(H)$. Equip $M(\Pi)$ with a symplectic bilinear form $M(\Pi) \times M(\Pi) \to \mathbb{F}_p$ defined by

$$(\bar{a},\bar{b}) = \varepsilon([a,b]) , \qquad a,\ b \in H .$$

If $\bar{a} = \bar{a}'$, then $a^{-1}a' \in Z(H)$ and $[a,b] = [a',b]$. Thus $(\bar{a},\bar{b})$ is well defined. This implies that the set $\{\bar{x}_1, \bar{y}_1,\ldots,\bar{x}_d, \bar{y}_d\}$ is a symplectic basis for the symplectic vector space $M(\Pi)$. Since $\bar{H}$ is abelian, $\bar{G}/\bar{H} = G/H \simeq Gal(F_1/F)$ acts on $\bar{H}$ by conjugation. Note that for each $a,\ b \in H$, $[a,b]$ is a scalar matrix. Therefore

$$[\bar{a}^g,\bar{b}^g) = \varepsilon([a^g,b^g]) = \varepsilon([a,b]^g) = \varepsilon([a,b]) = [\bar{a},\bar{b}] .$$

Thus $M(\Pi)$ is a symplectic $\mathbb{F}_p[T_F]$-module.

If M_0 is an isotropic submodule of $M(\Pi)$ (i. e., $(\bar{a},\bar{b}) = 0$ for each $\bar{a}, \bar{b} \in M_0$), then its preimage A_0 in $GL(n,\mathbb{C})$ is an abelian subgroup of H, normal in G. By a theorem of Rigby [59, Thm. 2.2] $A_0 \subseteq Z(G)$. Hence $A_0 \subseteq Z(H)$ and $M_0 = 0$. Thus $M(\Pi)$ is completely anisotropic.

Construction of Π from $M(\Pi)$.

Let M be a completely anisotropic symplectic $\mathbb{F}_p[T_F]$-module. We construct a primitive representation Π of G_F such that $M(\Pi)$ is projectively equivalent to M as $\mathbb{F}_p[T_F]$-modules. Koch reduces the general case to the case where M is primary. Then M is the direct sum of at most two irreducible modules. We restrict ourselves to the case where M is irreducible.

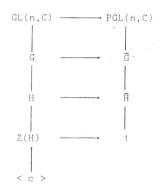

Let $n = 2d = \dim_{\mathbb{F}_p} M$. Construct a p-group H as in (1) such that $(H: Z(H)) = p^{2d}$. Let $x_1, y_1, \dots, x_d, y_d \in H$ as before and let $\bar{x}_1, \bar{y}_1, \dots, \bar{x}_d, \bar{y}_d$ be a symplectic basis for M. Then the map $x_i \longmapsto \bar{x}_i$, $y_i \longmapsto \bar{y}_i$, $i = 1, \dots, d$ extends to an epimorphism of H onto M with $Z(H)$ as the kernel

$$1 \to Z(H) \to H \to M \to 1 .$$

Also, H has several inequivalent faithful representations $\Pi: H \to GL(n,\mathbb{C})$ such that $\Pi(Z(H)) \leq Z(GL(n,\mathbb{C}))$. However they are all projectively equivalent. So we may identify M as a symplectic vector space with a subgroup of $PGL(n,\mathbb{C})$.

Denote the group of automorphisms of M that respect the symplectic structure by $Sp(M)$. Denote the normalizer of M in $PGL(n,\mathbb{C})$ by N. Koch proves that $N/M \cong Sp(M)$.

On the other hand, since the action of T_F on M is continuous it has an open kernel. Thus, M is a faithful $\mathbb{F}_p[Gal(E/F)]$-module, where E/F is a finite normal tamely ramified extension.

Let $E^{(p)}$ be the maximal extension of E such that $Gal(E^{(p)}/E)$ is an elementary p-group. The local reciprocity map gives an isomorphism

$$\theta: E^*/(E^*)^p \to Gal(E^{(p)}/E)$$

that respects the action of G_F (i. e., θ is a G_F-isomorphism) on both groups. Moreover, θ induces an isomorphism

$$\theta: U_E/U_E^p \to Gal(E_{wr}^{(p)}/E) ,$$

where U_E is the group of units of E and $E_{wr}^{(p)}$ is a maximal totally ramified (hence wildly ramified) extension of E in $E^{(p)}$ that is Galois over F. Iwasawa proved that U_E/U_E^p is isomorphic as an $\mathbb{F}_p[Gal(E/F)]$-module to $\mathbb{F}_p \oplus \mathbb{F}_p[Gal(E/F)]^m$, where $m = [F:\mathbb{Q}_p]$ and $Gal(E/F)$ acts trivially on $\mathbb{F}_p$. Let $e_1, \dots, e_m$ be a basis of $\mathbb{F}_p[Gal(E/F)]^m$ and let $a_1, \dots, a_m$ be any nonzero m-tuple of elements of M. Then the map $e_i \longmapsto a_i$, $i = 1, \dots, m$, extends to a homomorphism r of U_E/U_E^p onto a nonzero $\mathbb{F}_p[Gal(E/F)]$-submodule M_o of M. Since M is irreducible $M_o = M$ and r is surjective. This gives rise to a commutative diagram

$$\begin{array}{ccc} U_E/U_E^{(p)} & \xrightarrow{\ \theta\ } & \mathrm{Gal}(E_{wr}^{(p)}/E) \\ \downarrow{\scriptstyle r} & & \downarrow{\scriptstyle \mathrm{res}} \\ M & \xrightarrow{\ \bar{\theta}\ } & \mathrm{Gal}(L/E) \end{array}$$

where L/E is a totally ramified extension and $\bar{\theta}$ is an isomorphism that respects the action of $\mathrm{Gal}(E/F)$. In particular L/F is a Galois extension.

The action of $\mathrm{Gal}(E/F)$ on M respects the symplectic structure. Therefore $\mathrm{Gal}(E/F) \leq \mathrm{Sp}(M)$. Thus there is an intermediate group J, $M \leq J \leq N$ such that under the isomorphism $N/M \simeq \mathrm{Sp}(M)$ we have $J/N \simeq \mathrm{Gal}(E/F)$. This gives a commutative diagram

$$\begin{array}{ccccccccc} 1 & \longrightarrow & \mathrm{Gal}(L/E) & \longrightarrow & \mathrm{Gal}(L/F) & \longrightarrow & \mathrm{Gal}(E/F) & \longrightarrow & 1 \\ & & \downarrow{\scriptstyle \bar{\theta}^{-1}} & & & & \downarrow & & \\ 1 & \longrightarrow & M & \longrightarrow & J & \longrightarrow & J/M & \longrightarrow & 1 \end{array} \qquad (2)$$

where the vertical arrows are isomorphisms.

Koch proves that the $\mathrm{Gal}(E/F)$-module M is cohomologically trivial. Hence there exists an isomorphism $\gamma: \mathrm{Gal}(L/F) \to J$ that completes (2) to a commutative diagram

$$\begin{array}{ccccccccc} 1 & \longrightarrow & \mathrm{Gal}(L/E) & \longrightarrow & \mathrm{Gal}(L/F) & \longrightarrow & \mathrm{Gal}(E/F) & \longrightarrow & 1 \\ & & \downarrow{\scriptstyle \bar{\theta}^{-1}} & & \downarrow{\scriptstyle \alpha} & & \downarrow & & \\ 1 & \longrightarrow & M & \longrightarrow & J & \longrightarrow & J/M & \longrightarrow & 1 \end{array} .$$

The map $\bar{\Pi} = \alpha \circ \mathrm{res}_L : G_F \to J$ is a projective representation of G_F. It can be lifted to a representation $\Pi : G_F \to GL(n,\mathbb{C})$ (since $H^2(G_F,\mathbb{C}^\times) = 1$) and $M(\Pi)$ is isomorphic to M as $\mathbb{F}_p[T_F]$-modules. Finally, the assumption that M is completely anisotropic implies (Theorem 4.1 of [59]) that Π is primitive.

References

[59] H. Koch: Classification of the primitive representations of the Galois Group of local fields, Inventiones mathematicae 40 (1977), 195 - 216.

Contemporary Mathematics
Volume 86, 1989

ON THE NUMERICAL LOCAL LANGLANDS CONJECTURE

by M. Lorenz, Bonn

The purpose of this talk was to <u>state</u> the numerical local Langlands conjecture (NLLC) and to describe a few reductions. The material presented here is due to H. Koch and E.W. Zink and is based on the articles [63], [65]. These notes contain only one proof which simplifies Koch's original argument somewhat. In the last talk of this conference, G. Henniart announced and outlined his <u>solution</u> of the NLLC which does in fact use the reductions described in the following.

<u>Notations and Conventions.</u> Throughout, F will denote a local field, G_F the Galois group of a fixed separable closure F^{sep} of F, and $\hat{G}_F$ the set of equivalence classes of finite irreducible complex representations of G_F. Moreover, n will be a fixed positive integer, D will be a division F-algebra with $[D:F] = n^2$, and $\hat{D}^*$ will denote the set of finite irreducible complex representations of the multiplicative group D^* of D. Sets of representations will be denoted by upper case script letters, and the corresponding non-script letter will denote the cardinality of this set (e. g., card $\mathcal{R}_{n,j} = R_{n,j}$, etc.).

§ 1. Statement of the numerical local Langlands conjecture

The conjectural Langlands bijection

$$\Phi: \ \mathcal{R}_n := \{\Pi \in \hat{G}_F \mid \dim \Pi | n\} \xrightarrow{1-1} \hat{D}^*$$

should have the following properties (amongst others which are irrelevant for our purposes here): For any $\Pi \in \mathcal{R}_n$, one should have

(1) The Swan conductor of Π, $sw(\Pi)$, and the index of $\Phi(\Pi)$, $j(\Phi(\Pi))$, are related by

$$n_\Pi \cdot sw(\Pi) = j(\Phi(\Pi)) ,$$

where we have set $n_\Pi := n/\dim \Pi$.

Moreover, viewing 1-dimensional representations χ of G_F as characters of F^* via class field theory, the following should hold:

(2) $(\det \Pi)^{n_\Pi}$ is the central character of $\Phi(\Pi)$.

(3) $\Phi(\chi) = \chi \circ \mathrm{Nrd}_{D/F}$ ($\mathrm{Nrd}_{D/F}$ = reduced norm), and

$\Phi(\Pi \otimes \chi) = \Phi(\Pi) \otimes \Phi(\chi)$.

It follows from (1) and (2) that Φ will in particular yield bijections, for all $j \geq 0$,

$$\mathcal{R}_{n,j} := \{\Pi \in \mathcal{R}_n \mid n_\Pi \cdot sw(\Pi) \leq j \ , \ (\det \Pi)(\pi_F^{n_\Pi}) = 1\}$$

$$\xrightarrow{1-1} \mathcal{S}_{n,j} := \{\Sigma \in \hat{D}^* \mid j(\Sigma) \leq j \ , \ \Sigma(\pi_F) = 1\} \ ,$$

where π_F is a fixed prime element of F . The NLLC can now be stated as follows:

For all $j \geq 0$, one has $R_{n,j} = S_{n,j}$.

The right hand side of this equality is explicitely known ([61]; cf. Prof. Geyer's talk during the conference):

$$S_{n,j} = \sum_{f|n} \frac{n}{f^2} \sum_{d|f} \mu(\tfrac{f}{d})(q^d-1)q^{d[j/f]} \ ,$$

where $q = p^*$ is the cardinality of the residue field k of F and [ ] is the usual Gauss bracket.

Actually, a somewhat refined version of the above equality is usually considered. To explain this, let $G_{F,o} \subseteq G_F$ denote the inertia subgroup of G_F , let $U_D \subseteq D^*$ be the unit group of D^* , and put, for $f|n$,

$$\mathcal{R}_{n,j,f} := \{\Pi \in \mathcal{R}_{n,j} \mid \Pi|_{G_{F,o}} \text{ has (composition) length } f\}$$

and

$$\mathcal{S}_{n,j,f} := \{\Sigma \in \mathcal{S}_{n,j} \mid \Sigma|_{U_D} \text{ has length } f\} \ .$$

A slight variation of the arguments in [2, proof of Satz 19] and very little additional work yields the following formula [ ]:

$$S_{n,j,f} = \frac{n}{f^2} \sum_{d|f} \mu(\tfrac{f}{d})(q^d-1)q^{d[j/f]} ,$$

which of course implies the above formula for $S_{n,j}$. Furthermore, using fairly standard arguments from Clifford-Mackey theory for cyclic group extensions one easily establishes the following

Proposition. Suppose that $\Psi: \mathcal{R}_n \to \widehat{D^*}$ is injective and satisfies (3). Then, for all $\Pi \in \mathcal{R}_n$, length $(\Pi|_{G_{F,o}})$ = length $(\Psi(\Pi)_{U_D})$.

In particular, the conjectural Langlands bijection Φ should map $\mathcal{R}_{n,j,f}$ onto $S_{n,j,f}$. So one is lead to consider the following refined version of the NLLC (for all F, n, j, f as above):

(*)
$$\boxed{R_{n,j,f} = \frac{n}{f^2} \sum_{d|f} \mu(\tfrac{f}{d})(q^d-1)q^{d[j/f]}} .$$

For $f = 1$ (the case where $\Pi|_{G_{F,o}}$ is irreducible), this specializes to

(**)
$$\boxed{R_{n,j,1} = n\cdot(q-1)\cdot q^j} .$$

The main result to be explained below reduces the proof of (*) to the proof of (**), where one can also assume that n is a power of p = char k .

Examples. (a) The case $j = 0$: Letting $G_{F,1} \subseteq G_{F,o}$ denote the wild ramification group of G_F, one has

$$\mathcal{R}_{n,o,1} = \{\Pi \in \mathcal{R}_n \mid \Pi|_{G_{F,o}} \text{ is irreducible, } \Pi|_{G_{F,1}} \text{ is trivial,}$$
$$(\det \Pi)(\pi_F^{n_\Pi}) = 1\} .$$

Such Π's must be 1-dimensional, since $G_{F,o}/G_{F,1}$ is topologically cyclic, and so class field theory identifies $\mathcal{R}_{n,o,1}$ with $F^*/\langle U_F^1, \pi_F^n\rangle \simeq \mathbb{Z}/(q-1)\mathbb{Z} \times \mathbb{Z}/n\mathbb{Z}$ (U_F^1 = one-unit group of F^*). Thus $R_{n,o,1} = n\cdot(q-1)$, as required.

(b) The case $n = 1$: Again by class field theory, one can identify $\mathcal{R}_{1,j,1}$ with the set of characters χ of F^* such that $\chi|_{U_F^{j+1}}$ is trivial and $\chi(\pi_F) = 1$. Thus $\mathcal{R}_{1,j,1} \xleftrightarrow{1-1} F^*/< U_F^{j+1}, \pi_F >$ and so $R_{1,j,1} = (q-1)q^j$.

Remarks. (a) Put $\mathcal{C}_n := \{\sigma \in \hat{G}_F \mid \sigma$ is trivial on $G_{F,o}$ and $\pi_F^n\}$. Then $\mathcal{C}_n$ is a cyclic group of order n which operates on $\mathcal{R}_{n,j,1}$ via $\Pi \longmapsto \Pi \otimes \sigma$. All orbits in $\mathcal{R}_{n,j,1}$ have length n , and restriction of representations from G_F to $G_{F,o}$ yields a bijection of $\mathcal{R}_{n,j,1}/\mathcal{C}_n$ onto

$$\mathcal{T}_{n,j} := \{\Gamma \in \hat{G}_{F,o} \mid \dim \Gamma | n , \ \Gamma \text{ is } G_F\text{-invariant}, \ \Gamma|_{G_F^{(j/n)+\varepsilon}} \text{ is trivial for all } \varepsilon > 0\}$$

Here, as usual, G_F^j denotes the j-th ramification group of G_F . In particular, $R_{n,j,1} = n \cdot T_{n,j}$ and (**) is equivalent with $T_{n,j} = (q-1) \cdot q^j$.

(b) Similarly, if $\mathcal{D}$ denotes the set of G_F-invariant characters σ of $G_{F,o}$ which are trivial on $G_{F,1}$, then $\mathcal{D}$ is a cyclic group of order $q-1$ which acts on $\mathcal{T}_{n,j}$ via $\Gamma \longmapsto \Gamma \otimes \sigma$. In the case where n is a p-power (the crucial case) all orbits have length $q-1$ and restriction from $G_{F,o}$ to $G_{F,1}$ identifies $\mathcal{T}_{n,j}/\mathcal{D}$ with

$$\mathcal{U}_{n,j} := \{\Delta \in \hat{G}_{F,1} \mid \dim \Delta | n , \ \Delta \text{ is } G_F\text{-invariant}, \ \Delta|_{G_F^{(j/n)+\epsilon}} \text{ is trivial for all } \varepsilon > 0\} .$$

Therefore, $R_{n,j,1} = n \cdot (q-1) \cdot U_{n,j}$ and (**) becomes $U_{n,j} = q^j$.

§ 2. Finiteness and Some Reductions

Here we describe the main results of [63], which in particular reduce the proof of (*) to the proof of (**) for p-powers n .

For a given $\alpha \in F^*$ with $v_F(\alpha) \neq 0$ and a root of unity $\zeta \in \mathbb{C}^*$, put

$$\mathcal{R}_{n,j}(\alpha,\zeta) = \{\Pi \in \mathcal{R}_n \mid n_\Pi \cdot sw(\Pi) \le j , \quad \det \Pi(\alpha^{n_\Pi}) = \zeta\} ,$$

$$\mathcal{R}_{n,j,f}(\alpha,\zeta) = \{\Pi \in \mathcal{R}_{n,j}(\alpha,\zeta) \mid \Pi|_{G_{F,o}} \text{ has length } f\} .$$

Thus $\mathcal{R}_{n,j} = \mathcal{R}_{n,j}(\pi_F, 1)$ and similarly for $\mathcal{R}_{n,j,f}$. Part (b) of the following result ensures the independence of $R_{n,j,f}$ of the choice of the uniformizer $\pi_F \in F$.

<u>**Theorem 1.**</u> (a) <u>$\mathcal{R}_{n,j}(\alpha,\zeta)$ is finite.</u>

(b) $R_{n,j,f}(\alpha,\zeta) = |v_F(\alpha)| \cdot R_{n,j,f}$.

The proof of (b) is relatively straightforward and, of course, part (a) is now clear in view of Henniart's solution of the NLLC. Nevertheless, we will sketch an easy proof of (a) in § 3. In the statement of the next result, due to Koch, our previous notations will be extended in an obvious manner so as to make the ground field in question explicit.

<u>**Theorem 2.**</u> (a) $R_{n,j,f}(F) = \frac{1}{f} \sum_{d|f} \mu(\frac{f}{d}) \cdot R_{n/f,[j/f],1}(F_d)$,

<u>where</u> $F_d/F \subseteq F^{sep}/F$ <u>is the unramified extension of degree</u> d.

(b) <u>Write</u> $n = n_p \cdot m$ <u>with</u> n_p <u>a p-power and</u> $(p,m) = 1$. <u>Then</u>

$$R_{n,j,1}(F) = \sum_L R_{n_p,j,1}(L) \quad (= m \cdot R_{n_p,j,1}(F)) ,$$

<u>where</u> L <u>runs over the totally ramified extensions of degree</u> m <u>of</u> F <u>(inside</u> F^{sep}).

Part (a) reduces (*) to (**), and (b) reduces the general case of (**) to the case where n is a p-power (there are m such L). The proofs use Clifford-Mackey theory for the group extensions

$$1 \to G_{F,o} \to G_F \to G_F/G_{F,o} \simeq \hat{\mathbb{Z}} \to 0 \quad \text{(part (a))}$$

$$1 \to G_{F,1} \to G_F \to G_F/G_{F,1} \to 1 \quad \text{(part (b))} .$$

Since $R_{1,j,1}$ is of the desired form, by Example (b), Theorem 2 proves the NLLC for all n with $p \nmid n$.

§ 3. A Proof of Theorem 1(a).

We first note some standard facts (e. g., [64]). For any finite field extension M/L with $F \subseteq L \subseteq M \subseteq F^{sep}$, put

$$sw(M/L) := d(M/L) - e(M/L) + 1 ,$$

where $d(\cdot)$ denotes the exponent of the different. Then $sw(M/L) \geq 0$, and equality holds precisely when M/L is tamely ramified.
The following formulas will be implicitely used below:

(a) For $\Pi \in \hat{G}_M$ and $\alpha \in L^*$,

$$sw(\mathrm{Ind}_{G_M}^{G_L} \Pi) = f(M/L)\cdot(sw(M/L)\cdot\dim \Pi + sw(\Pi)) ,$$

$$\det(\mathrm{Ind}_{G_M}^{G_L} \Pi)(\alpha) = \det(\mathrm{Ind}_{G_M}^{G_L} 1)(\alpha)^{\dim \Pi}\cdot\det \Pi(\alpha) .$$

(b) For $\Pi \in \hat{G}_L$ and M/L tamely ramified,

$$sw(\Pi|_{G_M}) = e(M/L)\cdot sw(\Pi) ,$$

$$\det(\Pi|_{G_M})(\beta) = \det \Pi(N_{M/L}\, \beta) \qquad (\beta \in M^*) .$$

Now let n, j, α and ζ be given, as above. The proof of Theorem 1(a) proceeds by induction on $\dim \Pi$, for $\Pi \in \mathcal{R}_{n,j}(\alpha,\zeta)$. Those Π with $\dim \Pi = 1$ correspond, by class field theory, to characters of F^* which factor over $F^*/< U_F^{[j/n]+1}, \alpha^{n\cdot t} >$, where t denotes the order of ζ. As $v_F(\alpha) \neq 0$, the latter group is finite and so there are only finitely many such Π's. Suppose now that Π is induced, say $\Pi = \mathrm{Ind}_{G_L}^{G_F} \Pi'$ for $\Pi' \in \hat{G}_L$, L/F a finite subextension of F^{sep}/F. Then $[L:F] \leq n$ and $sw(L/F) \leq j$, whence only finitely many such L do exist (Krasner, Serre [100]). Moreover,

$$\Pi' \in \mathcal{R}_{n/[L:F],j}(\alpha, \pm\zeta)(L) .$$

If $L \neq F$, then our induction hypothesis implies that there are only finitely many possibilities for Π', hence for Π. In particular, we see that $\mathcal{R}_{n,j}(\alpha,\zeta)$ contains only finitely many monomial representations, and we can concentrate on the case where Π is primitive.

Put $G := \Pi(G_F) \subseteq GL_n(\mathbb{C})$. Then G is a finite linear group of the form $G = G_{E/F}$ for E/F finite Galois. We let $G_0 \subseteq G$ denote the inertia group of G and $G_1 \subseteq G_0$ the wild ramification group, a finite p-group. Then G_1 is contained in the Fitting subgroup $N := \mathrm{Fitt}(G)$ of G, and so G/N is metacyclic and N corresponds to a tame Galois extension H/F. There are upper bounds for the index $[G:\mathrm{Fitt}(G)]$ in terms of n (for example, Jordan's theorem yields a rough bound). So, again by the result of Krasner-Serre quoted above, there are only finitely many possibilities for H. Now we use the following generalization of Blichfeldt's theorem (cf.[53 , Theorem 6.22] for an even more general result):

Lemma. Let G be a finite group and let U be a normal subgroup of G such that G/U is supersolvable. Then, for each $\Pi \in \hat{G}$, there exists a subgroup $H \leq G$ with $U \subseteq H$ and a $\Sigma \in \hat{H}$ with $\Sigma|_U$ irreducible such that $\Pi = \mathrm{Ind}_H^G \Sigma$.

Inasmuch as Π is primitive, we conclude from the lemma that $(\Pi|_N = \Pi|_{G_H})$ is irreducible. Hence the same lemma implies that $\Pi|_N = \Pi|_{G_H}$ is monomial. Moreover, $\Pi|_{G_H} \in \mathcal{R}_{n,j\,e(H/F)}(\alpha, \zeta^{[H:F]})(H)$ and so, by our above remark about monomial representations, there are only finitely many possibilities for $\Pi|_{G_H}$. Since the different Π's corresponding to one $\Pi|_{G_H}$ only differ by a character of $G_{H/F}$, the number of possible Π's is bounded. This completes the proof.

References

[53] I.M. Isaacs: Character Theory of Finite Groups. Academic Press, New York 1976.

[61] H.Koch: Eisensteinsche Polynomfolgen und Arithmetik in Divisionsalgebren über lokalen Körpern; Math. Nachr. 104 (1981), 239 - 251.

[63] H. Koch: Bemerkungen zur numerischen lokalen Langlands-Vermutung; Trudy Math. Inst. Steklov 163 (1984), 108 - 114.

[64] H.Koch, E.-W. Zink: Zur Korrespondenz von Darstellungen der Galoisgruppen und der zentralen Divisionsalgebren über lokalen Körpern (der zahme Fall); Math. Nachr. 98 (1980), 83 - 119.

[100] J.P. Serre: Une "formule de masse" pour les extensions totalement ramifiées de degreé donné d'un corps local; C.R. Acad. Sci. Paris, Serie A, 286 (1978), 1031 - 1036.

Contemporary Mathematics
Volume 86, 1989

RAMIFICATIONS OF WEIL-REPRESENTATIONS OF LOCAL GALOIS GROUPS

by H. Opolka, Göttingen

Introduction

This is a report on the ramification properties of a certain class of representations of the absolute Galois group $G_F = Gal(F/F)$ of a local number field $F/\mathbb{Q}_p$. The representations in question are Weil-representations with respect to the wild ramification subgroup $P_F \leq G_F$ which means that their corresponding projective representations are irreducible and abelian when restricted to P_F. They occur in the work of H. Koch [59] who shows that every primitive representation of G_F is a Weil-representation with respect to P_F.

In the first section we briefly decribe their classification according to [116] which follows the method in [59]. In the second section we consider the problem of computing the (exponent of the) conductor of a projective Weil-representation which is defined to be the minimal conductor of its leftings. A complete answer is known only in special cases, for instance for Weil-representations of degree p. Here we follow [8], chapter II, and also [41], [117]. This problem leads to a certain filtration in the alternating square of the multiplicative group of a local number field which is induced by the upper ramification group filtration in its maximal nilpotent class two extension. A main part of W. W. Zink's work on local Galois representations is concerned with this filtration [117], [119]. In particular, he computes the jumps in this filtration in special case. These results are described in section 3.

Convention

A linear or projective finite dimensional complex representation of a profinite group is supposed to be continuous, so its kernel is of finite index.

§ 1. Weil-representations of local Galois groups

The basic references for this sections are [59] and [116].
Let U be a profinite group. According to H. Koch a projective representation $\Pi: U \to PGL(n,\mathbb{C})$ is called a Weil-representation if there is a closed normal subgroup

$A \trianglelefteq U$ such that the restriction Π_A of Π to A is irreducible and has abelian image. A linear representation $\Delta: U \to GL(n,\mathbb{C})$ is called a Weil-representation if the corresponding projective representation

$$\overline{\Delta}: U \xrightarrow{\Delta} GL(n,\mathbb{C}) \longrightarrow PGL(n,\mathbb{C})$$

is a Weil-representation. For instance, using Clifford's tensor product theorem one can easily show that every primitive (projective) representation with solvable image is isomorphic to a tensor product of Weil-representations. However, there are profinite groups with nonsolvable - primitive and imprimitive - Weil-representations. Weil-representations occur also in connection with the classification of primitive projective groups, see [7]. (Recall that a projective representation Π of a profinite group U is called irreducible resp. primitive if every lifting of Π to a linear representation of a central group extension of U is irreducible resp. primitive, i.e. irreducible and not induced by a representation of a proper subgroup.) If $A \trianglelefteq U$ is a closed normal subgroup we denote by $\mathcal{W}(U,A)$ the set of isomorphism classes (Π) of projective Weil-representations Π of U with respect to A.

Let $F/\mathbb{Q}_p$ be a finite extension, let $\overline{F}$ be an algebraic closure of F and let F_1/F be the maximal tamely ramified extension in F. Denote by $G_F = \mathrm{Gal}(\overline{F}/F)$ the absolute Galois group of F. The ramification subgroup $P_F \leq G_F$ is the fix group of F_1. Let $T_F := \mathrm{Gal}(F_1/F)$. A basic observation of H. Koch compare [59] says that the set of isomorphism classes of primitive projective representations of G_F, which we denote by $\mathcal{PP}(G_F)$, is contained in the set of isomorphism classes of Weil-representations $\mathcal{W}(G_F,P_F)$ of G_F with respect to P_F. Our first task is to give a description of both sets in terms of symplectic T_F-modules. The principal units U_V^1 of the finite normal tamely ramified extensions V/F constitute a projective system with respect to the norm maps $\mathrm{Norm}_{V/V'}: U_V^1 \to U_{V'}^1$, $V' \subseteq V$. Denote by $U^1 = \varprojlim U_V^1$, V/F, the corresponding projective limit. U^1 is a continuous T_F-module. Let $\mathrm{Alt}(U^1)$ denote the set of continuous alternating pairings $\omega: U^1 \times U^1 \to \mathbb{C}^*$; here "continuous" means that the radical $R(\omega)$ is of finite index in U^1. The action of T_F on U^1 induces an action of T_F on $\mathrm{Alt}(U^1)$. Let $\mathrm{Alt}(U^1)^{T_F}$ be the corresponding set of T_F-invariant alternating pairings on U^1. Then one can construct a map

$$\mathcal{W}(G_F,P_F) \xrightarrow{\rho} \mathrm{Alt}(U^1)^{T_F}. \tag{1.1}$$

The main properties of ρ are collected in the following theorem.

(1.2) Theorem. (a) If $p \neq 2$ then ρ is surjective.

(b) If $(\Pi) \in \mathcal{W}(G_F, P_F)$ and if $\omega = \rho((\Pi))$ is the corresponding T_F-invariant alternating pairing on U^1 then $\deg(\Pi)^2 = (U^1 : R(\omega))$.

(c) For $\omega \in \mathrm{Alt}(U^1)^{T_F}$ the preimage $\rho^{-1}(\omega)$ is a principal homogeneous space over $H^1(T_F, U^1/R(\omega))$.

(d) Let $(\Pi) \in \mathcal{W}(G_F, P_F)$ with $\rho((\Pi)) = \omega$. Then P is imprimitive if and only if $U^1/R(\omega)$ contains a isotropic T_F-submodule.

For the following assume $p \neq 2$.

(e) Let $(\Pi) \in \mathcal{W}(G_F, P_F)$ with $\rho((\Pi)) = \omega$. Then the kernel field of Π is an abelian fully ramified p-extension of the kernel field V of the action of T_F on $U^1/R(\omega)$ and the corresponding norm subgroup N in V^* has the form $N = R(\omega)\cdot C_V$ where $\omega = \rho((\Pi))$ is viewed as a pairing on U_V^1 and where C_V is a $\mathrm{Gal}(V/F)$-submodule in V^* such that $V^* \cong C_V \times U_V^1$.

(f) Let $\omega \in \mathrm{Alt}(U_1)^{T_F}$ and let V/F denote the kernel field of the action of T_F on $U^1/R(\omega)$. Then the representations in the preimage $\rho^{-1}(\omega) \in \mathcal{W}(G_F, P_F)$ define exactly $(U_F^1 : U_F^1 \cap R(\omega))$ different kernel fields with Galois groups isomorphic to the semidirect product $U_V^1/R(\omega) \rtimes \mathrm{Gal}(V/F)$. The subset of all $(\Pi) \in \rho^{-1}(\omega)$ with a fixed kernel field K can be identified with $H^1(\mathrm{Gal}(V/F), U_V^1/R(\omega))$.

A pairing $\omega \in \mathrm{Alt}(U^1)$ is called anisotropic if $U^1/R(\omega)$ is an anisotropic symplectic T_F-module; in this case, for group theoretical reasons, $U^1/R(\omega)$ is p-elementary. Denote by $\mathrm{Alt}_o(U_1)$ the set of all $\omega \in \mathrm{Alt}(U^1)$ which are anisotropic. Then by (1.2), (d), the map (1.1) yields a map

$$\mathcal{PP}(G_F) \xrightarrow{\rho_o} \mathrm{Alt}_o(U^1)^{T_F}. \tag{1.3}$$

The main properties of ρ_o are as follows.

(1.4) Theorem. (a) ρ_o is bijective.

(b) Let $\omega \in \mathrm{Alt}_o(U^1)^{T_F}$ and let V be the kernel field of the action of T_F on $U^1/R(\omega)$. View ω as a pairing on U_V^1. Then $U_F^1 \leq R(\omega)$, and the corresponding norm subgroup $R(\omega)\cdot C_V$ in V^* is independent of the chosen complementary $\mathrm{Gal}(V/F)$-module C_V.

(c) Let d be a natural number. Then for all $\omega \in \mathrm{Alt}_o(U^1)^{T_F}$ with

$(U^1:R(\omega)) = p^{2d}$ the kernel field of the action of T_F on $U^1/R(\omega)$ is contained in a finite extension V/F with ramification index $e_{V/F} = \mathrm{kgV}\ \{p^i+1 \mid i = 1,\dots,d\}$.

Note that (1.4) is true without the restriction $p \neq 2$.

We shall briefly explain the map ρ mentioned under (1.1).

Let U be a finite group, let $A \trianglelefteq U$ be an abelian normal subgroup and let $\mathcal{W}(U,A)$ be the set of isomorphism classes of projective Weil-representations of U with respect to A. Let $\mathcal{PI}(A)$ be the set of isomorphism classes of irreducible projective representations of A. The restriction induces a map res: $\mathcal{W}(U,A) \to \mathcal{PI}(A)$. The natural action of $G := U/A$ on A induces an action of G on $\mathcal{PI}(A)$, and by Clifford's theory every irreducible projective representation Π of A which is the restriction of a projective representation of U satisfies $\Pi^\sigma = \Pi$ for all $\sigma \in G$, i.e. $(\Pi) \in \mathcal{PI}(A)^G$. So we have a map res: $\mathcal{W}(U,A) \to \mathcal{PI}(A)^G$. It is well known that there is a bijective correspondence between $\mathcal{PI}(A)$ and the set $\mathrm{Alt}(A) = (A \wedge A)^\wedge$ of alternating (symplectic) pairings $\omega: A \times A \to \mathbb{C}^*$ which is obtained as follows: For an irreducible projective representation $\Pi: A \to PGL(n,\mathbb{C})$ choose a section $\tilde{\Pi}: A \to GL(n,\mathbb{C})$, i.e. $\tilde{\Pi}$ is a map such that its composition with the natural map $GL(n,\mathbb{C}) \to PGL(n,\mathbb{C})$ yields Π, and put $\omega_{(\Pi)}(x,y)\cdot \mathrm{Id} := \tilde{\Pi}(x)\tilde{\Pi}(y)\tilde{\Pi}(x)^{-1}\tilde{\Pi}(y)^{-1}$, $x,y \in A$. Schur's lemma implies that the commutator on the right is a scalar which obviously depends only on the isomorphism class of Π. Moreover, the correspondence $\mathcal{PI}(A) \ni (\Pi) \longmapsto \omega_{(\Pi)} \in \mathrm{Alt}(A)$ is compatible with the action of G, i.e. we have $\omega_{(\Pi^\sigma)}(x,y) = \omega_{(\Pi)}(\sigma(x), \sigma(y))$, $\sigma \in G$, $x,y \in A$. So we have a composition of maps

$$\Omega : \mathcal{W}(U,A) \xrightarrow{\text{res}} \mathcal{PI}(A)^G \xrightarrow{\sim} \mathrm{Alt}(A)^G . \tag{1.5}$$

By Schur's lemma the kernel of $(\Pi) \in \mathcal{PI}(A)$ is equal to the radical $R(\omega)$ of the corresponding alternating pairing $\omega = \omega_{(\Pi)}$, and for $\omega \in \mathrm{Alt}(A)^G$ the factor group $A/R(\omega)$ is a nondegenerate symplectic G-module. The following statement summarizes the main properties of the map Ω.

(1.6) (a) <u>If the group extension</u> $1 \to A \to U \to G \to 1$ <u>splits then the map</u> Ω <u>is surjective.</u>

(b) <u>If</u> $\omega \in \mathrm{Alt}(A)^G$ <u>is in the image of the map</u> (1.5) <u>then the preimage</u> $\Omega^{-1}(\omega)$ <u>is a principal homogeneous space over</u> $H^1(G, A/R(\omega))$.

(c) <u>If</u> $(\Pi) \in \mathcal{W}(U,A)$ <u>and</u> $\omega = \Omega((\Pi))$ <u>then</u> $\deg(\Pi)^2 = (A:R(\omega))$.

(d) <u>Let</u> $(\Pi) \in \mathcal{W}(U,A)$ <u>and</u> $\omega = \Omega((\Pi))$. <u>Then</u> P <u>is imprimitive if and only if</u>

$A/R(\omega)$ contains an isotropic G-submodule. If (Π) is primitive then the primary components of $A/R(\omega)$ are elementary abelian.

For a full treatment of (1.6) see [116]. We explain statement (a):
Take $\omega \in \mathrm{Alt}(A)^G$ and let $\Pi: A \to PGL(n,\mathbb{C})$ be an irreducible projective representation of A which corresponds to ω. Then $\Pi^\sigma \cong \Pi$ for all $\sigma \in G$ which implies that there is a projective representation $\Gamma: G \to PGL(n,\mathbb{C})$ such that $\Gamma(\sigma)\Pi(x)\Gamma(\sigma)^{-1} = \Pi(\sigma(x))$ for all $\sigma \in G$, $x \in A$. We get an irreducible projective representation $\Lambda: U = A \rtimes G \to PGL(n,\mathbb{C})$ such that $\Omega((\Lambda)) = \omega$ by defining $\Lambda((x,\sigma)) := \Pi(x)\Gamma(\sigma)$, $x \in A$, $\sigma \in G$.

We remark that (1.6) is also true for profinite U and A.

Now identify U^1 with the factor commutator group P_F^{ab} as follows: For a finite normal tamely ramified extension V/F let V_{ab}/V denote the maximal abelian extension contained in $\overline{F}$. Let $\tilde{V}^* := \varprojlim_n V^*/V^{*n}$ be the compactification of the multiplicative group of V. The local reciprocity map yields an exact sequence $1 \to \tilde{V}^* \to \mathrm{Gal}(V_{ab}/F) \to \mathrm{Gal}(V/F) \to 1$. Since V/F is tamely ramified we can identify U_V^1 with the ramification subgroup $\mathrm{Gal}(V_{ab}/F)_1$ in a way which is compatible with subextensions $V'/F \subseteq V/F$. Since P_F^{ab} is the projective limit of the groups $\mathrm{Gal}(V_{ab}/F)_1$ with respect to the restriction maps we have $U^1 \xrightarrow{\sim} P_F^{ab}$. This isomorphism is equivariant with respect to the continuos T_F-action. Therefore we have an isomorphism of the corresponding sets of T_F-invariant continuous alternating pairings:

$$\mathrm{Alt}(U^1)^{T_F} \xrightarrow{\sim} \mathrm{Alt}(P_F^{ab})^{T_F} . \tag{1.7}$$

$\mathrm{Alt}(U^1)^{T_F}$ is the direct limit of the sets $\mathrm{Alt}(U_V^1)^{\mathrm{Gal}(V/F)}$ for tamely ramified finite normal extensions V/F with respect to the maps which are induced by the norm and $\mathrm{Alt}(P_F^{ab})^{T_F}$ is the direct limit of the sets $\mathrm{Alt}(\mathrm{Gal}(V_{ab}/F)_1)^{\mathrm{Gal}(V/F)}$ with respect to the maps which are induced by restriction.
These considerations in mind we can use the maps (1.5) and (1.7) to construct the maps

$$\text{(1.8)} \qquad \rho: \mathcal{W}(G_F,P_F) \to \mathrm{Alt}(U^1)^{T_F} \quad \text{and} \quad \rho_o: \mathcal{PP}(G_F) \to \mathrm{Alt}_o(U^1)^{T_F} . \qquad \text{(1.8)}$$

In view of (1.6), (a), the surjectivity of these maps, which was asserted in statement (a) of (1.2) and (1.4), follows in the case $p \neq 2$ from the well known fact that the group extension $1 \to P_F/[P_F,P_F] \to G_F/[P_F,P_F] \to T_F \to 1$ splits. The injectivity of the map ρ_o, which was asserted in statement (a) of (1.4), is by (1.6), (b), equivalent to the statement $H^1(T_F,U^1/R(\omega)) = 0$ for every

$\omega \in Alt_o(U^1)^{T_F}$, which is indeed true. Full proofs for the results in this section are contained in [59] and [116].

§ 2. Lifting of local Galois representations

The basic references for this section are [8], chapter I; [41]; [117]. Let G be a profinite group. A linear representation $\Delta: G \to GL(n,\mathbb{C})$ is called a lifting of the projective representation $\Pi: G \to PGL(n,\mathbb{C})$ if Δ composed with the natural map $GL(n,\mathbb{C}) \to PGL(n,\mathbb{C})$ yields Π. If Δ is a lifting of Π then $\lambda \otimes \Delta$ is also a lifting of Π for every character of finite order $\lambda: G \to \mathbb{C}^*$, and every lifting of Π is of this form. The exact sequence $1 \to \mathbb{C}^* \to GL(n,\mathbb{C}) \to PGL(n,\mathbb{C}) \to 1$ yields an exact sequence of cohomology sets

$$\ldots \to Hom(G,GL(n,\mathbb{C})) \to Hom(G,PGL(n,\mathbb{C}) \xrightarrow{\delta} H^2(G,\mathbb{C}^*) ,$$

and (Π) determines a cocycle class $\delta(\Pi) \in H^2(G,\mathbb{C}^*)$. We have

(2.1) The projective representation Π of G has a lifting if and only if its cocycle class $\delta(\Pi) \in H^2(G,\mathbb{C}^*)$ is trivial.

Let G_o be an open normal subgroup of G and put $G = G/G_o$. Then the Hochschild-Serre spectral sequence yields the following exact sequence

$$0 \to \hat{G} \xrightarrow{inf} \hat{G} \xrightarrow{res} \hat{G}_o^G \xrightarrow{\tau} H^2(G,\mathbb{C}^*) \xrightarrow{inf} H^2(G,\mathbb{C}^*) ; \qquad (2.2)$$

here and in the following for any abelian locally compact topological group C we denote by $\hat{C}$ its Pontrjagin-dual; τ is the so called transgression map given by

$$\tau(\chi)(\sigma,\tau) = \chi(c(\sigma,\tau)) , \quad \chi \in \hat{G}_o^G , \quad \sigma,\tau \in G ;$$

where c is a cocycle corresponding to the group extension $1 \to G_o \to G \to G \to 1$. Denote by $I(G,G_o)$ the set of isomorphism classes of irreducible linear representations Δ of G such that the restriction Δ_{G_0} is a scalar representation and let $P(G,G_o)$ denote the set of isomorphism classes of irreducible projective representations Π of G such that $G_o \leq Ker(\Pi)$.
We get maps

$$\nu : I(G,G_o) \to P(G,G_o) \quad \text{and} \quad res: I(G,G_o) \to \hat{G}_o^G .$$

We have also the cocycle map $\delta: P(G,G_o) \to H^2(G,\mathbb{C}^*)$ and the transgression map $\tau: \hat{G}_o^G \to H^2(G,\mathbb{C}^*)$ mentioned above. All these maps fit together in the following commutative diagram, see [114].

$$\begin{array}{ccc} I(G,G_o) & \xrightarrow{\ \nu\ } & P(G,G_o) \\ {\scriptstyle \text{res}}\Big\downarrow & & \Big\downarrow{\scriptstyle \delta} \\ \hat{G}_o^G & \xrightarrow{\ \tau\ } & H^2(G,\mathbb{C}^*) \end{array} \qquad (2.3)$$

It is easy to see that res and δ are surjective. And ν is surjective if and only if τ is surjective. From (2.1) and (2.2) we deduce

(2.4) <u>The maps ν and τ in the above commutative diagram are surjective for every open normal subgroup $G_o \trianglelefteq G$ if and only if the group $H^2(G,\mathbb{C}^*)$ is trivial.</u>

Let $F/\mathbb{Q}_p$ be a finite extension with absolute Galois group $G_F = \mathrm{Gal}(\bar{F}/F)$ and let K/F be a finite normal subextension of $\bar{F}/F$ with $G_K = \mathrm{Gal}(\bar{F}/K)$ and $G = \mathrm{Gal}(K/F)$. In the situation above put $G = G_F$, $G_o = G_K$. The group extension $1 \to G_K \to G_F \to G \to 1$ determines a cocycle class $(u_o) \in H^2(G,G_K^{ab})$; it is the image of the local fundamental class $(u) \in H^2(G,K^*)$ under the map $H^2(G,K^*) \to H^2(G,G_K^{ab})$ which is induced by the G-equivariant local reciprocity map $K^* \to G_K^{ab}$. So the transgression map $\tau: \hat{G}_K^G \to H^2(G,\mathbb{C}^*)$ is given by

$$\tau(\chi)(\sigma,\tau) := \chi(u_o(\sigma,\tau)), \quad \chi \in \hat{G}_K^G, \quad \sigma,\tau \in G .$$

Now we show

$$H^2(G_F,\mathbb{C}^*) = 0 . \qquad (2.5)$$

Proof (see [8], chapter I): By (2.4) this is equivalent to the statement that for every finite factor group $G = \mathrm{Gal}(K/F)$ of G_F the transgression map $\tau: \hat{G}_K^G \to H^2(G,\mathbb{C}^*)$ is surjective. The local reciprocity map induces a G-equivariant isomorphism $\hat{G}_K \xrightarrow{\sim} \hat{K}^*$. Furthermore, $H^2(G,\mathbb{C}^*) \cong H^3(G,\mathbb{Z})$ is dual to $H^{-3}(G,\mathbb{Z})$, and by the local duality theorem the cup product with the fundamental class induces an isomorphism $H^{-3}(G,\mathbb{Z}) \cong H^{-1}(G,K^*) = K^{*N}/K^{*I}$, where $K^{*N} = \{x \in K^* \mid \mathrm{Norm}_{K/k}(x) = 1\}$ and K^{*I} is generated by all elements x^σ/x, $x \in K^*$, $\sigma \in G$. Finally we have a restriction map $\hat{K}^{*G} \to H^{-1}(G,K^*)^\wedge$.

A straight forward computation shows that all maps fit together in the following commutative diagram.

$$\begin{array}{ccc} \hat{G}_K^G & \xrightarrow{\ \tau\ } & H^2(G,\mathbb{C}^*) \\ \text{"Artin"}\ \big\downarrow & & \big\downarrow\ \text{"cup"} \\ \hat{K}^{*G} & \xrightarrow{\ res\ } & H^{-1}(G,K^*)^{\wedge} \end{array} \qquad (2.6)$$

It shows in particular that τ is surjective if and only if every character on the compact group K^{*N} which is trivial on K^{*I} can be extended to a G-invariant character on K^*. This last statement is easily deduced from the following well known extension lemma in the theory of topological groups, applied to $C = K^{*N}$ and $B = 1$.

(2.7) Extension lemma. Let H be an abelian locally compact topological group, let B be a closed subgroup, let C be a compact subgroup and let χ' be a character of C. Then χ' can be extended to a continuous homomorphism $\chi: H \to C^*$ with $B \leq Ker(\chi)$ if and only if $B \cap C \leq Ker(\chi')$.

The idea to use the extension lemma in this and similar contexts is contained in [8], chapter I.

Let $\Pi: G_F \to PGL(n,\mathbb{C})$ be a projective representation with image $G = Gal(K/F)$ and let $\Delta: G_F \to GL(n,\mathbb{C})$ be a lifting of Π. Then the restriction of Δ to G_K is a scalar representation and therefore determines uniquely a G-invariant character $\psi: G_K \to \mathbb{C}^*$ which is called the centric character of Δ. A G-invariant character of G_K which is centric for some lifting of Π is called centric for Π. If Δ, Δ' are two liftings of Π with the same centric character ψ then $\Delta' = \lambda \otimes \Delta$ for some character $\lambda: G \to \mathbb{C}^*$. Thus the centric character determines the lifting up to the operation of twisting with characters of G. A G-invariant $\chi: K^* \to \mathbb{C}^*$ is centric for a lifting Δ of Π or for Π if the character $\psi: G_K \to \mathbb{C}^*$ which corresponds to χ under the local reciprocity map is centric for Δ or Π. If $\chi: K^* \to \mathbb{C}^*$ is a centric character for Π then the restriction of χ to K^{*N} is uniquely determined by Π and is denoted by χ_Π; in the diagram (2.6) χ_Π-corresponds to $\delta(\Pi)$.
The following statement is another useful application of the extension lemma, see [8], chapter I.

(2.8) Let $\Pi: G_F \to PGL(n,\mathbb{C})$ be a projective representation with image $G = Gal(K/F)$ and let $(f) = \delta(\Pi)$ be the corresponding cocycle class in $H^2(G,\mathbb{C}^*)$. Let E/F be an extension such that $E \subset K$. Then there is a character $\psi: E \to \mathbb{C}^*$ such that $\psi \circ Norm_{K/E}$ is centric for Π if and only if the restriction of (f) to $Gal(K/E)$ is trivial.

We have to introduce some notation, see [99]. If M/L is a finite Galois extension of local number fields, let $b_o(M/L)$ be the largest integer m such that $Gal(M/L)_m \neq 1$, let $e_{M/L}$ be the ramification index, $f_{M/L}$ the residue class degree, $\mathcal{P}_L$ the prime ideal in the ring of integers of L, $\mathcal{D}_{M/L}$ the different, v_L the valuation on L such that $v_L(L^*) = \mathbb{Z}$, U_L^m the group of units of the ring of integers of L such that $v_L(x-1) \geq m$ and $\varphi_{M/L}$ the Herbrand function which connects the upper and lower ramification groups: $Gal(M/L)^{\varphi(u)} = Gal(M/L)_u$; and $\psi_{M/L}$ the inverse of $\varphi_{M/L}$: $Gal(M/L)^v = Gal(M/L)_{\psi(v)}$.

If $\Delta: G_F \to GL(n,\mathbb{C})$ is a Galois representation then we denote by $a(\Delta) \in \mathbb{Z}$ the (exponent of the) Artin conductor of Δ. If Δ is one-dimensional then Δ corresponds by local class field theory to a character $\lambda: F^* \to \mathbb{C}^*$ and $a(\Delta)$ is the conductor of λ, i.e. the smallest integer m such that λ is trivial on U_F^m. In general one can define $a(\Delta)$ by requiring that $a(\Delta_1 \oplus \Delta_2) = a(\Delta_1) + a(\Delta_2)$ and that $a(\Delta) = (1 + \sup\{m \mid \Delta(G_F^m) \neq 1\})/n$ for irreducible Δ. If Π is a projective representation of G_F then its Artin conductor $a(\Pi)$ is defined to be the smallest possible value of $a(\Delta)$ as Δ ranges over all liftings of Π.

We shall also need the following obvious application of the extension lemma.

(2.9) Let $\Pi: G_F \to PGL(n,\mathbb{C})$ be a projective representation with image $G = Gal(K/F)$. It is possible to find a character $\chi: K^* \to \mathbb{C}^*$ which is centric for Π and of conductor m if and only if

$$(*) \qquad U_K^m \cap K^{*N} \leq K^{*I},$$

i.e. if and only if the map $H^{-1}(G,U_K^m) \to H^{-1}(G,K^*)$ is trivial. Of course, (*) is true for sufficiently large m since for large enough m the group U_K^m is contained in a cohomologically trivial subgroup of K^*.

We denote by $b_{-1}(K/F)$ the largest integer m that does not satisfy (*). Thus the minimal value of the conductor $a(\chi)$ for χ a character centric for Π is $b_{-1}(K/F) + 1$.

The following basic lemma is important for computing the conductor of a projective Weil-representation.

(2.10) **Lemma.** Let $\Delta: G_F \to GL(n,\mathbb{C})$ be an irreducible linear representation with image $\tilde{G} = Gal(L/F)$ and let $r = b_0(L/F)$. Assume that there is a subgroup $H \leq \tilde{G}_r$ such that the restriction Δ_H is a nontrivial scalar representation. Let E/F be a subextension of L/F which is fixed by H . Then

$$e_{E/F}a(\Delta) = n\cdot v_E(\mathcal{D}_{E/F}) + a(\Delta_H) .$$

(2.11) **Remark.** If the restriction of Δ to the ramification group $\tilde{G}_1$ is irreducible, then $\tilde{G}_r = H$ works.

Proof (see [8], chapter II): Let $\chi_\Delta: \tilde{G} \to \mathbb{C}$ be the character of Δ and let $\theta_{\tilde{G}}$ be the character of the Artin representation of $\tilde{G}$, see [99], p. 107. Then $\theta_{\tilde{G}} = a(\Delta)\cdot\chi_\Delta$ + further terms not containing χ_Δ , i.e.

$$a(\Delta) = (\chi_\Delta, \theta_{\tilde{G}}) = \frac{1}{\#(\tilde{G})} \sum_{g\in\tilde{G}} \chi_\Delta(g)\bar{\theta}_{\tilde{G}}(g) .$$

Since

$$\chi_\Delta(xy) = trace(\Delta(xy)) = \chi(x)\chi_\Delta(y) , \quad x \in H , \quad y \in G ,$$

where χ is the centric character for Δ , and since $\theta_{\tilde{G}}$ is constant on nontrivial cosets $C = gH$, $g \notin H$, we have

$$\sum_{g\in C} \chi_\Delta(g)\bar{\theta}_{\tilde{G}}(g) = \text{constant} \cdot \sum_{x\in H} \chi(x) = 0$$

and therefore

$$\sum_{g\in C} \chi_\Delta(g)\bar{\theta}_{\tilde{G}}(g) = \sum_{g\in H'} \chi_\Delta(g)\bar{\theta}_{\tilde{G}}(g)$$

for any subgroup H' which contains H . If we take $H' = \tilde{G}$ in this formula and use the well known fact [99], p. 108,

$$\theta_{\tilde{G}|H} = f_{E/F}(v_E(\mathcal{D}_{E/F})\rho_H + \theta_H)$$

where ρ_H is the character of the regular representation of H , then a straightforward computation gives the asserted formula.

(2.12) Corollary. With the notation of (2.10) assume that the restriction $\Delta_{\tilde{G}_1}$ is irreducible. Then

$$e_{F_1/F}a(\Delta) = n(e_{F_1/F} - 1) + a(\Delta_{\tilde{G}_1}) .$$

Proof. By (2.11) we can take E to be the maximal tamely ramified subextension F_1/F in L/F . It is well known that $v_{F_1}(\mathcal{D}_{F_1/F}) = e_{F_1/F} - 1$.

(2.13) Corollary. With the notation of (2.10) assume that the restriction $\Delta_{\tilde{G}_1}$ is irreducible. Let $\bar{\Delta}:G_F \to PGL(n,\mathbb{C})$ be the projective representation corresponding to Δ with image $G = Gal(K/F)$. Let $\chi:K^* \to \mathbb{C}^*$ be a centric character for Δ . Then

$$e_{K/F}a(\Delta) = n\cdot v_K(\mathcal{D}_{K/F}) + a(\chi)$$

Proof. Apply (2.10) with $E = K$ and $H = \tilde{G}_r$ which is possible by (2.11).

The following lemma, which is well known in genus theory, see e.g. [102], § 1, is useful for computing the minimal conductor of a centric character.

(2.14) Let F'/F be the maximal tamely ramified subextension in K/F . Then $b_{-1}(K/F) \leq \max \{b_o(K/F) , b_{-1}(K/F')\}$.

We are ready to compute the Artin-conductor of a p-dimensional primitive projective representation of G_F .

(2.15) Theorem. ([8], [41]) Let $\Pi: G_F \to PGL(p,\mathbb{C})$ be a primitive projective Galois representation of degree p (= residue characteristic of F) with image $G = Gal(K/F)$. Then

$$a(P) = p + (\frac{p+1}{t})\, b_o(K/F) ,$$

where $t = e_{F'/F}$ is the tame ramification index of K/F .

If $\Delta: G_F \to GL(p,\mathbb{C})$ is any lifting of Π and if $\chi: K^* \to \mathbb{C}^*$ is a centric character for Δ, then (2.14) gives

$$e_{K/F}a(\Delta) = p\cdot v_K(\mathcal{D}_{K/F}) + a(\chi) .$$

Since $e_{K/F} = p^2\cdot t$ and $v_K(\mathcal{D}_{K/F}) = \sum_{i=0}^{\infty} \#(G_i) - 1 = p^2\cdot t - 1 + r(p^2-1)$, $r = b_0(K/F)$, we obtain

$$p\cdot t\cdot a(\Delta) = rp^2 + t\cdot p^2 - r - 1 + a(\chi) .$$

So (2.15) is a consequence of the following statement about centric characters for Π :

$$\operatorname*{Min}_{\chi} a(\chi) = (p+1)b_0(K/F) + 1 , \tag{2.16}$$

the minimum being taken over all characters χ which are centric for Π, which follows from the following two assertions

$$b_{-1}(K/F) = b_{-1}(K/F') \tag{2.17}$$

$$b_{-1}(K/F') = (p+1)b_0(K/F') ; \tag{2.18}$$

F'/F denotes the maximal tamely ramified subextension of K/F. In view of (2.12) and (2.13) these assertions are easily seen to be equivalent to the following claims: Let Π' denote the restriction of Π to G_1. Then

$$t\cdot a(\Pi) = p(t-1) + a(\Pi') \tag{2.19}$$

$$a(\Pi') = p + (p+1)b_0(K/F') . \tag{2.20}$$

First we prove (2.17) using (2.18).
It follows from the structure of primitive Galois representations of degree p that the restriction map res: $H^2(G,\mathbb{C}^*) \to H^2(G_1,\mathbb{C}^*)$ is an isomorphism, see [59]. Therefore by local duality the corestriction map cor: $H^{-1}(G_1,K^*) \to H^{-1}(G,K^*)$ is an isomorphism. This implies: $K^{*\,\mathrm{Norm}_{K/F'}} \cap K^{*\,I_G} = K^{*\,I_{G_1}}$, i.e. $b_{-1}(K/F) \geq b_{-1}(K/F')$. The reverse inequality follows from (2.18) and (2.14).
So we are left to prove (2.20). Let E/F' be an extension of degree p in K. By (2.8) a centric character for Π' is of the form $\psi \circ \mathrm{Norm}_{K/E}$, and it is easy to see that $\mathrm{Ind}(\psi)$ is a lifting of Π and that any lifting is of this form. Assertion (2.20) follows as a special case of the following crucial theorem by

taking K/E to be of degree p and assuming that there is only one break in the filtration of $Gal(K/F)$. It can be shown that this last assumption implies that $b_0(K/F) = b_0(K/E) = b_0(E/F)$ so that in the notation of the theorem $r' = r + 1$.

(2.21) **Theorem.** _Let E/F be a Galois extension of degree p. Let $\psi_0: E^* \to \mathbb{C}^*$ be a wildly ramified character such that $\Pi = Ind(\psi_0)$ is an irreducible p-dimensional representation of G_F. Let K be the fixed field of the kernel of the corresponding projective representation $\bar{\Pi}$ of G_F. Define r and r' by $r = 0$ if E/F is unramified and $r = b_0(E/F)$ if E/F is ramified, and let r' be the least integer m such that $U_E^m \leq Norm_{K/E}(K^*)$. Then the minimal possible value of the exponent of the Artin conductor $a(\psi)$ for $\psi: E^* \to \mathbb{C}^*$ a character such that $Ind(\psi)$ lifts $\bar{\Pi}$ is $r + r'$._

Proof. (see [8], chapter II) The case in which E/F is unramified follows the lines of the argument below; the proof is simpler and will be omitted.

If $\lambda: F^* \to \mathbb{C}^*$ is a character of F^* then $\lambda_E = \lambda \circ Norm_{E/F}$ is the character of E^* obtained by composing with the norm map.

If $\psi: E^* \to \mathbb{C}^*$ is a character such that $Ind(\psi)$ lifts Π then $Ind(\psi) \cong \lambda \otimes Ind(\psi_0) = Ind(\lambda_E \otimes \psi_0)$ for some character $\lambda: F^* \to \mathbb{C}^*$. Since E/F is normal the character of $Ind(\psi)$ vanishes outside G_E and the restriction of $Ind(\psi)$ to G_E is of the form $\sum_g \psi^g$ where the summation is over all elements g of the cyclic group $Gal(E/F)$. Hence $\underline{Ind(\psi)} = \underline{Ind(\psi_0)}$ if and only if $\psi = \lambda_E(\psi_0)^g$ for some $g \in Gal(E/F)$. Since $a(\psi) = a(\psi^g)$ it follows that the proposition is reduced to the following assertion:

$$\inf a(\lambda_E \psi_0) = r + r'$$

where the infinum is over all characters λ of F^*. Let s be a generator of $Gal(E/F)$ and let $\varphi: E^* \to \mathbb{C}^*$ be defined by $\varphi(x) := \psi_0(x^s/x) = \psi_0^{s-1}(x)$. Since K is the fixed field of the kernel of $\bar{\Pi} = \overline{Ind(\psi_0)}$ it follows that $Norm_{K/E}(K^*)$ is the intersection of the kernels of the conjugates of φ under $Gal(E/F)$. Thus the conductor of φ is r'.
Note that $r' \not\equiv 1 \bmod p$ because $F^* \leq Ker(\varphi)$.
Since $Gal(E/F)$ is cyclic a character $\psi: E^* \to \mathbb{C}^*$ is of the form $\lambda_E \psi_0$ if and only if $\psi^{s-1} = \varphi = \psi_0^{s-1}$, i.e. ψ and ψ_0 agree on $Ker(Norm_{E/F}) = E^{*s-1}$. Thus we have a fixed character α on E^{*s-1} defined by $\alpha(x^{s-1}) := \varphi(x)$ and we have to find an extension of α to E^* with the smallest possible conductor. In order to apply the extension lemma the following technical lemma is necessary.

Lemma. Let M/L be a wildly ramified Galois extension of local fields of residue characteristic p such that $Gal(M/L)$ is cyclic of order p. Let s be a generator of $Gal(M/L)$ and let $r = b_0(M/L)$. Any $y \in M^*$ can be written in the form $y = z\pi^m y_1 y_2 y_3 \ldots$ where $z \in L^*$, $0 \leq m < p$, π is a uniformizing parameter in M, $y_i = 1$ if $i \equiv 0 \bmod p$, and for $i \not\equiv 0 \bmod p$ either $y_i = 1$ or $y_i \in U_M^i - U_M^{i+1}$. If y has this form then $v_M(y^{s-1} - 1) = r$ if $m \neq 0$ and $= r + i$, if $m = 0$ and i is the smallest index for which $y_i \neq 1$.

Proof. It is easy to see that y can be represented in the above form. If we apply s to this expression for y and then divide by y we get $y^{s-1} = (\pi^{s-1})^m \prod_i y_i^{s-1}$. By the assumptions above π^{s-1} must have the form $1 + a$ where $v_M(a) = r$. It is a standard fact that if $y = 1 + b$ where $v_M(b) = j > 0$ then $y^{s-1} \equiv 1 + jab \bmod \pi^{j+r+1}$, and the last assertion of the lemma follows.

Now we return to the proof of (2.20). We want to find the minimal conductor of a character ψ that extends α. According to the previous lemma we have $U_E^{r+r'} \cap E^{*s-1} \leq Ker(\alpha)$ because of the definition of α and the fact that F^* and $U_E^{r'} \leq Ker(\varphi)$. By the extension lemma (2.7) we can therefore choose a ψ extending α in such a way that $a(\psi) \leq r + r'$.
Choose y in $U_E^{r'-1}$ with $\varphi(y) \neq 1$, so that $\alpha(y^{s-1}) \neq 1$. By the previous lemma $y^{s-1} \in U^{r+r'-1}$ because $r' \not\equiv 1 \bmod p$. It follows that $a(\psi) \geq r + r'$ for any ψ extending α. This shows that $r + r'$ is the smallest possible value of $a(\psi)$ for a ψ such that $Ind(\psi) = Ind(\psi_0)$.

§ 3. The Zink-filtration

The basic references for this section are [17], [19].

As mentioned earlier the set of isomorphism classes $\mathcal{PI}(A)$ of irreducible projective representations of a profinite abelian group A is in bijective correspondence with the set of continuous alternating pairings $\mathrm{Alt}(A)$. If $(\Pi) \in \mathcal{PI}(A)$ corresponds to $\omega \in \mathrm{Alt}(A)$ then

$$\mathrm{Ker}(\Pi) = R(\omega) \quad \text{and} \quad (A:R(\omega)) = \deg(\Pi)^2 .$$

Let $F/\mathbb{Q}_F$ be a finite extension with absolute Galois group $G_F = \mathrm{Gal}(\bar{F}/F)$ and factor commutator group $A_F = G_F^{ab}$. Define

$$FF^* := \varprojlim_N (F^*/N \wedge F^*/N)$$

where N runs over all open normal subgroups of finite index in F^*. Denote by $\mathrm{Alt}(F^*)$ the Pontrjagin-dual of FF^*, i.e. the set of alternating pairings $\omega : F^* \times F^* \to \mathbb{C}^*$ such that $(F^*: R(\omega)) < \infty$. Then the local reciprocity map, which yields an isomorphism between A_F and the profinite completion F^* of F^*, induces a natural bijection

$$\mathcal{PI}(A_F) \xrightarrow{\sim} \mathrm{Alt}(F^*) . \tag{3.1}$$

A main part of E.W. Zink's work is concerned with a certain filtration on FF^* which is closely related to the conductor of an irreducible projective representation of A_F. To introduce this filtration let F_2/F be the maximal Galois extension of F in $\bar{F}$ such that $\mathrm{Gal}(F_2/F)$ is nilpotent of class 2. Put $H_F = \mathrm{Gal}(F_2/F)$. Then we have a central group extension

$$1 \to H_F^c \to H_F \to A_F \to 1$$

where H_F^c is the closed commutator subgroup of H_F. The commutator pairing induces an isomorphism $A_F \wedge A_F \xrightarrow{\sim} H_F^c$, and therefore the local reciprocity map yields an isomorphism

$$c: FF^* \xrightarrow{\sim} H_F^c .$$

Now let H_F^i, $i \in \mathbb{R}_+$, denote the filtration given by the upper ramification groups in H_F . Then the <u>Zink-filtration</u> $UU_F^i \leq FF^*$, $i \in \mathbb{R}_+$, is defined by

$$c(UU_F^i) = H_F^c \cap H_F^i .$$

Since $H_F^c \leq H_F^o$ we have $UU_F^o = FF^*$. So for $\omega \in \mathrm{Alt}(F^*)$ we define the index $j(\omega) = j_F(\omega)$ by

$$j_F(\omega) := \begin{cases} 0 , & \text{if } \omega \text{ is trivial} \\ \limsup \{i \mid UU_F^i \leq \mathrm{Ker}(\omega)\} , & \text{otherwise.} \end{cases} \tag{3.2}$$

The following statement relates the conductor of an irreducible projective representation of A_F to the index of the corresponding alternating pairing.

<u>If</u> $(\Pi) \in PI(A_F)$ <u>corresponds to</u> $\omega \in \mathrm{Alt}(F^*)$ <u>then we have</u> **(3.3)**

$$sw_F(P)/\deg(\Pi) = j_F(\omega) .$$

For technical reasons we use here and in the following the (exponent of the) Swan conductor $sw_F(\Delta)$ of a linear representation Δ of G_F ; it is related to the (exponent of the) Artin conductor $a(\Delta)$ as follows: $a(\Delta) = \dim V - \dim V_o + sw_F(\Delta)$ where V is a representation space for Δ and V_o is the subspace of V which is elementwise fixed by the inertia group.

<u>Proof.</u> Assume $\omega_{UU_F^i} \neq 1$ and let Δ be a linear representation of H_F which lifts Π . Then the restriction of Δ to H_F^c contains a unique character ψ of degree 1 such that $\psi \circ c = \omega$; especially ψ is nontrivial on $c(UU_F^i) = H_F^c \cap H_F^i$, hence $\Delta_{H_F^i} \neq 1$. Therefore $j_F(\omega) \geq i$ implies $sw_F(\Delta)/\deg(\Delta) \geq i$ for every lifting Δ of Π , and this shows $sw_F(\Delta)/\deg \Pi \geq j_F(\omega)$.
On the other hand, assume $\omega_{UU_F^i} = 1$. Consider the exact sequence $1 \to FF^*/UU_F^i \xrightarrow{c} H_F/H_F^i \to A_F/A_F^i \to 1$. By (2.5) Π can be lifted to an irreducible representation Δ of H_F/H_F^i which, when restricted to $H_F^c \cdot H_F^i/H_F^i$ yields a unique character ψ of degree 1 such that $\psi \circ c = \omega$. Therefore, if $j_F(\omega) < i$ then there is a lifting Δ of Π such that $sw_F(\Delta)/\deg(\Delta) < i$, i.e. $sw_F(\Delta)/\deg(\Pi) \leq j_F(\omega)$.

$v \in \mathbb{R}_+$ is called a <u>jump</u> in the filtration $\{UU_F^i\}$, $i \in \mathbb{R}_+$, if $UU_F^{v+\varepsilon} \neq UU_F^v$ for all $\varepsilon > 0$. In view of (3.3) the jumps yield information about the conductors of irreducible projective representations of A_F. Recall that in the abelian situation the local reciprocity map gives isomorphisms

$$A_F \cong \tilde{F}^* \quad \text{and} \quad A_F^i \cong U_F^i \quad \text{for all} \quad i \in \mathbb{R}_+ .$$

Here U_F^i, $i \in \mathbb{R}_+$, denotes the group of principal units U_F^w where w is the smallest integer such that $w \geq i$. Especially, the jumps of this filtration are all the natural numbers.

In the following we denote for subgroups $M, N \leq F^*$ by $M \wedge N$ the topological closure of $M \otimes N$ under the natural map $F^* \otimes F^* \to FF^*$. Some simple properties of the Zink filtration are as follows:

(i) $U_F^i \wedge F^* \leq UU_F^i$ <u>for all</u> $i \in \mathbb{R}_+$ **(3.4)**

(ii) UU_F^i <u>is of finite index in</u> FF^*

(iii) $UU_F^0 = U_F^0 \wedge F^* = FF^*$

(iv) $UU_F^1 = U_F^1 \wedge F^*$

(v) <u>The Zink-filtration has no jumps</u> v <u>such that</u> $0 < v < 1$.

In particular, in view of (3.3) statement (ii) above shows that all jumps of the Zink-filtration are rational.

Proof. (i): Since H_F^i is normal in H_F we obtain $c(U_F^i \wedge F^*) = [H_F^i, H_F] \leq H_F^c \cap H_F^i$, hence (i).

(ii): $FF^*/U_F^i \wedge F^* = F^*/U_F^i \wedge F^*/U_F^i$ is finite, so the assertion is obvious from (i).

(iii): obvious, since F^*/U_F^0 is cyclic.

(iv) and (v): For (v) it is sufficient to show that the factor group H_F/H_F^1 is the Galois group of a tamely ramified extension. From the definition of UU_F^1 we have an exact sequence $1 \to FF^*/UU_F^1 \to H_F/H_F^1 \to A_F/A_F^1 \to 1$. Because of (i) and since the order of $FF^*/U_F^1 \wedge F^* = F^*/U_F^1 \wedge F^*/U_F^1$ is prime to p, we find that the order of FF^*/UU_F^1 is prime to p. Therefore the assertion is obvious from the exact sequence. To see (iv) consider the exact sequence $1 \to FF^*/U_F^1 \wedge F^* \to H_F/c(U_F^1 \wedge F^*) \to A_F \to 1$. The order of the first term is prime to p and H_F^1 generates a pro-p-subgroup in the second term. Therefore $H_F^1 \cdot c(U_F^1 \wedge F^*) \cap H_F^c = c(U_F^1 \wedge F^*)$. The left side in this equation is

$(H_F^1 \cap H_F^c) \cdot c(U_F^1 \wedge F^*)$. Hence (iv).

The following lower bound for $j_F(\omega)$ is easily deduced from (3.4):

$$sw_F(R(\omega)) \leq j_F(\omega) ;$$

here $sw_F(R(\omega))$ denotes the Swan-conductor of the abelian extension of F defined by $R(\omega)$.
Recall that the Swan-conductor of an abelian extension is = usual conductor − 1 resp. = 0 , if the usual conductor vanishes.
The following upper bound is more difficult; it is based on Buhler's result (2.21) and on subtle reduction arguments:

$$j_F(\omega) < sw_F(R(\omega)) + e/(p-1) , \quad e = e_{F/\mathbb{Q}_p} .$$

We summarize furhter observations concerning the Zink-filtration:

(3.5) Let $e = e_{F/\mathbb{Q}_p}$ and let c denote the largest integer such that $c < e(p-1)$. Then

(i) $UU_F^i = U_F^i \wedge F^*$ for $i = 0,1,\ldots,p$.

(ii) $U_F^i \wedge F^* \leq UU_F^i \leq U_F^{i-c} \wedge F^*$, if $i > p$.

(iii) If $e \leq p-1$ then $UU_F^i = U_F^i \wedge F^*$ for all integers $i \geq 0$.

The next statement is proved by subtle local methods.

$$(UU_F^i)^p = UU_F^{i+e} \text{ for } i \text{ large enough.} \tag{3.6}$$

From this one can deduce the existence of a p-power p^m such that all jumps in the filtration $\{UU_F^i\}$, $i \in \mathbb{R}_+$, belong to $(1/p^m) \cdot \mathbb{Z}$, and Zink conjectures that one can take $m = [((F:\mathbb{Q}_p) + 1)/2]$.

Since U_F^1 is a pro-p-group and since U_F^0/U_F^1 is cyclic of order prime to p , we obtain

$$U_F^1 \wedge U_F^1 = U_F^0 \wedge U_F^0 \leq FF^* .$$

Moreover, for a fixed prime element π for F , one obtains

$$\text{(i)} \quad FF^* = U_F^1 \wedge U_F^1 \times U_F^0 \wedge < \pi > \quad . \tag{3.7}$$

$$\text{(ii)} \quad UU_F^i = (UU_F^i \cap U_F^1 \wedge U_F^1) \times (U_F^i \wedge < \pi >) \; .$$

These decompositions show that it is sufficient to study the filtration $\{UU_F^i \cap (U_F^1 \wedge U_F^1)\}$, $i \in \mathbb{R}_+$. Zink proves

(3.8) The jumps v in the filtration $\{UU_F^i \cap (U_F^1 \wedge U_F^1)\}$, $i \in \mathbb{R}_+$, are not integers and satisfy $v > 1$. The jumps in the filtration $\{UU_F^i \cap (U_F^0 \wedge < \pi >)\} = \{U_F^i \wedge < \pi >\}$, $i \in \mathbb{R}_+$, are integers.

A main result of E.W. Zink's work gives a description of the filtration $\{UU_F^i \cap (U_F^1 \wedge U_F^1)\}$, $i \in \mathbb{R}_+$, modulo $U_F^2 \wedge U_F^1 = UU_F^2 \cap (U_F^1 \wedge U_F^1)$; especially, it yields all jumps v in the filtration $\{UU_F^i\}$, $i \in \mathbb{R}_+$, such that $1 < v < 2$. (Note that there are no jumps between 0 and 1 .) In the following we explain this result.

Denote by k the residue field of F . Fixing a prime element π for F we get isomorphisms

$$k \xrightarrow{\sim} U_F^1/U_F^2 \; , \quad a \longmapsto 1 + a\cdot\pi$$

and

$$k \wedge k \xrightarrow{\sim} U_F^1/U_F^2 \wedge U_F^1/U_F^2 = U_F^1 \wedge U_F^1/U_F^2 \wedge U_F^1 \; ,$$

which will be used to identify the filtration

$$\overline{UU}_F^i := (UU_F^i \cap U_F^1 \wedge U_F^1)/U_F^2 \wedge U_F^1$$

in $k \wedge k$. Namely, consider

$$V_{f_0} = k \wedge k \; , \quad V_{i+1} = V_i \cap \operatorname{Ker}(L_i) \quad \text{for } i \geq f_0 \tag{3.9}$$

where

$$f_0 = \text{smallest integer } \geq f/2 \; , \quad f = (k : \mathbb{F}_p)$$

$$L_i : k \wedge k \to k \; , \quad L_i(a \wedge b) := a\cdot\tau^i(b) - \tau^i(a)\cdot b \; , \quad i \in \mathbb{Z}$$

τ = Frobenius-automorphism.

Then

(3.10) Theorem. (i) The jumps in the filtration $\{\overline{UU}_F^i\}$, $i \in \mathbb{R}_+$, are the numbers $s_w = 1 + 1/p^{f-w}$, $w = f_o, \ldots, f-1$.

(ii) Under the isomorphism $k \wedge k \cong \overline{UU}_F^1$ we have $V_w = \overline{UU}_F^{s_w}$ for all $w = f_o, \ldots, f-1$.

Remarks. (a) If $f = (k:\mathbb{F}_p) = 1$ then $k \wedge k = 0$. So we assume $f \geq 2$, i.e. $f_o \leq f-1$.

(b) We have $\tau^{f-i} \circ L_i = -L_{f-i}$, hence $\mathrm{Ker}(L_i) = \mathrm{Ker}(L_{f-i})$.

(c) The group $\overline{UU}_F^{s_v}$ determined under (ii) is independent of the chosen prime element π. For if $\pi' = h \cdot \pi$ is another prime element for F, then the isomorphisms $u, u': k \wedge k \xrightarrow{\sim} U_F^1/U_F^2 \wedge U_F^1/U_F^2$ determined by π and π' satisfy $u'(a \wedge b) = (1+a\pi') \wedge (1+b\pi') = u(ha \wedge hb)$. Considering the $\mathbb{F}_p$-space $k \wedge k$ as a k^*-module under the action $k^* \times (k \wedge k) \to k \wedge k$, $(h, a \wedge b) \to ha \wedge hb$, then for an arbitrary subspace $V \leq k \wedge k$ we have $u'(V) = u(h \circ V)$. Obviously the subspaces V_v defined above are stable under the action of k^*, i.e. $h \circ V_v = V_v$ for all $h \in k^*$. This proves our claim.

A basic idea is now to dualize the filtration $\{\overline{UU}_F^i\}$, i.e. we consider the filtration in

$$(k \wedge k)^* = \mathrm{Alt}(k, \mathbb{F}_p) = \text{bilinear alternating } \mathbb{F}_p\text{-pairings } \omega: k \times k \to \mathbb{F}_p ,$$

which is dual to the filtration $\{\overline{UU}_F^i\}$. Using a well known interpretation of $\mathrm{Alt}(k, \mathbb{F}_p)$ in terms of additive polynomials this new filtration is seen to occur in a natural way.

Recall that a polynomial $p \in k[X]$ is called additive if $p(X+Y) = p(X) + p(Y)$, i.e. $p(X) = \sum_{i=o}^{t} a_i X^{p^i}$. The set of additive polynomials $\mathrm{Pol}_+ \subseteq k[X]$ is a ring with respect to addition and substitution $(P_1 \circ P_2)(X) = P_1(P_2(X))$.

Let $\tau \in \mathrm{Gal}(\overline{\mathbb{F}}_p/\mathbb{F}_p)$ be the Frobenius-substitution. Consider $k\{\tau\}$, the non-commutative ring of polynomials $\Sigma a_i \tau^i$, $a_i \in k$, with respect to the commutation rule $\tau^i \cdot a = \tau^i(a) \cdot \tau^i$, $a \in k$. Then we get a ring isomorphism

$$k\{\tau\} \xrightarrow{\sim} \mathrm{Pol}_+ , \quad \Sigma a_i \tau^i \longmapsto \Sigma a_i X^{p^i} ,$$

such that the multiplication in $k\{\tau\}$ corresponds to the substitution in Pol_+. In the following we shall identifiy Pol_+ with $k\{\tau\}$. For $p(\tau) = \Sigma a_i \tau^i \in k\{\tau\}$ define

$$\deg(p) := \max \{i \mid a_i \neq 0\}$$

and denote by $\mathrm{Pol}_+(f)$ the set of all $p \in k\{\tau\}$ such that

$$\deg(p) < f = (k : \mathbb{F}_p) .$$

Then there is a natural isomorphism

$$\mathrm{Pol}_+(f) \xrightarrow{\sim} \mathrm{End}_{\mathbb{F}_p}(k)$$

which sends $a \in k$ to the translation $k \ni x \longmapsto a \cdot x \in k$ and τ to the linear operator of the $\mathbb{F}_p$-space k. The image of $p \in \mathrm{Pol}_+$ is a linear operator which is denoted by D_p. We have

$$\dim_{\mathbb{F}_p} \mathrm{Ker}(D_p) = \deg(p) \quad \text{for} \quad p \not\equiv 0 .$$

It follows that the map $\mathrm{Pol}_+(f) \to \mathrm{End}_{\mathbb{F}_p}(k)$, $p \longmapsto D_p$, is injective, and a dimension count shows that it is also surjective. So we shall identify $\mathrm{Pol}_+(f)$ and $\mathrm{End}_{\mathbb{F}_p}(k)$ and we put $p(a) := D_p(a)$, $a \in k$. Denote by $\mathrm{Bil}(k, \mathbb{F}_p)$ the space of $\mathbb{F}_p$-bilinear pairings $\omega: k \times k \to \mathbb{F}_p$. We use the trace pairing $k \times k \ni (a,b) \longmapsto \mathrm{trace}_{k/\mathbb{F}_p}(a \cdot b) \in k$ to identify

$$\mathrm{End}_{\mathbb{F}_p}(k) \xrightarrow{\sim} \mathrm{Bil}(k, \mathbb{F}_p) , \quad D \longmapsto \omega_D , \quad \omega_D(a,b) := \langle a, D(b) \rangle .$$

Consider the ring $\Lambda := k\{\tau, \tau^{-1}\}$ of additive polynomials in τ and τ^{-1} with the commutation rule $\tau^i \cdot a = \tau^i(a) \cdot \tau^i$, $i \in \mathbb{Z}$, and define a map

$$\Lambda \to \Lambda \quad \text{by} \quad p = \Sigma\, a_i \tau^i \longmapsto p^* = \Sigma\, \tau^{-1}(a_i)\tau^{-i} .$$

Then

$$\langle p(a), b \rangle = \langle a, p^*(b) \rangle .$$

Therefore p^* is called the adjoint of p. We have

(3.11) _Under the isomorphisms_

$$\mathrm{Pol}_+(f) \xrightarrow{\sim} \mathrm{End}_{\mathbb{F}_p}(k) \xrightarrow{\sim} \mathrm{Bil}(k,\mathbb{F}_p)$$

let $p \in \mathrm{Pol}_+(f)$ _correspond to_ $D \in \mathrm{End}_{\mathbb{F}_p}(k)$ _and_ $\omega \in \mathrm{Bil}(k,\mathbb{F}_p)$. _Then the following statements are equivalent._

(i) ω _is alternating._

(ii) $D^* + D = 0$ _where_ D^* _denotes the adjoint of_ D _with respect to the trace pairing and_ $< a,D(a) > = 0$ _for all_ $a \in k$.

(iii) $\tau^f \cdot p^* + p = 0$.

The polynomials $p \in \mathrm{Pol}_+(f)$ with the property $\tau^f \cdot p^* + p = 0$ are called alternating. Let $\mathrm{Alt}(f)$ denote the set of all alternating polynomials in $\mathrm{Pol}_+(f)$. Then (3.11) implies

(3.12) _The map_ $p \longmapsto \omega_p$, $\omega_p(a,b) = < a,b(b) >$, _yields an isomorphism_

$$\mathrm{Alt}(f) \xrightarrow{\sim} \mathrm{Alt}(k,\mathbb{F}_p) .$$

If $\mathrm{Ker}(p)$ denotes the $\mathbb{F}_p$-space of the set of zeros of p in the algebraic closure $\bar{\mathbb{F}}_p$, then $\mathrm{Ker}(p) \cap k = \mathrm{Ker}(D_p) = R(\omega)$. If $d = \deg(p)$, then $\dim_{\mathbb{F}_p} \mathrm{Ker}(p) = 2d - f$, hence $\dim_{\mathbb{F}_p}(R(\omega)) \leq 2d - f$.

On $(k \wedge k)^* = \mathrm{Alt}(k,\mathbb{F}_p)$ consider the filtration given by

$$W_i := \{\omega \in \mathrm{Alt}(k,\mathbb{F}_p) \mid \omega = \omega_p \text{ such that } \deg(p) \leq i\} .$$

For $p \in \mathrm{Alt}(f)$ one has $f/2 \leq \deg(p) < f$, hence

$$W_i = 0 \text{ for } i < f/2 \text{ and } W_{f-1} = \mathrm{Alt}(k,\mathbb{F}_p) .$$

The natural pairing

$$(k \wedge k)^* \times (k \wedge k) \to \mathbb{F}_p$$

induces a dual filtration

$$0 = W_{f-1}^{\perp} \leq \dots \leq W_i^{\perp} = k \wedge k \text{ for } i < f/2 ,$$

such that

$$V_i = W_{i-1}^{\perp} \quad \text{for all} \quad i = f_o, \dots, f \; ; \tag{3.13}$$

where V_i is the filtration defined under (3.9). This filtration gives an equivalent formulation of (3.10): The isomorphism $k \wedge k = U_F^1 \wedge U_F^1/U_F^2 \wedge U_F^1$ yields a dual isomorphism $\mathrm{Alt}(U_F^1/U_F^2) \xrightarrow{\sim} \mathrm{Alt}(k)$. Fixing a primitive p-th root of unity one has $\mathrm{Alt}(k) \xrightarrow{\sim} \mathrm{Alt}(k, \mathbb{F}_p)$ .So one has isomorphisms

$$\mathrm{Alt}(U_F^1/U_F^2) \xrightarrow{\sim} \mathrm{Alt}(k, \mathbb{F}_p) \xrightarrow{\sim} \mathrm{Alt}(f) \; .$$

Let $\omega \in \mathrm{Alt}(U_F^1/U_F^2)$ correspond to $p \in \mathrm{Alt}(f)$ and view ω as an alternating pairing on F^* , comp. (3.7). Then

$$j_F(\omega) = 1 + 1/p^{f-d} \; , \quad d = \deg(p) \; , \quad f/2 \leq d < f \; . \tag{3.14}$$

By (3.13) ω determines an irreducible projective representation Π of A_F . Using formula (3.3) for the Swan-conductor we obtain

$$sw_F(\Pi) = (k\colon R(\omega))^{1/2} j(\omega) \; ,$$

and since $R(\omega)$ is equal to the kernel of the endomorphism $D_p \in \mathrm{End}_{\mathbb{F}_p}(k)$, we have

$$sw_F(\Pi) = p^{n(\Pi)} + p^{n(\Pi)+\deg(\Pi)-f}$$

where $n(\Pi) = \mathrm{codim}_{\mathbb{F}_p}(\mathrm{Ker}(D_p))/2$.

The first approach to the Zink-filtration uses the duality $H^3(G,\mathbb{Z})^\wedge = H^{-1}(G,K^*) = K^{*N}/K^{*I}$ where K/F is a finite Galois extension with Galois group G . On $H^{-1}(G,K^*)$ one considers the filtration

$$U_K^i \cap H^{-1}(G,K^*) := U_K^i \cap K^{*N}/U_K^i \cap K^{*I} \; .$$

If K/F is abelian and $W = W(K/F)$ is the corresponding relative Weil-group (corresponding to the fundamental class in $H^2(G,K^*)$) we get a central group extension $1 \to K^*/K^{*I} \to W/K^{*I} \to G \to 1$. The commutator in W/K^{*I} induces isomorphisms

$$UU_F^i/(N \wedge F^*) \xrightarrow{c} U_K^{\phi(i)} \cap H^{-1}(G,K^*)$$

where N is the norm subgroup of F^* corresponding to K/F and where $\psi=\psi_{K/F}$ is the Herbrand function of K/F. One can show that

$$UU_F^i = \varprojlim_{K/F} (U_K^{\psi_{K/F}(i)} \quad H^{-1}(g,K^*))$$

where K/F runs over all finite abelian extensions and the projective limit is taken with respect to the norm maps.

Remarks. (1) As remarked by J. Ritter the conductor formula in (2.15) could also be deduced from his recent explicit Brauer induction formula for local Galois characters [96].

(2) The ramification theory of nilpotent class two extensions of the local field F could probably be treated in the framework of Kirillov's orbit theory. For a first algebraic step which might be useful in this extent see [170].

Refercences

[7] R. Brauer: On primitive projective groups, in : Contributions to Algebra, ed. by H. Bass, P.J. Cassidy, J. Kovacic; Ac. Press, 1977, 63 - 82.

[8] J. Buhler: Icosahedral Galois representations; Lecture Notes in Mathematics 654, Springer Verlag, Berlin-Heidelberg-New York 1978.

[41] G. Henniart: Représentations du groupe de Weil d'un corps local; L'Ens. Math., IIe série, 26 (1980), 155 - 172.

[47] R. Howe: Kirillov theory for compact p-adic groups: Pac. J. Math. 73 (1977), 365 - 382.

[59] H. Koch: Classification of the primitive representations of the Galois group of a local field; Invent. math. 40 (1977), 195 - 216.

[99] J.P. Serre: Corps locaux, 2ième éd., Hermann, Paris, 1968.

[102] S. Shirai: On the central class field mod m of an algebraic number field; Nagoya-Math. J. 71 (1978), 61 - 85.

[114] K. Yamazaki: On projective representations and ring extensions of finite groups; J. Fac. Sc. Univ. Tokyo, Sect. I, 10 1963/64, 147 - 195.

[116] E.W. Zink: Weil-Darstellungen und lokale Galoistheorie; Math. Nachr. 92 (1979), 265 - 288.

[117] E.W. Zink: Lokale projektive Klassenkörpertheorie - Grundbegriffe und erste Resultate; Preprint R-Math-1/82, Akad. d. Wiss. d. DDR, Berlin, 1982.

[119] E.W. Zink: Lokale projective Klassenkörpertheorie II; Math. Nachr. 114 (1983), 123 - 150.

[120] E.W. Zink: Über das Kirillow-Dual endlicher nilpotenter Ringe; Preprint R-Math-03/85, Akad.d.Wiss.d. DDR, Berlin, 1985.

Contemporary Mathematics
Volume 86, 1989

REPRESENTATIONS OF CERTAIN GROUP EXTENSIONS

by W. Willems, Mainz

This article is a condensed and slightly different form of paragraph 8 in the book of Bushnell and Fröhlich [11]. Using a lemma on characters of certain extensions of extraspecial p-groups for which Dade has given a very elegant proof ([52], Chap. V, 17.13), some of the cumbersome calculations in [11] can be avoided.

The reader who is interested in more information on metric spaces may have a look in paper [93]. Most of the notation in the following is standard and taken from Huppert's book [52].

§ 1. The Problem

1.1. Notation. Let (G,N,χ) denote a triple where G is a group (not necessarily finite), N a normal subgroup of G, $V = G/N$ a finite elementary abelian p-group and

$$\chi: N \longrightarrow \mathbb{C}^*$$

a character on which we impose the following three conditions.

(1) $|N/\mathrm{Kern}\ \chi| < \infty$.

(2) χ is stable under G, i.e.
$\chi(n^g) = \chi(n)$ for all $n \in \mathbb{N}$ and all $g \in G$.

In particular we have

(*) $[G,N] \leq \mathrm{Kern}\ \chi \trianglelefteq G$ and

(**) $\chi([g_1,g_2])^p = \chi([g_1,g_2]^p) = \chi([g_1^p,g_2]) = 1$,

hence $\chi(G') \leq \zeta_p$ where ζ_p denotes the group of p-th roots of unity.

(3) We define a form

$$h_\chi : V \times V \longrightarrow \zeta_p$$
by $\quad h_\chi(g_1N, g_2N) = \chi([g_1, g_2])$.

This form is well-defined, bilinear and alternating. (To see this use (*) and (**) of (2)).

In what follows, we require that h_χ is non-singular.

1.2 Remark. $\bar{G} = G/\mathrm{Kern}\ \chi$ is an extraspecial p-group, i. e. $\Phi(\bar{G}) = Z(\bar{G}) = \bar{G}'$ is cyclic of order p .

For our purpose the kernel of χ doesn't play any role. Thus for most of the proofs we may assume that $\mathrm{Kern}\ \chi = \langle 1 \rangle$. A triple (G,N,χ) is therefore nothing but an extraspecial p-group G with centre N and a non-trivial linear character χ of N .

The representation theory of extraspecial p-groups yields

1.3 Proposition. Given a triple as in 1.1. Then there is (up to equivalence) exactly one irreducible representation Π_χ of G over $\mathbb{C}$ such that $\Pi_\chi|_N$ contains χ as a constituent.
Moreover, by Clifford's Theorem $\Pi_\chi|_N = e\chi$ with $e \in \mathbb{N}$ and $\dim \Pi_\chi = |G : N|^{1/2}$.

Proof. ([52], Chap.V, 16.14) or ([11]), (8.3.3).

1.4 Notation. We extend the situation of 1.1 as follows. Again, let (G,N,χ) be a triple as in 1.1 and let T be a finite p'-group acting on G . Let $H = T \ltimes G$ be a semidirect product of G by T and suppose that χ is also stable under H . In particular, $\mathrm{Kern}\ \chi \trianglelefteq H$. Thus modulo $\mathrm{Kern}\ \chi$ we have an extraspecial p-group extended by a group whose order is prime to p .
Since $\gcd(|G/\mathrm{Kern}\ \chi|, |T|) = 1$ and χ is stable under T , the representation Π_χ of G constructed in 1.3 has an extension to H and all the extensions are well known. More presicely, we have

1.5 Proposition. a) There exists an irreducible representation $\tilde{\Omega}_\chi$ of H with $\tilde{\Omega}_\chi|_G = \Pi_\chi$ and the determinant of $\Omega_\chi = \tilde{\Omega}_\chi|_T$ is the trivial representation of T .

b) The properties of $\tilde{\Omega}_\chi$ in a) determine $\tilde{\Omega}_\chi$ uniquely to within equivalence.

c) If $\tilde{\Omega}$ is a representation of H with $\tilde{\Omega}|_G = \Pi_\chi$ then $\tilde{\Omega} = \tilde{\Omega}_\chi \otimes \alpha$ where α is the inflation of a linear character of T to H .

Proof. ([52], Chap. V, 17.12) or ([11], (8.4.1)).

1.6 The Problem. For computing root numbers, we have to know the character values $\mathrm{tr}\,\Omega_\chi(x)$ $(x \in T)$. (Here tr denotes the usual trace map.) Thus we are interested in an explicit formula to compute these values.

1.7 Remark. If we write $\tilde{\Omega}_{T,\chi}$ (resp. $\Omega_{T,\chi}$) for $\tilde{\Omega}_\chi$ (resp. Ω_χ), the uniqueness property of $\tilde{\Omega}_\chi$ yields

(a) $\Omega_{U,\chi} = \Omega_{T,\chi}|_U$ for all $U \leq T$.

(b) $\Omega_{T,\chi} = \mathrm{infl}\,\Omega_{\overline{T},\chi}$ in case that the action of T on G factors through $\overline{T}$.

Thus we may assume in the sequel that $T = \langle t \rangle$ is a cyclic group and that t acts faithfully on $V = G/N$.

Next we consider the form

$$h_\chi : V \times V \longrightarrow \zeta_p$$

$$(g_1N, g_2N) \longmapsto \chi([g_1,g_2]) .$$

Since χ is T-invariant, we obtain

$$h_\chi((g_1N)^x.(g_2N)^x) = \chi([g_1^x,g_2^x]) = \chi([g_1,g_2]^x) = \chi([g_1,g_2]) = h_\chi(g_1N,g_2N)$$

for all $x \in T$. This means that h_χ is T-invariant or equivalently, (V,h_χ) is an alternating $\mathbb{F}_pT$-space.

1.8 Proposition. Suppose that

$$(V,h_\chi) = (V^{(1)},h_\chi^{(1)}) \perp (V^{(2)},h_\chi^{(2)})$$

is the orthogonal sum of the two (alternating $\mathbb{F}_pT$ - spaces $(V^{(i)},h_\chi^{(i)})$. Let $G^{(i)}$ denote the complete pre-image of $V^{(i)}$ in G and let $\Omega_\chi^{(i)}$ be the representation of T corresponding to the triple $(G^{(i)},N,\chi)$. Then $\Omega_\chi = \Omega_\chi^{(1)} \otimes \Omega_\chi^{(2)}$.

Proof. ([11], (8.4.3)).

As a consequence of 1.8, Problem 1.6 reduces to the following

1.9 **Reduced Problem.** Decompose

$$(V,h_\chi) = \perp_{i=1}^{s} (V^{(i)},h_\chi^{(i)})$$

with orthogonal indecomposable (alternating) $\mathbb{F}_pT$-spaces $(V^{(i)},h_\chi^{(i)})$, and compute explicitly $\operatorname{tr} \Omega_\chi^{(i)}$.

Since T acts completely reducible on V by Maschke's theorem, it is easy to see that the orthogonal indecomposable $\mathbb{F}_pT$-spaces are of the following types.

1) $V^{(i)}$ is an irreducible $\mathbb{F}_pT$-module

or

2) $V^{(i)} = W_1^{(i)} \oplus W_2^{(i)}$ where

$W_2^{(i)} \cong W_1^{(i)*} = \operatorname{Hom}_{\mathbb{F}_pT}(W_1^{(i)}, \mathbb{F}_p)$ is the dual of $W_1^{(i)}$, $W_j^{(i)}$ is irreducible and $h_\chi|_{W_j^{(i)} \times W_j^{(i)}} \equiv 0$. $V^{(i)}$ is called a hyperbolic space.

For more information see 3.1.

§ 2 A formula for $\Omega_\chi^{(i)}$.

We assume the situation as in 1.9. If the trivial module $k = \mathbb{F}_p$ occurs as a composition factor in V, then k belongs to a component

$$V^{(i)} = k \oplus k$$

since k doesn't carry a non-singular alternating form.
But then we have

$$\operatorname{tr} \Omega_\chi^{(i)}(t) = |V^{(i)}|^{1/2} .$$

Thus it remains to deal with such components $V^{(i)}$ on which t does not act trivially. By 1.7 we may even assume that t operates fixed point freely on $V^{(i)}$. Now, this case can be handled by the following result.

2.1 Proposition. Let G be an extraspecial p-group of order p^{2m+1}. Let $T = \langle t \rangle$ be a subgroup of $\mathrm{Aut}(G)$ with $p \nmid |T|$. We assume that t centralizes $Z(G)$ and acts fixed point freely on $V = G/Z(G)$.
Then the irreducible characters of $H = T \ltimes G$ are as follows.

(a) The irreducible characters of the Frobenius group $H/Z(G)$.

(b) For each irreducible character $\chi \neq 1$ of $Z(G)$ and each linear character μ of H there is an irreducible character $\psi_{\chi,\mu}$ of H of degree p^m with

$$\psi_{\chi,\mu}(g) = \begin{cases} p^m\chi(g) & | \quad g \in Z(G) \\ 0 & | \quad g \in G - Z(G) \\ \delta\chi(z)\mu(x) & | \quad g = zx^y, \quad z \in Z(G), \quad 1 \neq x \in T, \quad y \in G. \end{cases}$$

Moreover $\delta \in \{\pm 1\}$, $|T| \mid p^m - \delta$ and

$$\psi_{\chi,\mu}|_T = \frac{p^m - \delta}{|T|} \mathrm{reg}_T + \delta\mu$$ where reg_T denotes the regular character of T.

Proof. ([52], Chap. VII, 17.13)

To avoid indices we assume that (V, h_χ) is orthogonal indecomposable. By 2.1, we have $\mathrm{tr}\ \tilde{\Omega}_\chi = \psi_{\chi,\mu}$ for a special pair (δ,μ).
According to 1.9 we must consider the following two cases.

(1') V is an irreducible alternating $\mathbb{F}_p T$-space on which T acts fixed point freely

or

(2') $V = V_1 \oplus V_2$, $h_\chi|_{V_i \times V_i} \equiv 0$ and T acts fixed point freely on the irreducible $\mathbb{F}_p T$-module V_1 (hence also on V).

Case (1').

By ([52], Chap. II, 9.23) the order $|T|$ of T divides $|V|^{1/2} + 1$. Hence $\delta = -1$ and therefore

$$(*) \qquad \mathrm{tr}\ \Omega_\chi = \frac{|V|^{1/2} + 1}{|T|} \mathrm{reg}_T - \mu .$$

Thus it remains to compute μ. To do this, we consider the Wedderburn decomposition

$$\mathbb{F}_p T = \bigoplus_{j=1}^{t} S_j$$

where the S_j are extension fields of $\mathbb{F}_p$ and irreducible $\mathbb{F}_p T$-modules.
Let $V = S = S_j$ and let

$$\theta : T \hookrightarrow S = \mathbb{F}_p(\theta(T))$$

be the natural injection.
If $^-$ denotes the involution on $\mathbb{F}_p T$ which carries x to x^{-1} for $x \in T$, then $S = \bar{S}$ since S is an alternating $\mathbb{F}_p T$-space. As S is different from (the trivial module) $\mathbb{F}_p$, the involution $^-$ induces a non-trivial automorphism on S.
Let S_o be the fixed field of S under $^-$ and let

$$U(S) = \{s \mid s \in S\ ,\ s\bar{s} = 1\} = \text{Kern Norm}_{S^x \to S_o^x} .$$

Obviously, $\operatorname{Im} \theta \leq U(S)$ and

$$|U(S)| = |S|^{1/2} + 1 = |V|^{1/2} + 1 .$$

By (*), we have

$$1 = \det \Omega_\chi(t) = (\operatorname{sgn}_{\mathbb{C}T}(t)^{(|V|^{1/2}+1)/|T|})\mu(t)^{-1} ,$$

and therefore

$$\operatorname{tr} \Omega_\chi = \begin{cases} |U(S) : \operatorname{Im} \theta| \operatorname{reg}_T - 1_T & \text{if } \operatorname{Im} \theta \leq U(S)^2 \\ |U(S) : \operatorname{Im} \theta| \operatorname{reg}_T - \mu & \text{if } \operatorname{Im} \theta \nleq U(S)^2 \end{cases}$$

where $\mu^2 = 1_T$, $\mu \neq 1_T$.
This gives the formula

$$\operatorname{tr} \Omega_\chi(t) = \begin{cases} -1 & \text{if } \theta(t) \in U(S)^2 \\ -\mu(t) = 1 & \text{if } \theta(t) \notin U(S)^2 . \end{cases}$$

<u>Case (2').</u>

Now $|T| \mid |V|^{1/2} - 1$, hence $\delta = 1$.
By 2.1, we get

$$1 = \det \Omega_\chi(t) = (\operatorname{sgn}_{\mathbb{C}T}(t)^{(|V|^{1/2} - 1)/|T|})\mu(t) .$$

This implies the formula

$$\operatorname{tr}\Omega_\chi(t) = \mu(t) = \operatorname{sgn}_{\mathbb{C}T}(t)^{(|V|^{1/2}-1)/|T|} = \operatorname{sgn}_{V_1}(t) .$$

§ 3 Remarks.

In [11] Bushnell and Fröhlich obtained a full classification of the indecomposable alternating $\mathbb{F}_pT$-spaces which is as follows.

3.1 Theorem. Let $T = \langle t \rangle$ and let $p \nmid |T|$. If (V,h) is an alternating $\mathbb{F}_pT$-space, then we have

a) The decomposition

$$(V,h) = \perp_{i=1}^{s} (V_i,h_i) \quad \text{with orthogonal indecomposable } \mathbb{F}_pT\text{-spaces}$$

(V_i,h_i) is unique up to isometry. (We say that the metric Krull-Schmidt Theorem holds.)

b) The isometry class of an indecomposable alternating $\mathbb{F}_pT$-space is uniquely determined by the isomorphism class of the corresponding $\mathbb{F}_pT$-module.

c) The following is a complete list of non-isometric representatives of the isometry classes of indecomposable alternating $\mathbb{F}_pT$-spaces.

(i) For each simple component $S \subseteq \mathbb{F}_pT$ with $S = \bar{S}$, $\bar{\ } \neq 1_S$ there is a unique space (S,h). The form h is isometric to the form

$$(x,y) = \operatorname{Trace}_{S \to \mathbb{F}_p}(xa\bar{y}) \qquad (x,y \in S)$$

where $0 \neq a = -\bar{a} \in S$.

(ii) For each simple component $S \subseteq \mathbb{F}_pT$ with $S \neq \bar{S}$ (i.e. $S \not\cong S^*$) the hyperbolic space $HS = S \oplus S^*$ with the form

$$(s_1+s_1^* , s_2+s_2^*) = s_2^*(s_1) - s_1^*(s_2) \qquad (s_i \in S , s_i^* \in S^*) .$$

(iii) For each simple component $S \subseteq \mathbb{F}_p$ with $S = \bar{S}$ and $\bar{\ } = 1_S$ the hyperbolic space HS (in this case $S = \mathbb{F}_p$).

3.2 Remark. The metric Krull-Schmidt Theorem doesn't hold for arbitrary groups in general. It holds if the field F is algebraically closed, the characteristic p of F is different from 2 and the form is alternating or symmetric (see for instance [93]).

3.3 Remark. Assume that S is a simple component of $\mathbb{F}_pT$ with $S = \bar{S}$ and non-trivial on S. If T acts fixed point freely on S then by the results of § 2 and 3.1 there are two ways to compute $\operatorname{tr}\,\Omega_\chi(t)$ for $(V,h_\chi) \cong HS$.

Since (V,h_χ) is a hyperbolic space we have

$$\operatorname{tr}\,\Omega_\chi(t) = \operatorname{sgn}_S(t) .$$

On the other hand, since HS is decomposable as an alternating space we also have

$$\operatorname{tr}\,\Omega_\chi(t) = \begin{cases} (-1)^2 = 1 & \text{if } \theta(t) \in U(S)^2 \\ 1^2 & \text{if } \theta(t) \notin U(S)^2 . \end{cases}$$

That the two results coincide can be seen as follows. If $p = 2$, then the order of t is odd and therefore $\operatorname{sgn}_S(t) = 1$. If p is odd, then $\theta(t) \in S^{\times 2}$, hence again $\operatorname{sgn}_S(t) = 1$.

3.4 Remark. If $(V,h_\chi) = HW$ is any hyperbolic $\mathbb{F}_pT$-space (not necessarily indecomposable as an alternating space) and t acts without non-trivial fixed points on W, then

$$\operatorname{tr}\,\Omega_\chi(t) = \operatorname{sgn}_W(t) .$$

Proof. If $W = W_1 \oplus W_2$ as an $\mathbb{F}_pT$-module, then

$$\operatorname{sgn}_W(t) = (\operatorname{sgn}_{W_1}(t))^{|W_2|}(\operatorname{sgn}_{W_2}(t))^{|W_1|} .$$

If p is odd, then $|W_1|$ and $|W_2|$ are odd. If $p = 2$, then $|W_1|$ and $|W_2|$ are even and the order of t is odd. Thus we obtain by induction

$$\operatorname{sgn}_W(t) = \operatorname{sgn}_{W_1}(t)\operatorname{sgn}_{W_2}(t) = (\operatorname{tr}\,\Omega_\chi^{(1)}(t))(\operatorname{tr}\,\Omega_\chi^{(2)}(t)) =$$

$$= \operatorname{tr}(\Omega_\chi^{(1)}(t) \otimes \Omega_\chi^{(2)}(t)) = \operatorname{tr}\,\Omega_\chi(t) .$$

References

[11] J. Bushnell, A. Fröhlich: Gauss sums and p-adic division algebras; Lecture Notes in Mathematics 987, Springer Verlag, Berlin-Heidelberg-New York 1983.

[52] B. Huppert: Endliche Gruppen. Springer Verlag, Berlin-Heidelberg-New York 1967.

[93] H.G. Quebbemann, W. Scharlau, M. Schulte: Quadratic and hermitian forms in additive and abelian categories; J.Alg. 59 (1979), 264 - 289.

Contemporary Mathematics
Volume 86, 1989

TRACE CALCULATIONS

by J. Brinkhuis, Rotterdam

0. Introduction

The most critical test for the correspondence $Irr_n(G_F) \simeq Irf(D)$ which has been constructed in [11] is whether it respects rootnumbers. The answer will be given in [E] by a formula which relates the rootnumbers on both sides and which involves moreover some harmless invariants and one rather tricky sign (± 1). This sign will there turn up as the trace of a certain group representation evaluated at a particular element. By the theoretical results in [Wm] its calculation can be reduced to a great number of exercises in combinatorial manipulation involving finite fields. The resulting table which gives the rule for the sign is all that will be needed in later chapters and therefore let us start with it.

1. The table.

The following set of parameters for the table keeps the amount of notation to be introduced minimal.

F	a finite field extension of the p-adic rational field $\mathbb{Q}_p$
n	a natural number not divisible by p
D	a central F-division algebra of finite dimension n^2
E	a subfield of D containing F
$(E/F,c)$	a primordial pair, so in particular $E = F(c)$
$q = \|k\|$	the cardinality of k, the residue field of F
$e = e(E/F)$	the ramification index
$f = f(E/F)$	the residue class degree
$m = \frac{n}{ef}$	(of course $ef\|n$)
$\mathcal{O}_E$	the valuation ring of E
$\mathcal{P}_E$	the maximal ideal of $\mathcal{O}_E$
$\mathcal{D}_E$	the absolute different of E
s	the rational integer defined by $c\,\mathcal{O}_E = \mathcal{P}_E^{1+s}\mathcal{D}_E$

a_c — the root of unity in E of order prime to p which is $\equiv c^e (\mathrm{mod}\ F^*U_1(E))$ where $U_1(E) = 1 + \mathcal{P}_E$

$(\frac{u}{v})$ — for u and v two non-zero relatively prime rational integers is defined to be the Legendre symbol for odd primes v and is extended by multiplicativity in v.

$(\frac{a_c}{\mathcal{P}_E})$ — the usual quadratic residue symbol in the case $p \neq 2$.

We can now give the table

$$\zeta_D(E/F,c) = \begin{cases} \left. \begin{array}{ll} (\frac{-1}{q})^{mf/2} & \text{if } ms \equiv 0 \ (\mathrm{mod}\ 2) \\ -(\frac{a_c}{\mathcal{P}_E})(\frac{-1}{q})^{f/2} & \text{if } ms \equiv 1 \ (\mathrm{mod}\ 2) \end{array} \right\} & \text{if } e \equiv 0 \ (\mathrm{mod}\ 2) \\ \left. \begin{array}{ll} (\frac{q}{e})^{mf} & \text{if } ms \equiv 0 \ (\mathrm{mod}\ 2) \\ -(\frac{a_c}{\mathcal{P}_E}) & \text{if } ms \equiv 1 \ (\mathrm{mod}\ 2) \end{array} \right\} & \text{if } e \equiv 1 \ (\mathrm{mod}\ 2) \end{cases}$$

Moreover $\zeta_D(E/F,c)$ is only defined if $fms \equiv 0 \ (\mathrm{mod}\ 2)$ and $(e,s) = 1$.

Remark. all symbols in the table are well-defined: if $p = 2$, then necessarily $e \equiv 1 \ (\mathrm{mod}\ 2)$ and $ms \equiv 0 \ (\mathrm{mod}\ 2)$.

The thing to remember from this table is that it does give an explicit rule for our sign, but that this rule is not particularly simple. The latter fact should be seen as a shortcoming of the constructed bijection $\mathrm{Irr}_n(G_F) \simeq \mathrm{Irf}(D)$. The details of the table might be helpful in attempts to construct a more canonical bijection (In [11] section 12 a bijection is constructed by a suitable twisting where no sign occurs in the comparison of rootnumbers (see [11],(12.1.4. VI)).

2. First definition of the sign.

In this section we will give a definition of the sign which reflects the way in which it will actually arise in [E]

Let A be the centralizer of E in D. Let $\varphi : A^* \to \mathbb{C}^*$ be a homomorphism with finite image and open kernel ("a one-dimensional finite admissible representation of A^*"). Let $sw(\varphi)$ be its Swan conductor and define the integer j by $sw(\Psi)\,\mathcal{O}_D = \mathcal{P}_D^j$; assume that j is even and ≥ 1. Set $j = 2i$. By a simple procedure which is given in detail in [E], one constructs from φ a character $\tilde{\varphi}$ of U_1V_{i+1} where

$$V_k = 1 + P_D^k \qquad \text{for all } k \geq 1$$
$$U_k = V_k \cap A$$

The "induction" construction from [Wm] can be applied with

$G = U_1V_i$

$N = U_1V_{i+1}$

$\chi = \tilde{\varphi}$

$\Gamma = C_A/C_F$ (C_A and C_F are the appropriate complementary sub-groups).

(Indeed C_A acts by conjugation on U_1V_i and this action stabilizes U_1V_{i+1} and the character $\tilde{\varphi}$; moreover C_F acts trivially and C_A/C_F is finite of order prime to p). As a result one gets a representation Ω_φ of $C_A/C_F \ltimes U_1V_i$; its inflation to AV_i is denoted again by Ω_φ .

Finally we evaluate Ω_φ at the element c and take the trace $tr(\Omega_\psi(c))$. This is the invariant to be considered. That it takes values in $\{\pm 1\}$ is not a priori obvious (to me).

3. Second definition of the sign.

In this section we give a second definition, which is "context-free" and which is very convenient for the purpose of calculating the sign. The proof of the equivalence of the two definitions is given in [Wm]. Looking at it in another way, $tr(\Omega_\varphi(c))$ has been "calculated in theory" in [Wm], the function t which will be introduced in this section is just a bookkeeping gadget which summarizes the results.

Let $\gamma = c \pmod{C_F}$, $\Gamma = <\gamma>$, $X = V_i/V_{i+1}$. Then the order of Γ is prime to p , so the ring $\mathbb{F}_p\Gamma$ is semisimple; Γ acts on X by conjugation ("$c^{-1}uc$") . Thus γ acts on X/X^Γ without fixed points. Our sign will be defined in terms of this action only.
Let θ be a homomorphism $\Gamma \to \bar{\mathbb{F}}_p^*$ of groups, and $S = \mathbb{F}_p(\theta(\Gamma))$ (That is, S is the component of the group ring $\mathbb{F}_p\Gamma$ corresponding to θ ; it is an irreducible $\mathbb{F}_p\Gamma$-module and one gets all such modules by varying θ). We assume $\theta \neq 1$. We define:

(a) t_{S+S} = the sign of the permutation of S given by the action of γ (via θ) which

$$\text{is} \begin{cases} 1 & \text{if } \theta(\gamma) \in S^{*2} \\ -1 & \text{if } \theta(\gamma) \notin S^{*2} \end{cases}$$

(b) if $\theta^{-1} = \theta^{\sigma}$ for some $\sigma \in \mathrm{Gal}(\bar{\mathbb{F}}_p/\mathbb{F}_p)$, so $S = \mathbb{F}_{\tilde{q}^2}$ with $\tilde{q}$ some power of p , $\sigma : x \to x^{\tilde{q}}$ and $\theta(\gamma) \in \mu_{\tilde{q}+1}(S)$, the group of $(\tilde{q}+1)$-st roots of unity, then

$$t_S = \begin{cases} +1 & \text{if } \theta(\gamma) \in \mu_{\tilde{q}+1}(S)^2 \\ -1 & \text{if } \theta(\gamma) \notin \mu_{\tilde{q}+1}(S)^2 . \end{cases}$$

Thus t_Y is defined for certain $\mathbb{F}_p\Gamma$-modules Y . This is then extended by "additivity" $(t_{Y+Z} = t_Y t_Z)$ to those $\mathbb{F}_p\Gamma$-modules which are a finite direct sum of such modules (they can be characterized as those $\mathbb{F}_p\Gamma$-modules on which γ acts fixed-point free and which can be given a symplectic structure). To see that t_- is well-defined we have to check that, in the case $\theta^{-1} = \theta^{\sigma}$, calculation of t_{S+S} via a) gives the same result as via b):

via b): $t_{S+S} = (t_S)^2 = 1$

via a): $\theta(\gamma) \in \mu_{\tilde{q}+1}(S)$ this is of course contained in S^{*2} , so $t_{S+S} = 1$.

One can endow the $\mathbb{F}_p\Gamma$-module X/X^{Γ} with a symplectic structure and therefore $t_{X/X^{\Gamma}} \in \{\pm 1\}$ is defined. We take this as the second definition of our invariant.

4. Some remarks.

To begin with we make an additional remark on the equivalence of the definitions given in sections 2 and 3. The results of [Wm] show that $\mathrm{tr}(\Omega_{\varphi}(c))$ is equal to $t_{V_i/U_i V_{i+1}}$ (with $U_i = V_i \cap A$) . The following proposition shows that this is indeed equal to $t_{X/X^{\Gamma}}$ and thus, fortunately, one does not have to deal with the group U_i .

Proposition. $U_i V_{i+1}/V_{i+1} = X^\Gamma$

Proof. $X = V_i/V_{i+1} \simeq P^i/P^{i+1} = R_D b \cup \{0\}$ where R_D is the group of roots of unity in D and where b is a generator of P^i in C_D (= complementary subgroup in D). Under this isomorphism $xb(x \in R_D)$ corresponds to the class of $1 + xb$.

- $\supset$: Assume that xb corresponds to an element in X^Γ, so $c^{-1} x bc = xb$, that is, xb and c commute, so xb lies in the centralizer of $F(c) = E$, which is by definition A.
- $\subset$: the argument above is reversible.

In the second place we point out that the theory of [Wm], seems to require for the calculation of our sign, the decomposition of X/X^Γ as an $\mathbb{F}_p\Gamma$-space that is, the decomposition must respect the symplectic structure. However, the marvelous thing is that the classification of the irreducible $\mathbb{F}_p\Gamma$-modules and of the irreducible $\mathbb{F}_p\Gamma$-spaces shows that the decomposition as a space is determined by the decomposition as a module. (with "decomposition" we mean here "the sequence of multiplicities of the irreducibles in a space (resp. module)"). Thus one does not have to bother about the symplectic structure, if one decomposes X/X^Γ.

5. Distinction of cases.

In order to give some indication of the labour involved in carrying out the calculations of the sign, it is maybe not a bad idea just to display the "tree" of cases one has to distinguish on the way.

We need some additional notation, where

g is defined by $|k| = p^g$

$u = a_c^{1-q^{f/2}} \in R_E$

h is defined by $|\mathbb{F}_p(u)| = p^h$

e_2 is the exact power of 2 dividing e.

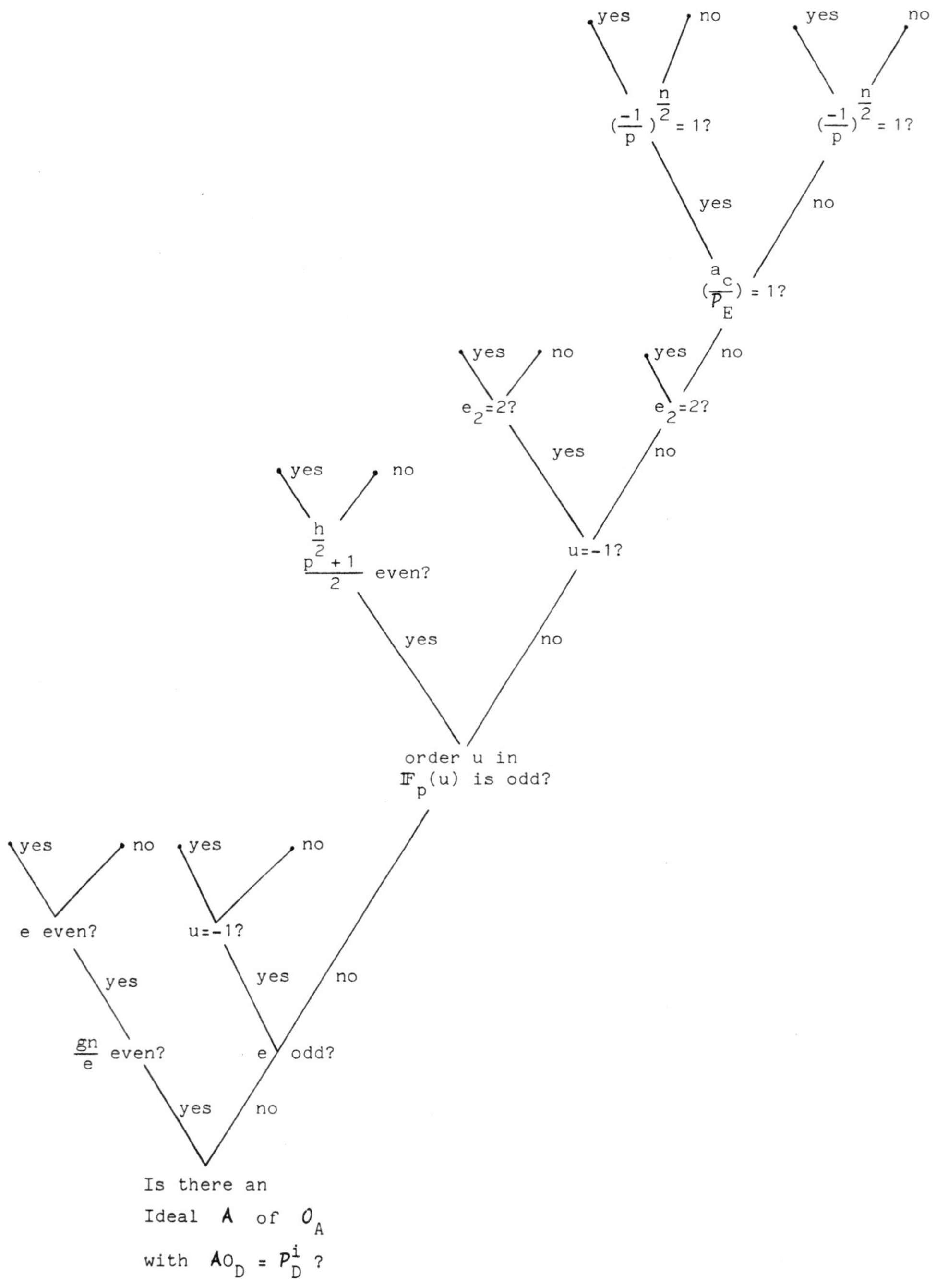
yes
no
yes
no
$(\frac{-1}{p})^{\frac{n}{2}} = 1$?
$(\frac{-1}{p})^{\frac{n}{2}} = 1$?
yes
no
$(\frac{a_c}{\mathcal{P}_E}) = 1$?
yes
no
yes
no
$e_2=2$?
$e_2=2$?
yes
no
yes
no
$\frac{p^{\frac{h}{2}}+1}{2}$ even?
u=-1?
yes
no
order u in $\mathbb{F}_p(u)$ is odd?
yes
no
yes
no
e even?
u=-1?
yes
yes
no
$\frac{gn}{e}$ even?
e odd?
yes
no
Is there an Ideal A of O_A with $AO_D = \mathcal{P}_D^i$?

It must have been a relief to the original explorer(s) of this tree that in the end at least it turned out to be finite.

6. Details of the calculation.

We will be very brief here. If the answer to the first alternative in the tree is yes, then X/X^{Γ} is the direct sum of $\frac{ng}{e}$ copies of the augmentation ideal of the group ring $\mathbb{F}_p(\Gamma/\Gamma^e)$. If it is no, then Γ acts on X with only the trivial fixed points, so $X/X^{\Gamma} = X = P^i/P^{i+1}$.

Thus one faces the following combinatorial problems. These modules have to be decomposed explicitly and the t-values of the components have to be calculated using the recipes a) and b) of 3. Maybe it should be pointed out that though some of the calculations are lengthy and rather involved, none of them requires something very subtle, in contrast to, for example, the classical sign determination of the quadratic Gauss sum.

One of the problems one encounters is that of determining the decomposition of $\mathbb{F}_p(u)[T]/(T^{e_2}-u)$. That is, one has to determine the decomposition of $T^{e_2} - u$ into irreducible factors over the field $\mathbb{F}_p(u)$. Simple as this may sound, this requires surprisingly enough about half of the work of the whole sign determination.

References

[11] C.J. Bushnell, A. Fröhlich: Gauss sums and p-adic division algebras; Lecture Notes in Mathematics 987, Springer Verlag, Berlin-Heidelberg-New York 1983.

[E] Everest, see this volume.

[Wm] Willems, see this volume.

Contemporary Mathematics
Volume 86, 1989

ROOT NUMBERS - THE TAME CASE

by G.R. Everest, Norwich

Introduction.

Suppose F is a local number field - a finite extension of $\mathbb{Q}_p$ - and D is a central F-division algebra of index n. The word tame in the title refers to the assumption that $p \nmid n$. This is not to be confused with the tameness of representations in the sequel.

Let $\bar{F}$ denote the algebraic closure of F and G_F the Galois group of $\bar{F}$ over F. Write $\mathrm{Irr}_n(G_F)$ for the set of equivalence classes of irreducible, continuous (complex) representations of G_F with dimension dividing n. On the other hand, write $\mathrm{Irf}(D^\times)$ for the set of equivalence classes of irreducible, admissible representations of $D^\times$ with finite image. In their book [11], Bushnell and Fröhlich describe a bijection

$$\mathrm{Irr}_n(G_F) \sim \mathrm{Irf}(D^\times) , \tag{1}$$

which satisfies the postulates of the Langlands conjectures. The bijection however, is not canonical. Indeed, the aim of the book is to highlight the role of the Gauss sum in Langlands philosophy.

The correspondence (1) is based upon an iterative procedure. We are going to describe one aspect of the division algebra side, namely, the comparison of root numbers. This is because it throws up an important inconsistancy. (It is by "twisting" away the inconsistancy that the correspondence (1) is obtained. However the twisting process renders the correspondence non-canonical.

Note: there are also inconsistancies on the Galois side.

1. Preliminaries.

Suppose D' is a subalgebra of D with centre F' and index m. We are going to define induction of representations and give a comparison of root numbers. There is some overlap here with Külshammer's talk [B]. We have tried to agree on notation.

Using the reduced trace $Tr_D: D \to F$ one obtains a decomposition of D :

$$D = D' \oplus D'^{\perp} , \tag{2}$$

where $D'^{\perp} = \{y \in D: Tr_D(D'y) = 0\}$. This decomposition even respects ideals,

$$P_D^i = (P_D^i \cap D') \oplus (P_D^i \cap D'^{\perp}) , \quad i \in \mathbb{Z} . \tag{3}$$

Write $V_i(D) = 1 + P_D^i$, to be abbreviated in the sequel as V_i . Define $U_i = V_i(D) \cap D'$ and $W_i = V_i(D) \cap (1 + D'^{\perp})$. For suffices $1 \leq i \leq j \leq 2i$, the decomposition (3) yields an isomorphism

$$V_i/V_j \cong U_i/U_j \times W_iV_j/V_j . \tag{4}$$

Twist (4) with $D'^{\times}$ to obtain the following identification,

$$D'^{\times}V_i/V_j = D'^{\times}/U_j \ltimes W_iV_j/V_j \quad \text{(semi-direct product).} \tag{5}$$

This is very important and we will refer to it many times. It gives the right framework for the definition of induction of representations.

Also, given a representation Π of $D^{\times}$ recall the definition of the <u>root number</u> $W(\Pi)$,

$$W(\Pi) = \frac{(-1)^n \tau(\Pi)}{Nf(\Pi)^{1/2}} . \tag{6}$$

2. Tame case.

Suppose Π is a tame character of $D'^{\times}$ and $(D/D',\Pi)$ is an admissible pair (meaning essentially, that Π cannot be factored non-trivially, through subextensions). Then Π is a representation of $D'^{\times}/U_1$. This group is canonically isomorphic to $D'^{\times}V_1/V_1$. Thus, inflate Π to a representation Π^* of $D'^{\times}V_1$ and define

$$\Pi' = \mathrm{Ind}_{D'^{\times}V_1}^{D} (\Pi^*) . \tag{7}$$

Theorem 1.

$$W(\Pi') = (-1)^{n-m} W(\Pi) .$$

Proof.

Here and throughout, write $A \sim B$ for two complex matrices, to mean $A = rB$ for some $r \in \mathbb{R}^+$ and say that A and B are equivalent.

First, choose c with $c \cdot O_D = P_D \cdot \mathcal{D}_D$. Form the operator

$$T(\Pi') = \sum_{x \in V_o/V_1} \Pi'(c^{-1}x)\phi_D(c^{-1}x) .$$

This is scalar, with eigenvalue $\tau(\Pi')$. Take the trace of both sides to obtain

$$\tau(\Pi') \sim \sum_{x \in V_o/V_1} \mathrm{tr}(\Pi'(c^{-1}x))\phi_D(c^{-1}x) .$$

Using the formula for the character of an induced representation, the right hand side becomes

$$\sum_{x \in V_o/V_1} \sum_{\substack{y \in D^\times/D'^\times V_1 \\ y^{-1}c^{-1}xy \in D'^\times V_1}} \mathrm{tr}(\Pi^*(y^{-1}c^{-1}xy))\phi_D(y^{-1}c^{-1}xy) .$$

Write $y^{-1}c^{-1}xy = c^{-1}u$. This forces $u \in V_o(D')V_1/V_1$. Given u and y, x is uniquely determined so reverse the order of summation

$$\tau(\Pi') \sim \sum_{y \in D^\times/D'^\times V_1} \sum_{u \in V_o(D')V_1/V_1} \mathrm{tr}(\Pi^*(c^{-1}u))\phi_D(c^{-1}u)$$

$$\sim \sum_{u \in V_o(D')V_1/V_1} \mathrm{tr}(\Pi^*(c^{-1}u))\phi_D(c^{-1}u) .$$

Observe that $V_o(d')V_1/V_1 \cong V_o(D')V_1(D')$ so we may take $u \in V_o(D')$. Then we can replace ϕ_D by $\phi_{D'}$ and Π^* by Π to obtain

$$\tau(\Pi') \sim \tau(\Pi) . \qquad \textbf{(8)}$$

It follows at once that

$$\tau(\check{\Pi}') \sim \tau(\check{\Pi}) , \qquad \textbf{(9)}$$

because induction commutes with contragredient. Theorem 1 follows from (9) because the Gauss sums differ by a positive, real number. Therefore $W(\Pi')/(-1)^{n-m}W(\Pi)$ is a positive, real number which must be 1 (take absolute values).□

3. Wild case.

This time Π is a wild character of $D'^{\times}$ and $(D/D',\Pi)$ is an admissible pair. Write $sw(\Pi)\mathcal{O}_D = \mathcal{P}_D^j$. There are two cases, differing greatly in difficulty.

Case (i): j odd. Let $2i = j+1$. Then Π is a representation of $D'^{\times}/U_{j+1}$, use the identification (5) with j replaced by $j+1$. Blow up Π to a representation of $D'^{\times}\cdot V_i/V_{j+1}$ by making it trivial on W_iV_{j+1}/V_{j+1}. Then inflate to Π^*, a representation of $D'^{\times}/V_i$. As before, define

$$\Pi' = \mathrm{Ind}_{D'^{\times}V_i}^{D^{\times}}(\Pi^*) .$$

Theorem 2.

$$W(\Pi') = (-1)^{n-m}W(\Pi) .$$

The proof follows very similar lines to that of Theorem 1.

Case (ii): j even. Set $2i = j$. The notion of complementary subgroup is very useful for this part. Of particular importance is the fact that, once a complementary subgroup C_F is fixed for F, the complementary subgroup C_D is determined to within conjugacy.

We have a wild representation Π of $D'^{\times}$ with $sw(\Pi) = \mathcal{P}_D^{2i}$.
Consider

$$\Pi \Big|_{1+sw(\Pi)} .$$

This contains the character α defined by

$$\alpha(1+x) = \psi_{D'}(-c^{-1}x) \qquad x \in sw(\Pi) ,$$

where c is an element of D which satisfies $c\mathcal{O}_D = \mathcal{D}_D f(\Pi)$. (see [11] p. 61

or [64] (1.4) for details)*. Now set α to be trivial on C_F, so we view α as a character of $F'^{\times}$. Write

$$\varphi = \alpha \circ \mathrm{Nrd}_{D'} ,$$

where $\mathrm{Nrd}_{D'}: D' \to F$ denotes the reduced norm. Then φ absorbs all of the conductor of Π and there is a factorisation

$$\Pi = \Pi_0 \otimes \varphi , \tag{10}$$

with $\quad sw(\Pi_0) \underset{\neq}{\mid} sw(\Pi)$.

Note: it is the presence of the character φ which causes the inconsistency referred to in the introduction - see the formula in Theorem 3.

Using (10) we will make a two-part definition of induction. Treat φ first. Restrict φ to U_1. Twisting (4) by U_1 allows us to identify

$$U_1 \cdot V_{i+1}/V_{j+1} = U_1/V_{j+1} \ltimes W_{i+1} \cdot V_{j+1}/V_{j+1} . \tag{11}$$

Now φ is trivial on U_{j+1} so it can be viewed as a character of U_1/U_{j+1}. Blow it up to a character $\tilde{\varphi}$ of $U_1 V_{i+1}/V_{j+1}$ by making it trivial on the other factor of (11). Then $\tilde{\varphi}$ lifts to $\Sigma_{\tilde{\varphi}}$, an irreducible representation of $U_1 V_i/V_{j+1}$ (uniquely determined by $\tilde{\varphi}$). Finally, use the decomposition

$$D'^{\times} V_i/V_{j+1} = C_{D'} \ltimes U_1 V_i/V_{j+1} ,$$

(which follows directly from the decomposition $D'^{\times} = C_{D'} \ltimes V_1(D')$) .
In his talk [ ], Willems shows that there is a unique representation Ω_φ of $D'^{\times} V_i/V_{j+1}$ such that:

(i) $\quad \Omega_\varphi \big|_{U_1 V_i/V_{j+1}} = \Sigma_\varphi$

(ii) $\quad \det \Omega_\varphi \big|_{C_{D'}} = 1$.

Secondly, treat Π_0 (as in (10)). This is a representation of $D'^{\times}/U_j$. Extend Π_0 to $D'^{\times} V_i/V_j$ using (5) and inflate to Π_0^* - a representation of $D'^{\times} V_i$.

---

* Footnote: The usual way of saying this is: (F'/F',c) is a <u>fundamental pair</u> and (F'/F,c) is a <u>primordial pair</u>.

Now set

$$\Pi^* = \Pi^*_o \otimes \Omega_\varphi ,$$

and define

$$\Pi' = \mathrm{Ind}_{D'^{\times}V_i}^{D^{\times}} (\Pi^*) .$$

Theorem 3.

$$W(\Pi') = (-1)^{n-m} W(\Pi) \mathrm{tr}(\Omega_\varphi(c)) .$$

Proof. (Sketch)

Set $\Phi' = \mathrm{Ind}_{D'^{\times}V_i}^{D^{\times}} (\Omega_\varphi)$. The following comparison is very important (and quite hard),

$$\tau(\varphi) \sim \tau(\Phi')\mathrm{tr}(\Omega_\varphi(c)) . \qquad (12)$$

Theorem 3 follows by setting

$$T^* = \sum_{x \in V_o(D')V_i/V_{j+1}} \Pi^*(c_1^{-1}x)\psi_D(c_1^{-1}x) \quad (\text{where} \quad c_1\mathcal{O}_D = f(\Pi')\mathcal{D}_D) ,$$

and observing that $\tau(\Pi') \sim \mathrm{tr}(T^*)$ by the method of Theorem 1. This (scalar) is equivalent to

$$\sum_{y \in U_1/U_j} \Pi(c^{-1}y)\psi_{D'}(c^{-1}y) \otimes \sum_{\substack{z \in W_iV_j/V_j \\ z \in W_i}} \Omega_\varphi(c^{-1}z) . \qquad (13)$$

Again, this follows by the method of Theorem 1 using the two steps

$$V_j/V_{j+1} = U_j/U_{j+1} \times W_jV_{j+1}/V_{j+1}$$

and

$$U_1V_i/V_j = U_1/U_j \times W_iV_j/V_j .$$

Thus both factors of (13) are scalar. The first is equivalent to $\tau(\Pi)$. The second is independent of Π and depends upon φ only. Using (12) it is easily calculated to be $tr(\Omega_\varphi(c))^{-1}$. It follows from Willems calculations [Wm] that this quantity behaves well under contragredients and the theorem follows at once.

The proof of formula (12) requires some work. We conclude by giving a broad outline of the argument. Let $\Gamma = < c >$ mod C_F (as in Brinkhuis' talk [Br]). Write

$$\Omega_\varphi\Big|_\Gamma = \sum_{\alpha\in\hat{\Gamma}} n_\alpha \alpha^{-1} .$$

Formula (12) is obtained by writing down two operators:

$$T_1 = \overline{\sum_{x\in V_o(D')V_i/V_{j+1}}} \Omega_\varphi(c^{-1}x)\psi_D(c^{-1}x) ,$$

a scalar operator with

$$tr(T_1) \sim \tau(\Phi') .$$

Also,

$$T_2 = \overline{\sum_{u\in U_1V_i/V_j}} (\Omega_\varphi \otimes \sum_{\alpha\in\Gamma} n_\alpha \alpha)(c^{-1}u)\psi_D(c^{-1}u) ,$$

a matrix with

$$tr(T_2) \sim \tau(\varphi) .$$

These are all related by the following chain of equivalences:

$$\tau(\varphi) \sim tr(T_2) = \sum_{\alpha\in\Gamma} n_\alpha tr(\alpha(c^{-1})T_1) = tr(T_1) \sum_\alpha n_\alpha \alpha(c^{-1}) \sim \tau(\varphi')tr(\Omega_\varphi(c)) .$$

Whence formula (12).

<u>**Note:**</u> See the talk by Brinkhuis ([Br]) for a table of values of $tr(\Omega_\varphi(c))$.

References

[11] C.J. Bushnell, A. Fröhlich: Gauss sums and p-adic division algebras; Lecture Notes in Mathematics 987, Springer Verlag, Berlin-Heidelberg-New York, 1983.

[64] H. Koch, E.-W. Zink: Zur Korrespondenz von Darstellungen der Galoisgruppen und der zentralen Divisionsalgebren über lokalen Körpern; Math. Nachr. 98 (1980), 83 - 119.

[Wm] Willems, see this volume.

[Br] Brinkhuis, see this volume.

[B] Becker/Külshammer, see this volume.

ntemporary Mathematics
lume **86**, 1989

REPRESENTATIONS OF LOCALLY PROFINITE GROUPS

by K. Wingberg, Regensburg

This article provides some basic informations on the representation theory of ocally profinite groups. The definitions and theorems stated here will be needed the subsequent paper of U. Jannsen presenting the results of Bernstein and elevinskii on the classification of smooth irreducible representations of $GL_n(F)$, a non-archimedian local field, by cuspidal representations. For a more detailed esentation of this theory the reader should look at the excellent articles of ernstein and Zelevinskii [4], Cartier [15] and Rodier [97] where also proofs of e following abstract can be found.

1. Locally profinite groups

.1 **Definition.** A topological group G is called locally profinite if the folowing equivalent conditions hold:

i) G is locally compact and totally disconnected.

ii) There exists a fundamental system of neighbourhoods of the unit element consisting of open compact subgroups.

xample. 1) Let F be a non-archimedean local field and $G = GL_n(F)$. Then $= GL_n(\mathcal{O})$ is a maximal compact open subgroup of G and the set $K_r = 1 + p^r M_n(\mathcal{O}), r \geq 1\}$ is a fundamental system of neighbourhoods of 1; here resp. p denotes the ring of integers of F resp. the maximal ideal in $\mathcal{O}$.

) Let F be the algebraic closure of a p-adic field F and let F_{nr} be the maximal unramified extension of F. Let $\varphi \in Gal(F_{nr}/F) \simeq \hat{\mathbb{Z}}$ be the Frobenius autoorphism, then the Weil group of F

$$W_F = \{\sigma \in Gal(\bar{F}/F) \mid \sigma_{|F_{nr}} = \varphi^n \quad \text{for} \quad n \in \mathbb{Z}\}$$

is a locally profinite group.

1.2 Remark. Closed subgroups, factor groups by closed normal subgroups, finite and restricted products of locally profinite groups are again locally profinite.

1.3 Definition. Let G be a locally profinite group. Then the Hecke algebra $\mathcal{H}(G)$ of G is defined as the $\mathbb{C}$-vector space consisting of all Schwartz functions on G, i. e. all locally constant functions $f\colon G \to \mathbb{C}$ with compact support. The multiplication on $\mathcal{H}(G)$ is defined by the convolution of $f_1, f_2 \in \mathcal{H}(G)$:

$$f_1 * f_2(g) = \int_G f_1(gg')f_2(g'^{-1})d\mu(g') \,, \quad g \in G \,.$$

Here μ denotes a left invariant Haar measure on G.
$\mathcal{H}(G)$ is an associative algebra without unit element (if G is not discrete).

Let $K \subset G$ be a compact open subgroup, then $\mathcal{H}(G,K)$ is defined as follows:

$$\mathcal{H}(G,K) = \{f \in \mathcal{H}(G) \mid f \text{ is constant at all double cosets } KgK \,,\ g \in G\} \,.$$

1.4 Proposition.

i) $\mathcal{H}(G) = \cup\, \mathcal{H}(G,K)$, <u>where the union is taken over all open compact subgroups of</u> G.

ii) $\mathcal{H}(G,K)$ <u>is a subalgebra of</u> $\mathcal{H}(G)$ <u>with the characteristic function</u> e_K <u>as unit element:</u>

$$e_K\colon G \to \mathbb{C} \,, \quad e_K(g) = \mu(K)^{-1} \quad \text{for } g \in K \text{ and } 0 \text{ otherwise.}$$

§ 2. Representations of locally profinite groups

2.1. Definition. A (complex) representation of a locally profinite group G is a pair (Π, V) consisting of a $\mathbb{C}$-vector space V (i. g. not of finite dimension) and a homomorphism

$$\Pi\colon G \to GL(V)$$

such that the following equivalent conditions hold:

i) For all $v \in V$ the stabilizer $I(v) = \{g \in G \mid \Pi(g)v = v\}$ is an open subgroup of G.

ii) $V = \cup V^K$, where the union is taken over all compact open subgroups of G and $V^K = \{v \in V \mid \Pi(g)v = v \text{ for all } g \in K\}$ is the subspace of V invariant under K.

iii) G acts via Π continuously on V (V supplied with the discrete topology).

Let Rep_G denote the category of the representations of G (the morphisms are given in the obvious way). Rep_G is an abelian category with arbitrary direct sums.

Let $V^* = Hom_{\mathbb{C}}(V,\mathbb{C})$ be the dual space. Then there is a canonical pairing:

$$V^* \times V \to \mathbb{C}, \quad (v^*,v) \longmapsto \langle v^*,v \rangle .$$

2.2 Definition. Let $\Pi \in Rep_G$. The function

$$\pi_{v^*,v} : G \to \mathbb{C}, \quad g \longmapsto \langle v^*,\Pi(g)v \rangle$$

is called matrix-coefficient of Π at (v^*,v). This function is locally constant, but has i. g. not a compact support.

2.3 Proposition. Let $\Pi \in Rep_G$.

i) The map $\mathcal{H}(G) \to End_G(V)$, $f \longmapsto \Pi(f)$, is a $\mathbb{C}$-algebra homomorphism, where $\Pi(f)$ is defined by

$$\langle v^*,\Pi(f)v \rangle = \int_G f(g)\pi_{v^*,v}(g)\,d\mu(g), \quad v^* \in V^*, \quad v \in V .$$

Hence V is supplied with a $\mathcal{H}(G)$-module structure.

ii) Let $K \subset G$ be a compact open subgroup, then $\Pi(e_K)$ is a projector onto V^K. In particular: V is a non-degenerated $\mathcal{H}(G)$-module, i. e. every $v \in V$ can be written as a sum $\Sigma h_i v_i$, $h_i \in \mathcal{H}(G)$, $v_i \in V$.

2.4 Proposition. The functor

$Rep_G \longrightarrow$ category of non-degenerate $\mathcal{H}(G)$-modules

$(\Pi,V) \longmapsto V$ as $\mathcal{H}(G)$-module

is an equivalence of categories.

2.5 Proposition. Let (Π,V) , $(\Pi',V') \in \mathrm{Rep}_G$ then

i) (Π,V) is irreducible if and only if for all compact open subgroups $K \subset G$ the $\mathcal{H}(G,K)$-module V^K is simple.

ii) $(\Pi,V) \simeq (\Pi',V')$ if and only if for one open compact subgroup $K \subset G$ the $\mathcal{H}(G,K)$-modules V^K and V'^K are non-zero and isomorphic.

2.6 Definition. A representation $\Pi \in \mathrm{Rep}_G$ is called admissible, if

$$\dim_{\mathbb{C}} V^K < \infty \quad \text{for all compact open subgroups} \quad K \subset G .$$

The admissible representations form a Serre subcategory of Rep_G .

2.7 Lemma. Let $(\Pi,V) \in \mathrm{Rep}_G$, then the following assertions are equivalent

i) (Π,V) is admissible.

ii) For all $f \in \mathcal{H}(G)$ the endomorphism $\Pi(f) : V \to V$ is of finite rank.

2.8 Definition. Let (Π,V) be an admissible representation of G , then the $\mathbb{C}$-linear map

$$\mathrm{tr}(\Pi) : \mathcal{H}(G) \to \mathbb{C} , \quad f \longmapsto \mathrm{tr}(\Pi(f))$$

is called character of (Π,V) .

2.9 Proposition. Let $\{(\Pi_i,V_i)\}$ be a family of pairwise non-isomorphic irreducible admissible representations of G . Then the family $\{\mathrm{tr}(\Pi_i)\}$ is linear independent over $\mathbb{C}$.

2.10 Corollary. Let (Π,V) , $(\Pi',V') \in \mathrm{Rep}_G$ be admissible and irreducible, then

$$(\Pi,V) \simeq (\Pi',V') \quad \text{if and only if} \quad \mathrm{tr}(\Pi) = \mathrm{tr}(\Pi') .$$

2.11 Definition. A representation $(\Pi,V) \in \mathrm{Rep}_G$ is called finitely generated, if V as $\mathcal{H}(G)$-module is finitely generated.

2.12 Remark. A finitely, generated representation $0 \neq (\Pi,V) \in \mathrm{Rep}_G$ has an irreducible quotient, since there exists a maximal $\mathcal{H}(G)$-submodule of V .

§ 3. Construction of representations

The contragredient representation.

Let $(\Pi, V) \in \mathrm{Rep}_G$; then the contragredient representation is defined by

$$\Pi^\vee : G \to GL(\tilde{V}) \ , \quad \tilde{V} = \cup \, V^{*K} = \{v^* \in V^* \mid I(v^*) \quad \text{open in} \quad G\} \ ,$$

$$< \Pi^\vee(g)v^*, v > := < v^*, \Pi(g^{-1})v > .$$

3.1 Proposition. Let $(\Pi, V) \in \mathrm{Rep}_G$ be admissible. Then

i) $(\Pi^\vee, \tilde{V})$ is admissible;

ii) the canonical morphism $(\Pi, V) \to (\Pi^{\vee\vee}, \tilde{\tilde{V}})$ is an isomorphism;

iii) $(\Pi^\vee, \tilde{V})$ is irreducible if and only if (Π, V) is irreducible.

The tensor product.

Let $G_1, \ldots, G_r$ be locally profinite groups and let $(\Pi_i, V_i) \in \mathrm{Rep}_{G_i}$, $1 \leq i \leq r$. Then the tensor product of (Π_i, V_i) is defined by

$$\Pi : G = G_1 \times \ldots \times G_r \to GL(\bigotimes_{i=1}^{r} V_i) \ , \quad \Pi(g_1, \ldots, g_r)(\otimes v_i) := \Pi_i(g_i)(v_i) \ .$$

3.2 Proposition. **i)** Let $(\Pi_i, V_i) \in \mathrm{Rep}_{G_i}$, $1 \leq i \leq r$ be admissible representations then $(\Pi, V) = \otimes(\Pi_i, V_i)$ is admissible. If in addition all (Π_i, V_i) are irreducible then this holds also for (Π, V) .

ii) Let $(\Pi, V) \in \mathrm{Rep}_G$ be an admissible irreducible representation. Then there is an isomorphism

$$(\Pi, V) \simeq \bigotimes_{i=1}^{r} (\Pi_i, V_i)$$

with (up to isomorphism) uniquely determined, admissible and irreducible representations $(\Pi_i, V_i) \in \mathrm{Rep}_{G_i}$, $1 \leq i \leq r$.

Induced representation.

Let H be a closed subgroup of G , then there is the restriction functor

$$\mathrm{Res}_H^G : \mathrm{Rep}_G \to \mathrm{Rep}_H , \quad (\Pi,V) \longmapsto (\Pi_{|H},V) .$$

3.3 **Theorem.** (Frobenius duality)

There exists a right-adjoint functor to Res_H^G : the induction functor

$$\mathrm{Ind}_H^G : \mathrm{Rep}_H \to \mathrm{Rep}_G ,$$

$$\mathrm{Hom}_G(\Pi,\mathrm{Ind}_H^G\Sigma) = \mathrm{Hom}_H(\mathrm{Res}_H^G\,\Pi,\delta_G\delta_H^{-1} \otimes \Sigma),$$

where δ_G is the square root of the modulus Δ_G of G, i. e.
$\delta_G = \Delta_G^{1/2} : G \to \mathbb{R}_+^*$, $\Delta_G(g)\mu_G = t(g)\mu_G$, $t(g)$ the right translation by $g \in G$.

Construction of $\mathrm{Ind}_H^G\Sigma$ for $(\Sigma,W) \in \mathrm{Rep}_H$:

Let V be the $\mathbb{C}$-vector space consisting of all maps $v: G \to W$ such that

$$v(hg) = \delta_G\delta_H^{-1}(h)\Sigma(h)v(g) , \quad h \in H , g \in G ,$$

$$v(gk) = v(g) , \quad g \in G , k \in K_v ,$$

for a compact open subgroup K_v of G .
Then $\Pi = \mathrm{Ind}_H^G\Sigma : G \to GL(V)$ is defined by right translation

$$\Pi(g)(v)(g') = v(g'g) , \quad g',g \in G , v \in V .$$

3.4 **Proposition.** The following assertions hold

i) The functor Ind is transitiv: $\mathrm{Ind}_H^G = \mathrm{Ind}_{H'}^G \circ \mathrm{Ind}_H^{H'}$ for closed subgroups $H \leq H' \leq G$.

ii) Ind is an exact functor.

iii) Let G modulo H be compact, then $\mathrm{Ind}_H^G\Sigma$ is admissible if this holds for Σ .

iv) The subspace $V_c = \{v \in V | v$ has compact support modulo $H\}$ is G-invariant. The corresponding representation

$$_c\mathrm{Ind}_H^G : G \to GL(V)_c$$

is called compact induction.

§ 4. Representation of $GL_n(F)$, F a non-archimedean local field

Let $G = GL_n(F)$ and let $\alpha = (n_1,\ldots,n_r)$ be a partition of n . Then there is a functor called parabolic induction

$$J = J_\alpha : \mathrm{Rep}_{GL_{n_1} \times \ldots \times GL_{n_r}} \to \mathrm{Rep}_G$$

which is defined as follows: Consider the parabolic subgroup P_α of G with the unipotent radical U_α

$$P_\alpha = \left\{ \begin{pmatrix} n_1 & & * \\ & \ddots & \\ 0 & & n_r \end{pmatrix} \right\} , \quad U_\alpha = \left\{ \begin{pmatrix} \begin{smallmatrix} 1 & 0 \\ & \ddots \\ 0 & 1 \end{smallmatrix} & & * \\ & \ddots & \\ 0 & & \begin{smallmatrix} 1 & 0 \\ & \ddots \\ 0 & 1 \end{smallmatrix} \end{pmatrix} \right\} ;$$

hence we have a semidirect product $P_\alpha = U_\alpha \times (GL_{n_1} \times \ldots \times GL_{n_r})$ of locally profinite groups. Let $\Pi_1 \otimes \ldots \otimes \Pi_r$ be a representation of $GL_{n_1} \times \ldots \times GL_{n_r}$ then

$$\Pi_1 \circ \ldots \circ \Pi_r := J_\alpha(\Pi_1 \otimes \ldots \otimes \Pi_r)$$

is given by extending $\otimes \Pi_i$ trivially to P_α: $(\otimes \Pi_i)(ug) = (\otimes \Pi_i)(g)$ for $u \in U_\alpha$, $g \in \prod GL_{n_i}$, and inducing this representation to G .

4.1 Proposition. <u>J_α is an exact functor and $J_\alpha(\Pi_1 \otimes \ldots \otimes \Pi_r)$ is admissible if all representations Π_i are admissible.</u>

There exists a right-adjoint functor to J_α . We define this functor in the following more general situation:
Let $P = U \ltimes M$ be a semi-direct product of locally profinite groups, where U is the union of all of its compact subgroups. Furthermore let $\theta: U \to \mathbb{C}$ be a character invariant under the action of M . For a representation $(\Pi, V) \in \mathrm{Rep}_P$ we set

$$V_{U,\theta} := V/(\Pi(u)v - \theta(u)v \mid u \in U , v \in V) ;$$

then Π factors through a representation

$$\Pi_{U,\theta} : P \to GL(V_{U,\theta}) .$$

The functor

$$r_{U,\theta} : \Pi \longmapsto r_{U,\theta}(\Pi) = \mathrm{Res}_M^P(\delta_P\delta_M^{-1} \otimes \Pi_{U,\theta}) : M \to GL(V_{U,\theta})$$

is called Jacquet-functor.

4.2 Proposition.

i) $r_{u,\theta}$ *is exact.*

ii) $\mathrm{Hom}_P(\Pi, \Sigma \otimes \theta) = \mathrm{Hom}_M(r_{U,\theta}(\Pi), \delta_P\delta_M^{-1} \otimes \Sigma)$ *for* $\Sigma \in \mathrm{Rep}_M$, $\Pi \in \mathrm{Rep}_P$.

Now let $P = P_\alpha$, $U = U_\alpha$, $\theta = 1$, then the exact functor

$$P_\alpha = r_{U,1} \circ \mathrm{Res}_{P_\alpha}^{GL_n}$$

is right-adjoint to J_α.

4.3 Theorem. (Jacquet) *If* $\Pi \in \mathrm{Rep}_{GL_n}$ *is admissible and finitely generated then* $R_\alpha(\Pi)$ *is admissible and finitely generated, respectively.*

4.4 Definition. An admissible representation $\Pi \in \mathrm{Rep}_{GL_n}$ is called cuspidal if for all proper partitions α of n

$$R_\alpha(\Pi) = 0$$

or equivalent

$$\mathrm{Hom}(\Pi, J_\alpha(\Sigma)) = 0$$

for all $\Sigma \in \mathrm{Rep}_{GL_{n_1} \times \ldots \times GL_{n_r}}$ and all proper partitions α of n.

4.5 Remark. Since R_α is exact all subquotients of a cuspidal representation are cuspidal.

4.6 Theorem. (Harish-Chandra) *Let* $\Pi \in \mathrm{Rep}_{GL_n}$ *be admissible. Then* Π *is cuspidal if and only if all matrix-coefficients* $\pi_{v^*,v}$ *have compact support modulo the center of* GL_n.

4.7 Theorem. Let $\Pi \in Rep_{GL_n}$ be admissible and irreducible. Then there exists a partition $\alpha = (n_1,\dots,n_r)$ of n and irreducible cuspidal representations $\Sigma_i \in Rep_{GL_{n_i}}$, $1 \leq i \leq r$, such that Π is a subrepresentation of $\Sigma_1 \circ \dots \circ \Sigma_r$.

Proof. Choose $\alpha = (n_1,\dots,n_r)$ such that $R_\alpha(\Pi) \neq 0$ and $R_\beta(\Pi) = 0$ for all finer partitions $\beta<\alpha$. Since Π is irreducible and admissible $R_\alpha(\Pi)$ is finitely generated and admissible (4.3). Hence $R_\alpha(\Pi)$ has and irreducible admissible quotient $\Sigma=\Sigma_1\otimes\dots\otimes\Sigma_r$ (2.12, 3.1.ii). The representations Σ_i are cuspidal by the choice of the partition α. Because of

$$\mathrm{Hom}(\Pi,\Sigma_1 \circ \dots \circ \Sigma_r) = \mathrm{Hom}(\Pi, J_\alpha(\Sigma_1 \otimes \dots \otimes \Sigma_r)) = \mathrm{Hom}(R_\alpha(\Pi),\Sigma) \neq 0$$

there exist a non-zero homomorphism

$$\Pi \to \Sigma_1 \circ \dots \circ \Sigma_r$$

which is injective, since Π is irreducible.

References

[4] I.N. Bernštein, A.V. Zelevinskii: Representations of the group GL(n,F) where F is a local non-archimedean field; Russion Math. Surveys 31 (1976), 1 - 68 (transl. from Uspekhi Math. Nauk 31, no.3 (1976), 5 - 70

[15] P. Cartier: Representations of p-adic groups: A survey; Proceedings of Symposia in pure Math. Vol. 33 (1979), part 1, 111 - 155.

[97] F. Rodier: Répresentations de GL(n,k) oú k un corps p-adique; Séminaire Bourbaki (1981/1982) 587,

Contemporary Mathematics
Volume **86**, 1989

THE THEOREMS OF BERNŠTEIN AND ZELEVINSKII

by U. Jannsen, Regensburg

This is a short survey of the work of I. N. Bernštein and A. V. Zelevinskii on the classification of irreducible representations of $GL_n(F)$ for a local field F in terms of the cuspidal ones. There is already a very good survey article by F. Rodier [97], so I shall restrict myself to presenting the results and discussing their meaning rather that say much about the proofs.

I shall freely use the notations and theorems of the previous article by K. Wingberg. There it was shown that every irreducible admissible representation Σ of $GL_n(F)$ can be embedded in a representation $\Sigma_1 \circ \ldots \circ \Sigma_r$ with cuspidal Σ_i. The idea for the classification is to do this embedding in a canonical way, by studying the composition factors of $\Sigma_1 \circ \ldots \circ \Sigma_r$. This is, of course, interesting in its own right.

1. Induced representations

Let F be a non-archimedean local field and $G_n = GL_n(F)$ for $n \geq 1$, $G_0 = \{1\}$. For a partition $\alpha = (n_1,\ldots,n_r)$ of n, i. e., $n_1 + \ldots + n_r = n$ let P_α be the standard parabolic described in the previous talk, with Levi part M_α isomorphic to $G_{n_1} \times \ldots \times G_{n_r}$.

1.1 Theorem. (Howe/Bernštein-Zelevinskii) <u>If</u> Σ_i <u>are irreducible cuspidal representations of</u> G_{n_i}, $i = 1,\ldots,r$, then $\Sigma_1 \circ \ldots \circ \Sigma_r := \mathrm{Ind}_{P_\alpha}^{G_n}(\Sigma_1 \otimes \ldots \otimes \Sigma_r)$ <u>is of finite length, at most</u> $r!$.

1.2 Theorem. (Bernštein-Zelevinskii) <u>Let</u> $\alpha = (m_1,\ldots,m_r)$ <u>and</u> $\beta = (n_1,\ldots,n_s)$ <u>be partitions of</u> n, <u>let</u> Φ_i <u>(resp.</u> Σ_j<u>) be irreducible cuspidal representations of</u> G_{m_i} <u>(resp.</u> G_{n_j}<u>) for</u> $i = 1,\ldots,r$ <u>(resp.</u> $j = 1,\ldots,s$). <u>Then the following statements are equivalent for the representations</u> $\Phi = \Phi_1 \circ \ldots \circ \Phi_r$ <u>and</u> $\Sigma = \Sigma_1 \circ \ldots \circ \Sigma_s$ <u>of</u> G_n:

i) Φ and Σ have a common subquotient.

ii) Φ and Σ have the same Jordan-Hölder series (with multiplicities).

iii) $\mathrm{Hom}(\Phi,\Sigma) \neq 0$.

iv) $r = s$ and there is a permutation $z \in S_r$ such that $m_i = n_{z(i)}$ and $\Phi_i = \Sigma_{z(i)}$ for all i .

Idea of the proof. Since

$$\mathrm{Hom}_{G_n}(\Phi,\Sigma) \simeq \mathrm{Hom}_{P_\beta}(\mathrm{Res}^{G_n}_{P_\beta}\Phi, \delta^{-1}_{P_\beta}\Sigma_1 \otimes \dots \otimes \Sigma_s) \simeq \mathrm{Hom}_{M_\beta}(R_\beta\Phi, \Sigma_1 \otimes \dots \otimes \Sigma_s) ,$$

it turns out that one has to study $R_\beta\Phi$ for both results. This is done by using the Bruhat decomposition

$$G_n = \bigcup_{w \in W_{\alpha,\beta}} P_\alpha\, w\, P_\beta ,$$

where $W_{\alpha,\beta}$ is a certain subset of the Weyl group of G_n , which is isomorphic to the symmetric group S_n .

This does not yet give the complete classification of irreducible representations, because $\Sigma_1 \circ \dots \circ \Sigma_r$ may have non-isomorphic subquotients. The idea is to order the Σ_i in a canonical way and then find canonical subrepresentations or quotients of $\Sigma_1 \circ \dots \circ \Sigma_r$.

2. The method of Gelfand and Kazhdan

It consists in studying the restriction to the subgroups

$$G_n \supseteq P_n = \left\{ \begin{pmatrix} & * & \\ 0 & 0 \dots 0 & 1 \end{pmatrix} \right\}$$

(note that this is not a parabolic).

2.1 Proposition. Let Π be a smooth representation of P_n . Then there exists a series of subrepresentations

$$\Pi = \Pi_0 \supset \Pi_1 \supset \dots \supset \Pi_n = 0$$

such that Π_{i-1}/Π_i is of the form

$$c\text{-}\mathrm{Ind}_{G_{n-i}\cdot U_n^i}^{P_n}(\delta^{-1}_{P_n}\cdot r_{U_n^i,\theta_n^i}(\Pi)\otimes\theta_n^i)$$

where

$$U_n^i = U_{(n-i,1,\ldots 1)} = \left\{\begin{pmatrix} 1_{n-i} & & & * & \\ & 1 & u_{n-i,\,n-i+2} & & \\ & & 1 & \ddots & \\ & 0 & & \ddots & u_{n-1,n} \\ & & & & 1 \end{pmatrix} = u\right\}$$

= unipotent radical of $P_{(n-i,1,\ldots,1)}$,

and θ_n^i is the character of U_n^i given by

$$\theta_n^i(u) = \psi(u_{n-i+1,n-i+2}) \cdots \psi(u_{n-1,n}) .$$

Here ψ is a non-trivial character of F .

Method of proof. One has $P_n = U_n^1 \rtimes G_{n-1}$, where $U_n^1 \simeq F^{n-1}$ is abelian. The characters of U_n^1 decompose into two orbits unter the action of G_{n-1} : one consisting of the trivial character and the other one containing the character $\theta_1 : U \longmapsto \psi(u_{n-1,n})$. By a Mackey-type argument one finds a subrepresentation Π_1 induced from P_{n-1} such that the quotient Π/Π_1 comes from G_{n-1} and may proceed by induction. More precisely one gets an exact sequence

$$0 \to c\text{-}\mathrm{Ind}_{P_{n-1}U_n^1}^{P_n}(\delta^{-1}_{P_n}\delta_{P_{n-1}U_n^1}\mathrm{Res}_{P_{n-1}}^{P_n}\Pi_{U_n^1,\theta_1}\otimes\theta_1) \to \Pi \to \Pi_{U_n^1,1} \to 0 ,$$

see [4] 5.12.

2.2 Example Let Π be the restriction to P_n of an irreducible cuspidal representation Π_o of G_n . Then

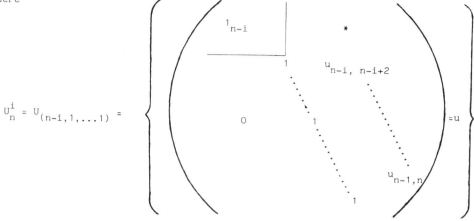

$$r_{U_n^i,\theta_n^i}(\Pi) = 0 \quad \text{for} \quad 1 \leq i \leq n-1 .$$

In fact, since θ_n^i is trivial on

$$U_{(n-i,i)} = \left\{ \begin{pmatrix} 1_{n-i} & * \\ 0 & 1_i \end{pmatrix} \right\},$$

$r_{U_n^i,\theta_n^i}(\Pi)$ is a quotient of $R_{(n-i,i)}(\Pi_o)$, which by definition is zero for cuspidal Π_o if $i \neq 0,n$.

On the other hand, by a result of Gelfand-Kazhdan one has

$$\dim r_{U_n^n,\theta_n^n}(\Pi) = 1$$

in our situation, see [32] thm. B and C.

In view of 2.1 we see that Π must be irreducible and $\Pi = {}_c\mathrm{Ind}_{U_n^n}^{P_n}(\theta_n^n)$.

3. Kirillov and Whittaker model

Whittaker or Kirillov models allow the realization of an abstract representation Π in a concrete space of functions on the group G_n or P_n, respectively (compare the definition of induced representations!). They are very important for the calculation of L- and ε-factors.

3.1 Definition. For a representation Π of P_n and $i \geq 1$ let $\Pi^{(i)} := r_{U_n^i,\theta_n^i}(\Pi)$, this is a representation of G_{n-i}. For a representation Π of G_n let $\Pi^{(i)} = (\mathrm{Res}_{P_n}^{G_n}(\Pi))^{(i)}$.

3.2 Theorem. (Zelevinskii) Let Π be an irreducible admissible representation of G_n and m be maximal with $\Pi^{(m)} \neq 0$ (this $\Pi^{(m)}$ is called the principal derivative).

a) Then $\Pi^{(m)}$ is irreducible.

b) If $\Sigma \neq 0$ is a subrepresentation of $\mathrm{Res}_{P_n}^{G_n}(\Pi)$, then also $\Sigma^{(m)} \neq 0$ (i.e.

$\mathrm{Res}_{P_n}^{G_n}(\Pi)$ <u>is homogeneous, see [115] 5.1)</u>

3.3 Corollary. $\mathrm{Res}_{P_n}^{G_n}(\Pi)$ <u>is uniquely realizable in a subspace</u>

$$K(\Pi) \subseteq \mathrm{Ind}_{G_{n-m}\cdot U_n^m}^{P_n}(\delta_{P_n}^{-1}\Pi^{(m)} \otimes \theta_n^m) .$$

This is called the degenerate Kirillov model.

Proof. By Frobenius reciprocity we have

$$\dim \mathrm{Hom}_{P_n}(\mathrm{Res}_{P_n}^{G_n}(\Pi), \mathrm{Ind}_{G_{n-m}\cdot U_n^m}^{P_n}(\delta_{P_n}^{-1}\Pi^{(m)} \otimes \theta_n^m))$$

$$= \dim \mathrm{Hom}_{G_{n-m}\cdot U_n^m}(\mathrm{Res}_{G_{n-m}\cdot U_n^m}^{G_n}(\Pi) , \delta^{-1}_{G_{n-m}\cdot U_n^m}\Pi^{(m)} \otimes \theta_n^m)$$

$$= \dim \mathrm{Hom}_{G_{n-m}}(\Pi^{(m)} = r_{U_n^m,\theta_n^m}(\Pi),\Pi^{(m)}) = 1 , \quad \text{by 3.2 a),}$$

so there is up to scalar multiples a unique non-trivial P_n-morphism

$$A_\Pi : \mathrm{Res}_{P_n}^{G_n}(\Pi) \to \mathrm{Ind}_{G_{n-m}\cdot U_n^m}^{P_n}(\delta_{P_n}^{-1}\Pi^{(m)} \otimes \theta_n^m) .$$

The point is to show that A_Π is injective (note that $\mathrm{Res}_{P_n}^{G_n}\Pi$ can be reducible), then we may set $K(\Pi) = \mathrm{Im}\, A_\Pi$. But by construction $(\mathrm{Ker}\, A_\Pi)^{(m)} = 0$, so by 3.2b) we must have $\mathrm{Ker}\, A_\Pi = 0$.

3.4 Corollary. <u>There is a character</u> θ' <u>of</u> U_n^n <u>such that</u> Π <u>is uniquely realizable in a space</u>

$$W(\Pi) \subseteq \mathrm{Ind}_{U_n^n}^{G_n}(\theta') .$$

This is called the degenerate Whittaker model.

Proof. Let $\lambda_o = 0$, $\Pi^{(o)} = \Pi$ and $\Pi^{(\lambda_i)}$ be the principal derivative of $\Pi^{(\lambda_{i-1})}$ for $i \geq 1$. For big k, $\Pi^{(\lambda_k)}$ must be a representation of $G_o = \{1\}$, and by 3.2a) it must be 1-dimensional. If θ' is the character of U_n^n defined by

$$\theta'(u) = \sum_{\substack{i \neq n-\lambda_1, \\ n-\lambda_1-\lambda_2, \\ \dots}} \psi(u_{i,i+1}) : \quad \begin{pmatrix} 1 & \ddots & & & & & & \\ & \ddots & \psi_1 & & & & & \\ & & \ddots & \psi & & & & \\ & & & \ddots & \ddots \;\Big\}\, \lambda_2 - 1 & & & \\ & & & & \ddots & \psi & & \\ & & & & & 1 & \psi & \\ & & & & & & \ddots & \ddots \;\Big\}\, \lambda_1 - 1 \\ & & & & & & & 1 \quad \psi \end{pmatrix},$$

it is therefore clear that $\dim r_{U_n^n,\theta'}(\Pi) = 1$. By Frobenius reciprocity there is up to scalars a unique non-trivial morphism

$$\Pi \to \mathrm{Ind}_{U_n^n,\theta'}^{G_n}(\theta'),$$

which must be injective, since Π is irreducible.

3.5 **Remarks.** **a)** If $\Pi^{(n)} \neq 0$, then Π is called non-degenerate. By definition this means, that there is a non-trivial functional $\ell: V \to \mathbb{C}$ on the representation space V of Π such that

$$\ell(\Pi(u)v) = \theta_n^n(u) \cdot \ell(v) \quad \text{for all} \quad v \in V \quad \text{and} \quad u \in U_n^n .$$

In this case, 3.2 is due to Bernštein and Zelevinskii ([4] 5.16 and 5.20), and one talks of Kirillov and Whittaker models. The existence of a Whittaker model is obviously equivalent to the non-degenerateness of a representation, while the existence of a Kirillov model for non-degenerate representations was conjectured before by Gelfand and Kazhdan [32].

Non-degenerate representations are also called generic, since for them we may take $\theta' = \theta_n^n$, which generates the "generic" orbit of characters of U_n^n under the action of $P_{(1,\dots,1)}$, containing all characters that do not vanish on any $u_{i,i+1}$.

b) In 2.2 we have seen that cuspidal representations Π of G_n are non-degenerate and for them $K(\Pi) = {}_c\mathrm{Ind}_{U_n^n}^{P_n}(\theta) \subseteq \mathrm{Ind}_{U_n^n}^{P_n}(\theta)$.

4. The Zelevinskii classification

Notation. For a representation Π of G_m and $s \in \mathbb{R}$ let $\Pi(s)$ be the representation of G_m defined by $\Pi(s)(g) = |\det g|^s \cdot \Pi(g)$ where $|\ |$ is the normalized absolute value of $F^\times$.

4.1 Theorem. (Bernšstein-Zelevinskii) *Let Φ_i be irreducible cuspidal representations of $G_{n_i}, i=1,\dots,r$. Then $\Phi_1 \circ \dots \circ \Phi_r$ is reducible if and only if there are i,j with $n_i = n_j$ and $\Phi_j = \Phi_i(1)$.*

By this theorem we obtain a lot of irreducible representations. To study the general situation, Zelevinskii first investigates the extreme case $\Phi \circ \Phi(1) \circ \dots \circ \Phi(r-1)$.

4.2 Definition. Call a set $\Delta = \{\Phi, \Phi(1), \dots, \Phi(r-1)\}$ of representations of G_m with Φ irreducible cuspidal a segment.

4.3 Theorem. (Zelevinskii) *Let $\Delta = \{\Phi, \dots, \Phi(r-1)\}$ be a segment.*

a) $\Phi \circ \Phi(1) \circ \dots \circ \Phi(r-1)$ *has a unique irreducible subrepresentation* $Z(\Delta)$.

b) $\mathrm{Res}_{P_n}^{G_n} Z(\Delta)$ *is irreducible, and one has*

$$R_{(m,\dots,m)}(Z(\Delta)) = \Phi \otimes \Phi(1) \otimes \dots \otimes \Phi(r-1)$$

and $Z(\Delta)^{(m)} = Z(\Delta^-)$, *where* $\Delta^- = \{\Phi, \dots, \Phi(r-2)\}$.

c) $\Phi \circ \dots \circ \Phi(r-1)$ *has a unique irreducible quotient* $L(\Delta)$. *This quotient is essentially square-integrable.*

Call segments $\Delta_1 = \{\Phi_1, \dots, \Phi_1'\}$ and $\Delta_2 = \{\Phi_2, \dots, \Phi_2'\}$ *linked*, if $\Delta_1 \not\subseteq \Delta_2$, $\Delta_2 \not\subseteq \Delta_1$ and $\Delta_1 \cup \Delta_2$ is a segment. In this case say Δ_1 preceeds Δ_2 if $\Phi_2 = \Phi_1(m)$ with $m > 0$.

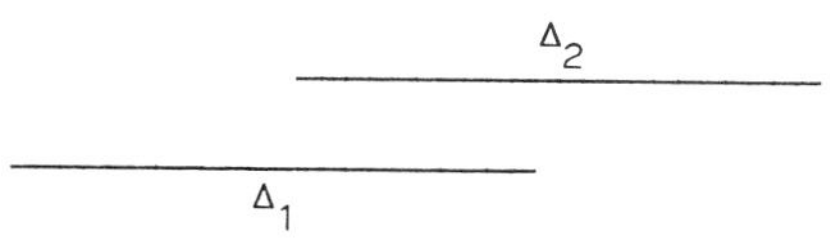

4.4 Theorem. (Zelevinskii) *Let $\Delta_1,\dots,\Delta_s$ be segments.*

a) *There is an ordering $\Delta_{\nu_1},\dots,\Delta_{\nu_s}$ such that no Δ_{ν_i} preceeds Δ_{ν_j} for $i < j$. In this case $Z(\Delta_{\nu_1}) \circ \dots \circ Z(\Delta_{\nu_s})$ has a unique irreducible subrepresentation $Z(\Delta_1,\dots,\Delta_s)$ (resp. irreducible quotient $L(\Delta_1,\dots,\Delta_s)$).*

b) *$Z(\Delta_1,\dots,\Delta_s) \simeq Z(\Delta_1',\dots,\Delta_t')$ (resp. $L(\Delta_1,\dots,\Delta_s) \simeq L(\Delta_1',\dots,\Delta_t')$) if and only if $t = s$ and $(\Delta_1,\dots,\Delta_s)$ and $(\Delta_1',\dots,\Delta_t')$ coincide up to ordering, in particular, $Z(\Delta_1,\dots,\Delta_s)$ and $L(\Delta_1,\dots,\Delta_s)$ do not depend on the particular choice of the ordering prescribed in a).*

c) *Every irreducible admissible representation Π of G_n is of the form $Z(\Delta_1,\dots,\Delta_s)$ (resp. $L(\Delta_1',\dots,\Delta_t')$) for some segments $\Delta_1,\dots,\Delta_s$ (resp. $\Delta_1',\dots,\Delta_t'$).*

By this theorem, there is a bijection between the set of irreducible admissible representation of G_m, $m \geq 1$ and the families $(\Delta_1,\dots,\Delta_r)$ of segments, provided that we do not regard the ordering in these families but do regard the multiplicities - this is correctly defined by introducing multisets, see [115].

4.5 Example. To illustrate the classification, we consider the classical case $n = 2$, i. e., representations of $GL_2(F)$, already considered in [55]. Let $\mu = \mu_1,\mu_2 : G_1 = F^\times \to \mathbb{C}^\times$ be quasi-characters.

a) By 4.1, $\mu_1 \circ \mu_2$ is irreducible for $\mu_1/\mu_2 \neq |\ |, |\ |^{-1}$.

b) $\mu \circ \mu(1)$ has two composition factors (compare 1.1): the 1-dimensional subrepresentation $Z(\Delta = \{\mu,\mu(1)\}) = \mu \det$ and the quotient $L(\{\mu,\mu(1)\})$ which is square-integrable - $L(\{1,|\ |\})$ is just the Steinberg representation.

c) $\mu(1) \circ \mu$ has the subrepresentation $Z(\Delta_1 = \{\mu(1)\},\Delta_2 = \{\mu\})$ and the quotient $L(\{\mu(1)\},\{\mu\})$, which is not square-integrable. In fact, by 1.2 or 4.4 we must have $L(\{\mu(1)\},\{\mu\}) = Z(\{\mu,\mu(1)\}) = \mu \det$ and $L(\{\mu,\mu(1)\}) = Z(\{\mu(1)\},\{\mu\})$.

4.6 Remark. $L(\Delta_1,\dots,\Delta_s)$ is connected with the Langlands classification of representations in terms of tempered ones, see [103]. The proof of this relies on an unpublished result by Bernštein, that the essentially square-integrable representations are exactly the $L(\Delta)$. The formulation of the classification in terms of $L(\Delta_1,\dots,\Delta_s)$ is due to Rodier, the statement in [115] is slightly weaker.

For the classification of the unitary representations of G_n see M. Tadić [104].

5. Connection with the Langlands correspondence

The local Langlands conjecture states that there is a canonical bijection between irreducible admissible representations of $GL_n(F)$ and n-dimensional complex representations of the Weil-Deligne group W'_F . Perhaps the appearing and definition of the latter is better understood, if one looks at ℓ-adic representations of the Weil group W_F of F , i. e., continuous morphisms

$$W_F \to GL(V)$$

where V is a finite-dimensional $\mathbb{Q}_\ell$-vector space (here ℓ is different from the residue characteristic p of F). Note also, that these really are the representations of W_F , which arise in a big number and most naturally - by the action of W_F on the ℓ-adic cohomology

$$V = H^i_{\acute{e}t}(X \times_F \bar{F}, \mathbb{Q}_\ell) \ , \quad i \geq 0 \ , \qquad \mathbf{(5.1)}$$

of each algebraic variety X over F .

This relation with W'_F is as follows. Using Grothendieck's construction of an "ℓ-adic monodromy", see [10] 4.2.2) , Deligne proves an equivalence of categories

$$\left\{ \begin{array}{l} \ell\text{-adic representations} \\ \\ W_F \to GL(V) \end{array} \right\} \longleftrightarrow \left\{ \begin{array}{l} \text{pairs } \Sigma' = (\Sigma, N) \text{ , where } \Sigma \\ \text{is a continuous \underline{semi-simple}} \\ \text{representation of } W_F \text{ on } V \text{ ,} \\ N: V \to V \text{ is a nilpotent endo-} \\ \text{morphism with} \\ \Sigma(w)N\Sigma(w)^{-1} = |w| \cdot N \text{ for } w \in W_F \end{array} \right\} .$$

Here $|\ |$ denotes the quasi-character $W_F \twoheadrightarrow W_F^{ab} = F^\times \xrightarrow{|\ |} \mathbb{Q}^\times$.
The same holds, if we replace $\mathbb{Q}_\ell$ by its algebraic closure $\bar{\mathbb{Q}}_\ell$, i. e., if V is a finite-dimensional $\bar{\mathbb{Q}}_\ell$-vector space on both sides.

Then the indecomposable objects on the right are of the form

$$\Sigma \otimes sp(r) \ , \quad r \in \mathbb{N} \ ,$$

where Σ is an irreducible representation of W_F (with $N = 0$) and $sp(r)$ has a basis $e_0, \ldots, e_{r-1}$ with

$$\rho(w)\, e_i = |w|^i\, e_i \quad \text{for} \quad w \in W_F$$

$$N\, e_i = e_{i+1} \quad (N\, e_{r-1} = 0)\, .$$

Since for a semi-simple ℓ-adic representation of W_F the restriction to the inertia group factorizes through a finite quotient, we again obtain a continuous representation of W_F, but this time a complex one, if we pass from $\bar{\mathbb{Q}}_\ell$ to the isomorphic field $\mathbb{C}$. Thus the description above shows, that the category on the right (but not the one on the left) is the "same" if we replace $\bar{\mathbb{Q}}_\ell$ by $\mathbb{C}$, i. e., if V now is a $\mathbb{C}$-vector space.
This category is just the category G_F of complex representations of the Weil-Deligne group W'_F, see [107] 4.1.2 (we may take this as a definition, since we do not need the definition of W'_F itself).

We have seen that there is an equivalence between continuous $\bar{\mathbb{Q}}_\ell$-representations of W_F and complex representations of W'_F. In particular, to any ℓ-adic representation V we can associate an element of G_F, and this is exactly how one defines the local L-and ε-factors for the representation described in 5.1, see [107] 4.2.4. The local Langlands correspondence would associate a canonical representation of $GL_n(F)$ to this, $n = \dim V$, and canonical means at least that it has the same L- and ε-factors. This is the necessary local step for the global "Langlands program" to relate L- and ε-functions of (compatible systems of) ℓ-adic representations - for example Hasse-Weil zeta functions - to automorphic L- and ε-functions.

5.2 Remark. The construction above depends on the choice of an abstract isomorphim between $\bar{\mathbb{Q}}_\ell$ and $\mathbb{C}$, which is not after everybody's taste and at least may infect the well-definedness of the L- and ε-factors for the representations 5.1. However, it is conjectured for these "motivic" representations, that the eigenvalues of a Frobenius element in W_F are algebraic numbers, so that the representation (Σ, N) can be defined over some algebraic number field (which we can embed in $\mathbb{C}$ without scruples), and that the isomorphy type of this representation is independent of $\ell (\nmid p)$, see [107] 4.2.4.

The Zelevinskii classification reduces the question of the Langlands correspondence to the study of cuspidal representations of $GL_n(F)$ on one side and irreducible complex representations of W_F on the other, in the following sense.

5.3 Theorem. <u>If</u> $\Sigma \to \Pi(\Sigma)$ <u>is an</u> L- <u>and</u> ε-<u>factor preserving bijection between irreducible</u> m-<u>dimensional representations of</u> W_F <u>and irreducible cuspidal representations of</u> $GL_m(F)$ <u>for all</u> m, <u>then</u>

$$\bigoplus_{i=1}^{s} \Sigma_i \otimes sp(r_i) \to L(\Delta_1,\dots,\Delta_s) ,$$

$$\Delta_i = \{\Pi(\Sigma_i),\Pi(\Sigma_i)(1),\dots,\Pi(\Sigma_i)(r_i-1)\}$$

is an L- and ε-factor preserving bijection between n-dimensional representations of W_F' and irreducible admissible representations of $GL_n(F)$ for all n .

One can show that this extension would preserve some more functorialities, like twist with characters or equality between central character of $\Pi(\Sigma)$ and determinant of Σ , see [97] 4.4. Although one does not know about the uniqueness of such an extension in general, the definition above seems to be a very natural one and coincides with the known correspondence for $GL_2(F)$ described in [97], compare the example 4.5.

References

[4] I.N. Bernštein, A.V. Zelevinskii: Representations of the group GL(n,F) where F is a local non-archimedean field; Russian Math. Surveys 31 (1976) , 1- 68 (trans. from Uspekhi Mat. Nauk 31, no. 3 (1976), 5 - 70).

[32] I.M. Gelfand, D.A. Kazhdan: Representations of the group GL(n,K) where K is a local field, in: Lie groups and their representations; J. Wiley and Sons, London 1975.

[55] H. Jacquet, R.P. Langlands: Automorphic Forms on GL(2); Lecture Notes in Mathematics no. 114, Springer Verlag, Berlin-Heidelberg-New York, 1970.

[97] F. Rodier: Représentations de GL(n,k) où k un corps p-adique; Séminaire Bourbaki (1981/1982) 587.

[103] A.J. Silberger: The Langlands Quotient Theorem for p-adic group; Math. Annalen 236 (1978), 95 - 104.

[104] M. Tadić: Solution of the unitarizablility problem for general linear groups (non-archimedean case); preprint Max-Planck-Institut für Mathematik Bonn 1985.

[107] J.T. Tate: Number theoretic background, in: Automorphic Forms, Representations, and L-Functions (Corvallis), Proc. Symp. Pure Math. 33, AMS, Providence 1979.

[115] A. Zelevinsky: Induced representations of reductive p-adic groups II: On irreducible representations of GL(n); Ann. Sci. ENS 13 (1980), 165 - 210.

Contemporary Mathematics
Volume **86**, 1989

PRINCIPAL ORDERS AND CONGRUENCE GAUß SUMS

by S.M.J. Wilson, Durham

Table of notation:

p is a prime integer;

F is a p-adic field, v its valuation (extended to $\bar{F}$);

$\mathcal{o}$ is its valuation ring and $\mathfrak{p} = (\pi)$ is its valuation ideal;

D is a division ring with centre F, $|D:F| = d^2$;

$\mathcal{O}$ is its valuation ring and $\mathcal{P} = (\pi_D)$ is its valuation ideal;

$k = \mathcal{O}/\mathcal{P}$.

$A \simeq \mathrm{Mat}_m(D)$, $n = dm$;

$\mathcal{A}$ is an $\mathcal{o}$-order in A and $\mathcal{J} = \mathrm{rad}(\mathcal{A})$;

$U_o = \mathcal{A}^\times$, $U_t = 1 + \mathcal{J}^t \subseteq \mathcal{A}^\times$ for $t > 0$;

$\mathrm{tr} : F \to \mathbb{Q}_p$ is the absolute trace;

$\mathrm{tr}_A : A \to \mathbb{Q}_p$ is the reduced trace followed by tr;

Let α be the standard character $\mathbb{Q}_p/\mathbb{Z}_p \to \mathbb{Z}[1/p]/\mathbb{Z} \to \mathbb{Q}/\mathbb{Z} \to \mathbb{C}^\times$ then

$\psi = \mathrm{tr} \circ \alpha$ and $\psi_A = \mathrm{tr}_A \circ \alpha$.

If M is a full $\mathbb{Z}_p$-lattice in F (resp. in A) then

$M^* = \{\lambda \in F \mid \mathrm{tr}(\lambda M) \subseteq \mathbb{Z}_p\}$ (resp. $M^* = \{a \in A \mid \mathbf{tr}_A(aM) \subseteq \mathbb{Z}_p\}$).

§ 1. Gauss Sums

If $\chi: F^\times \to \mathbb{C}^\times$ is a character with conductor $\mathfrak{f}(\chi)$, then the Gauss sum $\tau(\chi)$ of χ is defined to be

$$\tau(\chi) = \sum_{\bar{x} \in (\mathcal{O}/\mathfrak{f}(\chi))^\times} \chi(cx)\psi(cx)$$

where c generates $\mathfrak{f}(\chi)^*$.

We want to do the same thing for $\mathcal{A}$. Naively, take an admissible irreducible representation Π of $A^\times$ over $\mathbb{C}$. Say that Π has conductor $\mathfrak{f}(\Pi) = J^i$ if $\Pi|U_i$ is trivial but $\Pi|U_{i-1}$ is not. Choose $c \in A^\times$ such that $c \in \mathfrak{f}(\chi)^* \setminus \mathfrak{f}(\chi)^* J$. Put

$$T(\Pi)_c = \sum_{\bar{x} \in (\mathcal{A}/\mathfrak{f}(\Pi))^\times} \Pi(cx)\psi_A(cx)$$

There are several fatal problems with this approach.

(i) If $m \neq 1$ then the most interesting Π do not have conductors (indeed if $D = F$ then the only Π with conductors are the abelian ones and since these factor through the reduced norm we get, at best, our original Gauss sums back again).

(ii) It is by no means clear that $T(\Pi)_c$ is independent of c.

(iii) The vagueness in the choice of c probably means that $T(\Pi)_c = 0$ in many cases. (Recall that if, in the case of $\tau(\chi)$ above, we take $\mathfrak{f}(\chi)$ too small then the resulting Gauss sum will be zero).

(iv) If $T(\Pi)$ is non-zero we are still left with a rather nondescript endomorphism of the representation space V_Π which may well be infinite dimensional.

The following modifications suggest themselves.
Problem (i) is caused by the fact that if $m \neq 1$ the normalizer of $\mathcal{A}$ in $A^\times$ is a subgroup of infinite index. This suggests considering representations not of $A^\times$ but of some subgroup G of $A^\times$ (indeed of $N(\mathcal{A}) = \{a \in A^\times \mid a^{-1}\mathcal{A}a = \mathcal{A}\}$, the <u>normalizer</u> of $\mathcal{A}$). However if G is too small then the term $\Pi(cx)$ in the above sum may not be defined (since cx may not lie in G). We propose therefore:

<u>Modification (a):</u> Consider representations Π of some subgroup G of $A^\times$ such that

(1) $\mathcal{A}^\times \subseteq G \subseteq N(\mathcal{A})$

(2) $[\mathfrak{f}(\Pi)^* \setminus \mathfrak{f}(\Pi)^* J] \cap G \neq \emptyset$.

Problems (ii) and (iii) result from the freedom in the choice of c . Especially for problem (iii) it would be safest if we could choose c so that $c\mathcal{A} = \mathfrak{f}(\Pi)^*$. We propose therefore

Modification (b) Choose $\mathcal{A}$ and G so that characteristic ideals of $\mathcal{A}$ are principal and generated by elements of G .

1.1 Theorem. If we have (a) and (b) above and if c is chosen so that $c\mathcal{A} = \mathfrak{f}(\Pi)^*$ then $T(\Pi)$ is independent of the actual choice of c and commutes with $\Pi(G)$.

Proof. It is easily seen that

$$T(\Pi)_c = \sum_{\bar{y}} \Pi(y)\psi_A(y)$$

where $\bar{y} = y + \mathfrak{f}(\Pi)^*\mathfrak{f}(\Pi)$ runs over $\{\bar{y} \mid \bar{y}\mathcal{A} = \mathfrak{f}(\Pi)^*/\mathfrak{f}(\Pi)^*\mathfrak{f}(\Pi)\}$. So $T(\Pi)_c$ is independent of c .

Again $\mathfrak{f}(\Pi)$ and hence $\mathfrak{f}(\Pi)^*$ are G-invariant as is ψ_A. Hence we have for $g \in G$,

$$\Pi(g)T(\Pi)\Pi(g)^{-1} = \sum_{\bar{y}} \Pi(g^{-1}yg)\psi_A(y)$$

$$= \sum_{\bar{z}} \Pi(z)\psi_A(z) = T(\Pi) .$$

To dispose of (iv) we need to be able to apply Schur's lemma and hence to deduce from the centrality of $T(\Pi)$ that this is a scalar map, multiplication by some complex number $\tau(\Pi)$.

For this we need our representations to be finite dimensional and so we propose:

Modification (c): Choose G to be compact modulo $F^{\times}$. (= Compact mod centre = CMC). (It will emerge in §4 that (for more trivial reasons than are at present apparent) this condition is sufficient.)

Thus, with modifications (a), (b) and (c) we have Gauss sums $\tau(\Pi)$. These are referred to as non-abelian congruence Gauss sums.

These sums may, of course be zero and this aspect needs detailed investigation but an even more immediate question is whether the three modifications (a), (b) and

(c) can be made simultaneously without restricting to situations which are obviously too special to be generally useful. The following theorem is propitious:

1.2 Theorem.

(i) Every CMC subgroup of $A^\times$ is contained in a maximal one.

(ii) If G is a maximal CMC subgroup of $A^\times$ then $G = N(\mathsf{A})$ for some order A in A.

(iii) Moreover there is a unique maximal such A and this A satisfies the condition in (b), (i.e. A is a principal order).

This theorem will be proved in the next section.

§ 2. CMC subgroups, chains and principal orders

A is said to be a principal order if $J = g\mathsf{A}$ for some $g \in A^\times$.

This section will be devoted to establishing the correspondence between maximal CMC subgroups and principal orders. We start with a very basic result:

2.1 Lemma. For any A, $N(\mathsf{A})$ is CMC:

Proof. View A as an F-vector space and A as a full $\mathcal{O}$-lattice in A. Then the conjugation action of $A^\times$ on A makes $A^\times/F^\times$ a closed subgroup of $\mathrm{Aut}_F(A)$ and $N(\mathsf{A})/F^\times = A^\times/F^\times \cap \mathrm{Aut}_{\mathcal{O}}(\mathsf{A})$. Moreover $\mathrm{Aut}_{\mathcal{O}}(\mathsf{A})$ ($\simeq \mathrm{Mat}_{n^2}(\mathcal{O})$) is compact.

Let V be a left D-vector space of dimension m (and a right A-module). A chain $\underline{X}$ (of fundamental length r) in V is a doubly infinite sequence $\{X_i\}_{i\in\mathbb{Z}}$ of $\mathcal{O}$-lattices such that, for all i, $X_i \supseteq X_{i+1}$ and $X_{i+r} = PX_i$.

We view chains $\underline{X}$ and $\underline{X}'$ as the same (equivalent) if $\{X_i\} = \{X_i'\}$. If for some $g \in A^\times$ and $t \in \mathbb{Z}$ we have $X_i g = X_{i+t}$ for all i then we say that g is an automorphism of $\underline{X}$ of degree t. We write $\mathrm{Aut}(\underline{X})$ for the group of automorphisms of $\underline{X}$ of any degree and $\mathrm{Aut}_o(\underline{X})$ for the subgroup of those automorphisms of degree 0. We define $\mathcal{O}(X)$, the order of $\underline{X}$, to be $\{a \in A \mid X_i a \subseteq X_i \ \forall\ i\}$.

Now $X_{i+1} \supseteq \pi_D X_i$ so X_i/X_{i+1} is a k-vector space. We put $s_i = s_i(\underline{X})$ for its dimension. Note that

$$X_i/X_{i+1} \simeq \pi_D X_i/\pi_D X_{i+1} = X_{i+r}/X_{i+r+1} .$$

So $$s_{i+r} = s_i \tag{2.2}$$

Also $$\bigoplus_{i=1}^{r} X_i/X_{i+1} \simeq X_1/X_{r+1} = X_1/\pi_D X_1 .$$

So $$\sum_{i=1}^{r} s_i = m . \tag{2.3}$$

For each i, choose $B_i \subset X_i$ to reduce to a basis of X_i/X_{i+1} (so $|B_i| = s_i$) We may (and do) choose the B_i so that $\pi_D B_i = B_{i+r}$. By Nakayama's lemma,

$$\left.\begin{array}{ll} B = B_1 \cup B_2 \cup \ldots \cup B_{r-1} \cup B_r & \text{is an } \mathcal{O}\text{-basis for } X_1 \\ \pi_D B_1 \cup B_2 \cup \ldots \cup B_{r-1} \cup B_r & \text{is an } \mathcal{O}\text{-basis for } X_2 \\ \cdots\cdots\cdots & \\ \pi_D B_1 \cup \pi_D B_2 \cup \ldots \cup \pi_D B_{r-1} \cup B_r & \text{is an } \mathcal{O}\text{-basis for } X_r \end{array}\right\} \tag{2.4}$$

We represent $A \simeq \mathrm{Mat}_m(D)$ with respect to the basis B.

Notation: If the S_{ij} are sets and $t_1,\ldots,t_r, u_1,\ldots,u_o \in \mathbb{N}$ then

$$\begin{bmatrix} S_{11} & S_{12} & \cdots & S_{1s} \\ \vdots & \vdots & & \vdots \\ S_{r1} & S_{r2} & \cdots & S_{rs} \end{bmatrix} \quad (t_1,\ldots t_r)\times(u_1,\ldots,u_s)$$

denotes the set of matrices which are each an $r \times s$ array of blocks of which the i,j^{th} blocks is $t_i \times u_j$ and has entries lying in S_{ij}.

2.5 Propositions about chain orders

(i) $\mathcal{O}(\underline{X}) = \bigcap_{i=1}^{r} \mathcal{O}(X_i)$ where $\mathcal{O}(X_i) = \mathcal{O}_R(X_i) = \{a \in A \mid X_i a \subseteq X_i\}$.

(ii) $$\mathcal{O}(\underline{X}) = \begin{bmatrix} \mathcal{O} & \mathcal{O} & \cdots & \mathcal{O} & \mathcal{O} \\ P & \mathcal{O} & & \mathcal{O} & \mathcal{O} \\ \vdots & \vdots & & \vdots & \vdots \\ P & P & \cdots & P & \mathcal{O} \end{bmatrix} \quad (s_1,\ldots,s_r)\times(s_1,\ldots,s_r)$$

(i.e. block upper-triangular mod P).

(iii) $\mathrm{rad}(\mathcal{O}(\underline{X}))$ consists of those matrices in $\mathcal{O}(X)$ in which the diagonal blocks also are zero mod P. Moreover $\mathrm{rad}(\mathcal{O}(X) = \{a \in A \mid X_i a \subseteq X_{i+1} \forall i\}$.

(iv) $\mathcal{O}(\underline{X})/\mathrm{rad} = \bigoplus_{i=1}^{r} \mathrm{Mat}_{s_i}(k)$.

(v) The X_i are the only (non-zero) $\mathcal{O}(\underline{X})$-lattices in V.

(vi) $\mathcal{O}(\underline{X}) = \mathcal{O}(\underline{X}') \Leftrightarrow \underline{X}$ is equivalent to $\underline{X}'$.

(vii) $\mathrm{Aut}(\underline{X}) = N(\mathcal{O}(\underline{X}))$.

(viii) $\mathrm{Aut}_o(\underline{X}) = \mathcal{O}(\underline{X})^{\times}$.

(ix) $\mathcal{O}(\underline{X})$ is hereditary (every ideal is projective) and $\{X_1, \ldots, X_r\}$ is an irredundant set of indecomposable $\mathcal{O}(\underline{X})$-projectives.

Proof. **(i)** Clear. (Note that $\mathcal{O}(X_i) = \mathcal{O}(\pi_D X_i) = \mathcal{O}(X_{i+r})$.)

(ii) Follows from (i) and the bases given in 2.4.

(iii) If I is the proposed radical then clearly $I^r \subseteq (P)^{m\times m} \subseteq \mathrm{rad}\,(\mathcal{O}(\underline{X}))$. Hence $I \subseteq \mathrm{rad}(\mathcal{O}(\underline{X}))$. On the other hand $\mathcal{O}(\underline{X})/I$ is semisimple as described in (iv). The final part now follows from 2.4.

(iv) Clear.

(v) Suppose Y is an $\mathcal{O}(\underline{X})$-lattice in V and that $Y \notin \{X_i\}$. Choose i so that $X_i \supset Y \not\supseteq X_{i+1}$ and create a new chain $\underline{Z}$:

$$X_i \supset Y + X_{i+1} \supset X_{i+1} \ldots X_{i+r} \supset \pi_D(Y + X_{i+1}) \supset X_{i+r+1}$$

of fundamental length $r + 1$.

Clearly $\mathcal{O}(\underline{Z}) \supseteq \mathcal{O}(\underline{X})$ as all the lattices of Z are $\mathcal{O}(\underline{X})$-lattices. Moreover, since $\{Z_i\} \supseteq \{X_i\}$ we must have $\mathcal{O}(\underline{X}) \supseteq \mathcal{O}(\underline{Z})$. Hence $\mathcal{O}(\underline{X}) = \mathcal{O}(\underline{Z})$. But that cannot be the case as from (iv) we know that the order of a chain determines the fundamental length of that chain.

(vi) Clear from (v).

(vii) Clearly $\mathrm{Aut}(\underline{X}) \subseteq N(\mathcal{O}(\underline{X}))$. Also if $g \in N(\mathcal{O}(\underline{X}))$ then $\underline{X}g$ is a chain with order $g^{-1}\mathcal{O}(\underline{X})g = \mathcal{O}(\underline{X})$ so $\underline{X}g$ is (equivalent to) $\underline{X}$. Therefore, since $X_i \longmapsto X_i g$ preserves the ordering, $g \in \mathrm{Aut}(\underline{X})$.

(viii) Clear.

(ix) Clearly from (v), since $X_i \simeq X_{i+r}$, the lattices $X_1, \ldots, X_r$ represent all isomorphism classes of $\mathcal{O}(\underline{X})$ lattices in V .

Moreover if $X_i \simeq X_J$ then any isomorphism can be extended to an A-automorphism of V i.e. an element of $D^\times$. Such an element must be of the form $u\pi_D^t$ with $u \in \mathcal{O}_\times$ and then $X_j = \pi_D^t X_i = X_{i+rt}$. So $X_1,\dots,X_r$ are pairwise non-isomorphic.

Decomposing $\mathcal{O}(\underline{X})$ (as described in (ii)) into rows and comparing these with the bases for the X_i we see that $\mathcal{O}(\underline{X}) \simeq \oplus_i X_i^{(s_i)}$ and hence the X_i are projective. If I is a right ideal of $\mathcal{O}(\underline{X})$ choose $a \in V$ so that $aI \nsubseteq \{0\}$. Then $aI = X_i$ for some i , by (v). And then the epimorphism $I \to aI = X_i$ splits. So, by induction, $I \simeq \oplus X_i^{(?)}$ (and so, in particular, I is projective).

Hence $O(\underline{X})$ is hereditary with minimal projectives X_i .

2.6. Propositions about chain automorphisms

(i) The matrix $\pi_D I_m$ represents (with respect to B) an automorphism of $\underline{X}$ of degree r .

(ii) $\mathrm{Aut}(\underline{X})$ is CMC.

(iii) $\mathrm{Aut}_0(\underline{X}) = \{g \in \mathrm{Aut}(\underline{X}) \mid \mathrm{Nrd}(g) \in \mathcal{O}^\times\}$.

(iv) If $v \in X_k \setminus X_{k+1}$ then $\langle v\mathrm{Aut}_0(\underline{X})\rangle = X_k$.

(v) If $\mathrm{Aut}(\underline{X}) = \mathrm{Aut}(\underline{Y})$ then $\underline{X} = \underline{Y}$.

Proof. **(i)** is clear and **(ii)** follows from 2.5**(viii)** and 2.1.

(iii) The automorphisms of a lattice are its unimodular endomorphisms.

(iv) Renumber $\underline{X}$ so that $k = 1$ and choose the basis $\{v_i\} = B$ (see after 2.3) so that $v_1 = v$. If $w \in X_1$, define a,b in A by $v_1 a = w$, $v_1 b = v+w$ and $v_i a = v_i b = v_i$ for $i > 1$. Put $w = \Sigma q_i v_i$, $q_i \in \mathcal{O}$. If $q_1 \in \mathcal{O}^\times$ then, by 2.5.(ii), $a \in \mathcal{O}(\underline{X})^\times$; Otherwise $1 + q_1 \in \mathcal{O}^\times$ and $b \in \mathcal{O}(X)^\times$. In either case $w \in \langle v\mathcal{O}(X)^\times\rangle = \langle v\mathrm{Aut}_0(\underline{X})\rangle$.

(v) By (iv), $\underline{X} = \{\langle v\mathrm{Aut}_0(\underline{X})\rangle \mid v \in V\setminus\{0\}\}$ so $\mathrm{Aut}_0(\underline{X})$, and hence $\mathrm{Aut}(\underline{X})$, determines $\underline{X}$.

We say that the chain $\underline{X}$ is uniform if s_i is independent of i .

2.7 Theorem. The following are equivalent.

(i) $\underline{X}$ is uniform

(ii) X has an automorphism of degree 1 (and hence automorphisms of all degrees).

(iii) $\mathcal{O}(\underline{X})$ is a principal order.

Proof. **(i) ⇒ (ii).** Putting $s = s_i$ and

$$h = \left[\begin{array}{c|c} 0 & I_{(r-1)s} \\ \hline \pi_D I_s & 0 \end{array}\right]$$

in A, we find from the explicit bases (2.4) that $X_i h = X_{i+1}$.

(ii) ⇒ (iii). Let h be an automorphism of degree 1. Then

$$\begin{aligned} \mathrm{rad}(\mathcal{O}(\underline{X})) &= \{a \in \underline{A} \mid X_i a = X_{i+1}\} \\ &= \{a \in \underline{A} \mid X_i a = X_i h\} \\ &= \mathcal{O}(\underline{X})h = h\mathcal{O}(\underline{X}) \qquad (\text{since } h \in N(\mathcal{O}(\underline{X}))\ . \end{aligned}$$

(iii) ⇒ (i). Put $\mathcal{O}(\underline{X}) = \mathsf{A}$ and $J = h\mathsf{A}$. Of course J is two-sided so $J \supseteq \mathsf{A}h$. and since $\mathsf{A}/h\mathsf{A}$ and $\mathsf{A}/\mathsf{A}h$ have the same order we have $J = \mathsf{A}h$.

From 2.5 (iii), for all $i, X_{i+1} = X_i J = X_i h$ and so h induces an isomorphism $X_i/X_{i+1} \xrightarrow{\sim} X_{i+1}/X_{i+2}$. Hence $s_i = s_{i+1}$ .... .

Since our interest is in CMC subgroups of $A^\times$ which are as big as possible we shall assume that any CMC subgroup contains $F^\times$.

By the <u>eigenvalues</u> of an element $g \in A^\times$ we mean the roots of the reduced characteristic polynomial of g as an element of A.

2.8 **Lemma.** <u>Let</u> G <u>be</u> CMC.

(i) <u>If</u> $g \in G$ <u>then all the eigenvalues of</u> g <u>have the same valuations.</u>

(ii) <u>We have an exact sequence</u>

$$1 \longrightarrow G_0 \longrightarrow G \xrightarrow{v_A} \mathbb{Z}$$

<u>where</u> $v_A(g) = v(\mathrm{nrd}(g))$ <u>and</u>

$$G_0 = \{g \in G \text{ with all eigenvalues p-adic units}\}\ .$$

Proof. **(i)** Suppose for a contradiction that we have $g \in G$ with eigenvalues $\lambda_1, \ldots, \lambda_n$ of differing valuations and let $\omega = \min(v(\lambda_i))$. So $n\omega < v_A(g)$. Let $\lambda_1, \ldots, \lambda_t$ be the eigenvalues of g with $v(\lambda_i) = \omega$ and put $\sigma_t(g)$ for the tth symmetric function of the λ_i. We have

$$\sigma_t(g) = \lambda_1\lambda_2 \ldots \lambda_t + \text{smaller terms.}$$

So $v(\sigma_t(g)) = t\omega$ and, indeed, for $k > 0$, $v(\sigma_t(g^k)) = tk\omega$.

Put, for $a \in A^\times$,

$$f(a) = -v\left[\frac{\sigma_t(a)^n}{\mathrm{Nrd}(a)^t}\right]$$

then $f(F^\times) = \{0\}$. So, since G is compact mod $F^\times$ and f is continuous, $f(G)$ must be bounded. But $f(g^k) = tk(n\omega - v_A(g))$ and this tends to ∞ as k tends to ∞. So we have a contradiction.

(ii) This is now clear.

2.9 **Lemma.** Let G be CMC then

(i) G_0 is compact.

(ii) $G \subseteq N(\mathcal{A})$ for some $\mathcal{A}$ in A.

(iii) If $\mathcal{A}$ is maximal such that $G \subseteq N(\mathcal{A})$ then $\mathcal{A} = \mathcal{O}_R(\mathcal{J})$.

Proof. **(i)** Choose $g \in G$ so that $v_A(g)$ is least possible greater than zero. Thus by 2.8 (ii) $G = \langle G_0, g \rangle$ and $g^t \equiv \pi \bmod G_0$ where $t = n/v(g)$. So $G_0/\mathcal{O}^\times$ has index t in $G/F^\times$ and is therefore compact. But also $\mathcal{O}^\times$ is compact so G_0 is compact.

(ii) So if L is an $\mathcal{O}$-lattice in A then $\langle LG_0 \rangle$ is also a lattice and $\mathcal{B} = \mathcal{O}_R(\langle LG_0 \rangle)$ is G_0-invariant. Whence the order

$$\mathcal{A} = \bigcap_{i=0}^{t-1} g^i \mathcal{B} g^{-i} \quad \text{is } G\text{-invariant}$$

(since $g^t\mathcal{B}g^{-t} = \mathcal{B}$)

(iii) If $\mathcal{A}$ is maximal such that $G \subseteq N(\mathcal{A})$ then $g^{-1}\mathcal{J}g = \mathcal{J}$ for all $g \in G$ and consequently $\mathcal{O}_F(\mathcal{J})$ is G-invariant. But $\mathcal{O}_R(\mathcal{J}) \supseteq \mathcal{A}$ and hence $\mathcal{O}_R(\mathcal{J}) = \mathcal{A}$ by the maximality of $\mathcal{A}$.

We say that an order $\mathcal{A}$ is <u>extremal</u> if it is the only order $\mathcal{B}$ such that $\mathcal{B} \supseteq \mathcal{A}$ and $\mathrm{rad}(\mathcal{B}) \supseteq \mathrm{rad}(\mathcal{A})$.

2.10 Lemma. If $\mathcal{O}_R(J) = A$ then A is extremal.

Proof. Let B be an order in A and put $K = \mathrm{rad}(B)$. Suppose that $B \supseteq A$ and $K \supseteq J$.

Now $\pi^q K \subseteq J$ for some q (J is a full lattice) and K is nilpotent mod π so $K^t \subseteq J$ for some t. Hence $(K \cap A)^t \subseteq J$ whence $K \cap A \subseteq J$.

Choose t minimal such that $K^t \subseteq A$ and $t \geq 1$. We claim $t = 1$. For, by the above work, $K^t \subseteq J$ and so if $t > 1$ we have $JK^{t-1} \subseteq K^t \subseteq J$ whence $K^{t-1} \subseteq A$ since $\mathcal{O}_R(J) = A$ whence t is not minimal.

So $K \subseteq J$ and therefore $K = J$. And then $B \subseteq \mathcal{O}_R(K) = \mathcal{O}_R(J) = A$. So $B = A$.

2.11 Corollary. Any principal order is extremal.

Proof. This is clear.

2.12 Lemma. Let A be extremal then there is a chain $\underline{X} = \mathrm{ch}(A)$, unique up to equivalence such that $A = \mathcal{O}(\underline{X})$.

Proof. (The uniqueness follows from 2.5 (vi)).

Let L be an A-lattice in V and choose r least positive such that $LJ^r \subseteq \pi_D L$. Put $X_i = \pi_D^{[i/r]}(LJ^{(i-r[i/r])} + \pi_D L)$. Then $\underline{X} = \{X_i\}$ is a chain. Put $B = \mathcal{O}(\underline{X})$. Certainly $A \subseteq B$. Moreover, we see that $X_i J \subseteq X_{i+1}$. So, by 2.5 (iii), $J \subseteq \mathrm{rad}(B)$. But A is extremal. So $A = B = \mathcal{O}(\underline{X})$.

We now prove an expanded version of Theorem 1.2.

2.13 Theorem. **(i)** Every CMC subgroup G of $A^\times$ is contained in a maximal one.

(ii) If G is a maximal CMC subgroup of $A^\times$ then there is a unique maximal G-stable order A. Moreover A is a principal order and $G = N(A)$.

(iii) If A is a principal order then $\mathrm{ch}(A)$ is a uniform chain and $N(A) = \mathrm{Aut}(\mathrm{ch}(A))$ is a maximal CMC subgroup of $A^\times$.

(iv) If A is a principal order and $G = N(A)$ then every G-invariant two-sided ideal of A is a power of J and is generated by an element of G.

Proof. (i) By 2.9, G stabilizes an order $\mathcal{A}$. Since $\mathcal{A}$ has finite index in some maximal order we can assume that $\mathcal{A}$ is maximal subject to $G \subseteq N(\mathcal{A})$.

We claim that $N(\mathcal{A})$ is a maximal CMC subgroup. Indeed, if H is a CMC subgroup containing $N(\mathcal{A})$ we can choose an order $\mathcal{B}$ maximal subject to $H \subseteq N(\mathcal{B})$. By 2.9, 2.10 and 2.12 we can find chains $\underline{X} = ch(\mathcal{A})$ and $\underline{Y} = ch(\mathcal{B})$ such that $\mathcal{A} = \mathcal{O}(\underline{X})$ and $\mathcal{B} = \mathcal{O}(\underline{Y})$.
But then $Aut(\underline{X}) = N(\mathcal{A}) \subseteq H \subseteq N(\mathcal{B}) = Aut(\underline{Y})$ so by 2.6. (iv) $\mathcal{A} \subseteq \mathcal{B}$. But $G \subseteq H \subseteq N(\mathcal{B})$.
Hence, by the maximality property of $\mathcal{A}$, $\mathcal{A} = \mathcal{B}$ and so $H = N(\mathcal{A})$.
Hence, $N(\mathcal{A})$ is maximal CMC.

(ii) If G is maximal CMC then, choosing $\mathcal{A}$ and $\underline{X}$ as above, $G = N(\mathcal{A}) = Aut(\underline{X})$ so, by 2.6 (v), G determines $\underline{X}$ and hence $\mathcal{A}$.

It remains to show that $\mathcal{A}$ is a principal order.

Put $\underline{Z} = \{X_1 g \mid g \in G\} \subseteq \underline{X}$. By 2.6 (i) $\underline{Z}$ is a chain. Moreover $G \subseteq Aut(\underline{Z})$.
But $Aut(\underline{Z}) = N(\mathcal{O}(\underline{Z}))$ is CMC. Hence $G = Aut(\underline{Z})$ and so $\underline{X} = \underline{Z}$ by 2.6 (v).
Clearly $\underline{Z}$ possesses automorphisms of all degrees and so, by 2.7., $\underline{X} = \underline{Z}$ is uniform and $\mathcal{A} = \mathcal{O}(\underline{X})$ is principal.

(iii) If $\mathcal{A}$ is a principal order then by 2.11 and 2.12 we can form $\underline{X} = ch(\mathcal{A})$ and, by 2.7, $\underline{X}$ is uniform. By 2.5 (viii), $N(\mathcal{A}) = Aut(\underline{X})$.

Let $H \supseteq N(\mathcal{A})$ be a maximal CMC group. By the above work $H = Aut(\underline{Y})$ for some chain $\underline{Y}$. Now $Aut(\underline{X}) \subseteq Aut(\underline{Y})$ so by 2.6. (iv) $\underline{X} \supseteq \underline{Y}$. But, since $\underline{X}$ is uniform, $Aut(\underline{X})$ has elements of all degrees (2.7) and so
$\underline{X} = \{Y_1 g \mid g \in Aut(\underline{X})\} \subseteq \{Y_1 g \mid g \in Aut(\underline{Y})\} \subseteq \underline{Y}$. So $\underline{X} = \underline{Y}$ and
$N(\mathcal{A}) = Aut(\underline{Y}) = H$ is maximal CMC.

(iv) Put $\underline{X} = ch(\mathcal{A})$ and choose $h \in Aut(\underline{X}) = G$ of degree 1. Then (cf. the proof of 2.7) $\mathcal{J} = h\mathcal{A}$. Let the G-invariant two-sided ideal $\mathcal{I}$ of $\mathcal{A}$ be a maximal counter-example. Now, by 2.5 (v), $X_o\mathcal{I} = X_t$, for some $t \geq 0$, and then $X_i\mathcal{I} = X_o h^i \mathcal{I} = X_o \mathcal{I} h^i = X_{i+t}$ for all i. If $t \neq 0$ then $\mathcal{I} \subseteq \mathcal{J} = h\mathcal{A}$ and then $h^{-1}\mathcal{I} = \mathcal{J}^{-1}\mathcal{I}$ would also be a counter-example. So $t = 0$ and $X_i\mathcal{I} = X_i$ for all i. But by 2.5 (ix), $\mathcal{A} = \bigoplus_i X_i^{(s_i)}$ so $\mathcal{I} = \mathcal{A}\mathcal{I} = \mathcal{A}$ and $\mathcal{I}$ is not a counter-example.

§ 3. Results on principal orders and their normalizers

From this point onward G will denote a maximal CMC subgroup of $A^\times$ (containing $F^\times$). $\mathcal{A}$ will be the principal order such that $G = N(\mathcal{A})$ and $\underline{X}$ will be ch(A) . Since $\underline{X}$ is uniform $s_i = s_i(\underline{X})$ is independent of i so we write $s = s_i$. By 2.3 we have

$$rs = m \tag{3.1}$$

Again since $\underline{X}$ is uniform we can choose h in $G = \mathrm{Aut}(\underline{X})$ of degree 1 . And then, as emerges in the proof of 2.7,

$$J = hA \tag{3.2}$$

Let $I(A)$ be the set of two-sided G-invariant fractional ideals of A . From 2.13 (iv),

$$I(A) = \{J^i \mid i \in \mathbb{Z}\} = \{h^i A \mid i \in \mathbb{Z}\} \tag{3.3}$$

and is a group.

The results 2.5 (viii), 2.6 (iii), 2.8 (ii) and 2.13 (iv) may be summarized in the following commutative diagram in which the rows are exact.

$$\begin{array}{ccccccccc}
1 & \longrightarrow & A^{\times} & \longrightarrow & N(A) & \xrightarrow{\text{gen}} & I(A) & \longrightarrow & 1 \\
 & & \| & & \| & & \| & & \\
1 & \longrightarrow & \mathrm{Aut}_o(\underline{X}) & \longrightarrow & \mathrm{Aut}(\underline{X}) & \xrightarrow{\deg} & \mathbb{Z} & \longrightarrow & 1 \\
 & & \| & & \| & & \downarrow{\times s} & & \\
1 & \longrightarrow & G_o & \longrightarrow & G & \xrightarrow{v_A} & \mathbb{Z} & &
\end{array} \tag{3.4}$$

($\mathrm{gen}(g) = gA$) .

Now 2.5 (iv) gives

$$A/J \simeq \mathrm{Mat}_s(k)^{(r)} \tag{3.5}$$

Recall our definitions: $U_o = A^{\times}$, $U_t = 1 + J^t$ for $t > 0$. We have, in the usual way:

$$U_o/U_1 \simeq (A/J)^{\times} \quad (\simeq \mathrm{GL}_s(k)^{(r)}) \quad \text{by} \quad uU_1 \longmapsto u + J \tag{3.6}$$

and, for $t > 0$,

$$U_t/U_{t+1} \simeq J^t/J^{t+1} \simeq A/J$$

by
$$(1+h^t v)U_{t+1} \longmapsto h^t v + J^{t+1} \longmapsto v + J \tag{3.7}$$

All but the last of these isomorphisms respect the action of G. From 2.6 (i), we see that, with A represented as in 2.5 (ii), $J^r = x^r A = \pi_D A$, so, with $e = rd$,

$$J^e = \pi_D^d A = \pi A \tag{3.8}$$

With good reason, therefore, we refer to e as the ramification index $e(A/o)$ of A over o.

If a is a full ideal of A we put $N(a) = |A/a|$ in the usual way. If $a \in I(A)$ then $a = gA$ with $g \in G$ and $\deg(g)$ is $\deg(a)$ as defined in 3.4. Hence

$$N(a) = \operatorname{nrd}(g)^n = q^{sn\deg a} = q^{n^2\deg a/e} \tag{3.9}$$

where $q = |o/p|$ and so N extends to a multiplicative map on the whole of $I(A)$.

If T is a finite abelian group we write $T^\circ = \operatorname{Hom}(T,\mathbb{C}^\times)$. We define the <u>relative inverse different</u> of A over o by

$$\mathcal{D}(A/o)^{-1} = \{a \in A \mid \operatorname{tr}_A(aA) \subseteq o\} .$$

3.10 Theorem. **(i)** $\mathcal{D}(A/o)^{-1} = J^{1-e}$.

(ii) The absolute inverse different, A^* is $\mathcal{D}(A/o)^{-1}o^*$.

(iii) If $a \in I(A)$ then $a^* = a^{-1}A^*$.

(iv) If $a,b \in I(A)$ and $a \supseteq b$ then we have a natural isomorphism $b^*/a^* \simeq (a/b)^\circ$ by

$$b + a^* \longmapsto [: a + b \longmapsto \psi_A(ba)] .$$

Proof. (i) $\mathcal{D}(A/o)^{-1}$ is a G-invariant two-sided ideal and so must have the form J^{-t}. So we want the biggest t such that $\operatorname{tr}_A(J^{-t}) \subseteq o$. We see from 2.5 (ii) and (ii) that

$$\operatorname{tr}_A(J^{-e+1}) = \operatorname{tr}_A(p^{-1}J) = p^{-1}\operatorname{tr}_A(J) = p^{-1}p = o .$$

Whereas

$$\operatorname{tr}_A(J^{-e}) = \operatorname{tr}_A(p^{-1}A) = p^{-1}\operatorname{tr}_A(A) = p^{-1}o = p^{-1} .$$

So the required t is $e - 1$.

(ii) and **(iii)**. These are easy generalizations of standard results.

(iv) This is a standard duality result (basically, $\mathrm{Ext}(T,\mathbb{Z}) \simeq T^{\circ}$) using the non-singularity of the form $< a,b > = \mathrm{tr}_A(ab)$ on A.

§ 4. Non-Abelian Congruence Gauss sums

We take up now where we left off in § 1. With G and $\mathcal{A}$ as in § 3. we take an irreducible representation Π of G on a $\mathbb{C}$-vector space W. We will assume that Π is admissible. That is

(i) For all $x \in W$ the stabilizer G_x of x in G is open in G.

(ii) If H is an open subgroup of G then W^H has finite dimension.

4.1 **Theorem.** **(i)** $\ker(\Pi) \supseteq U_j$ *for some* j.

(ii) W *has finite dimension.*

Proof. Choose $x \in W \setminus \{0\}$. G_x is open so, for some t, $G_x \supseteq 1 + \pi^t A = 1 + J^{et} = U_{et}$. But U_{et} is a normal subgroup of G so its fixed space is a non-zero G-subspace of W and hence the whole of W since W is irreducible. Thus $\ker_\Pi \supseteq U_{et}$ and we have part (i). Part (ii) follows now by condition (ii) above.

Thus Π has a conductor $\mathfrak{f}(\Pi) = J^i$. We say that Π is unramified, tame or wild according as $j = 0$, $j = 1$ or $j > 1$. In any case we now have our Gauss sum $\tau(\Pi)$ as defined in § 1.

Since G/U_0 is isomorphic (under the degree map) to $\mathbb{Z}$ we have the following description of the unramified representations.

4.2 **Theorem.** *Let* Π *be an unramified representation of* G. *Then*

(i) Π *is 1-dimensional*

(ii) $\Pi = \varphi_\lambda$, *for some* $\lambda \in \mathbb{C}$, *where* $\varphi_\lambda(g) = \lambda(\deg(g))$.

(iii) Π *factors through* $I(\mathcal{A})$ *(via the map gen of 3.4).*

(iv) $\tau(\Pi) = \Pi(\mathcal{A}^*)$.

We say that Π is <u>non-degenerate</u> if $\tau(\Pi) \neq 0$ so certainly the unramified Π are non-degenerate.

If $a \in F^\times$ then, by Schur's lemma, we know that $\Pi(a)$ is a scalar multiplication. We define the character $\omega_\Pi: F^\times \to \mathbb{C}^\times$ by

$$\Pi(a) = \omega_\Pi(a)1_W \tag{4.3}$$

The action of G on W gives an action on $\mathrm{Hom}(W,\mathbb{C})$. This defines the contragredient representation, denoted $\check{\Pi}$. The natural way to investigate the degeneracy of Π is to examine $\tau(\Pi)\tau(\check{\Pi})$.

The following result will be discussed elsewhere.

4.3 Theorem. *If Π is tame then Π is non-degenerate. Indeed*

$$\tau(\Pi)\tau(\check{\Pi}) = \omega_\Pi(-1)p^k \quad \text{where} \quad p^k | Nf(\Pi) = |\mathcal{O}/\mathfrak{p}|^{sn}.$$

We confine ourselves to the wild case where degeneracy can occur. Suppose then that Π is wild and so $\mathfrak{f}(\Pi) = J^{t-1}$ for some $t \geq 1$. The restriction of Π to U_t is then a representation of the finite abelian group U_t/U_{t+1} and must therefore split into a sum of G-conjugate (since Π is irreducible) linear characters of which we choose one, χ say. Let W_χ be the subspace of W corresponding to χ.

By 3.7 and 3.10 (iv) we have G-isomorphisms

$$(U_t/U_{t+1})^\circ \simeq (J^{-1}\mathfrak{f}(\Pi)/\mathfrak{f}(\Pi))^\circ \simeq \mathfrak{f}(\Pi)^*/\mathfrak{f}(\Pi)^*J$$

so χ determines an element b in $\mathfrak{f}(\Pi)^*$ (mod $\mathfrak{f}(\Pi)^*J$) such that, for $1 + x \in U_t$,

$$\chi(1 + x) = \psi_A(xb) . \tag{4.4}$$

and Π itself determines the G-conjugacy class C_Π of b mod $\mathfrak{f}(\Pi)^*J$. Note that, since $U_1 \subseteq G$, if $C_\Pi \cap G \neq \emptyset$ then $C_\Pi \subset G$.

4.5 Theorem. *Suppose that Π is wild then*

i) *Π is degenerate if and only if $C_\Pi \not\subset G$.*

ii) *If $C_\Pi \subset G$ then $\tau(\Pi)\tau(\check{\Pi}) = \omega_\Pi(-1)N(\mathfrak{f}(\Pi))$.*

Proof. Choose c such that $cA = f(\Pi)^*$ then

$$\tau(\Pi)1_W = T(\Pi) = \sum_{\bar{u}\in U_0/U_{t+1}} \Pi(cu)\psi_A(cu) .$$

As $\bar{y}$ runs over U_0/U_t and $\bar{x}$ runs over J^t/J^{t+1}, $\overline{y(1+x)}$ will run over U_0/U_{t+1} so

$$\tau(\Pi)1_W = \sum_{\bar{y}\in U_0/U_t} \sum_{\bar{x}\in J^t/J^{t+1}} \Pi(cy(1+x))\psi_A(cy+cyx)$$

$$= \sum_{\bar{y}} \Pi(cy)\psi_A(cy) \sum_{\bar{x}} \Pi(1+x)\psi_A(cyx) .$$

Consider the internal sum restricted to W_χ. It becomes

$$\sum_{\bar{x}} \chi(1+x)\psi_A(cyx) = \sum_{\bar{x}} \psi_A(xb)\psi_A(cyx) \qquad \text{by 4.4}$$

$$= \sum_{\bar{x}} \psi_A(x(b+cy)) .$$

Now this sum vanishes unless $\bar{x} \longmapsto \psi_A(x(b+cy))$ is the trivial character on J^t and therefore unless $b + cy \in (J^t)^* = \mathfrak{f}(\Pi)^*J = cJ$. If this were the case then $b(\equiv -cy \bmod f(\Pi)^*J)$ would generate $f(\Pi)^*$ and therefore lie in G.

So if $C_\Pi \not\subset G$ the internal sum (and hence T_Π) vanishes on W_χ and therefore $\tau(\Pi) = 0$.

On the other hand, if $C_\Pi \subset G$ we can take $c = b$. The contragredient representation may (after the choice of a basis for W) be realized on W by putting $\check{\Pi}(g) = \text{transpose } (\Pi(g^{-1}))$. So we have

$$\tau(\check{\Pi})1_W = \text{transpose } (T(\check{\Pi})) = \sum_{\bar{v}\in U_0/U_{t+1}} \Pi(v^{-1}c^{-1})\psi_A(cv)$$

whence

$$\tau(\check{\Pi})\tau(\Pi)1_W = \sum_{\bar{u},\bar{v}\in U_0/U_{t+1}} \Pi(v^{-1}c^{-1})\Pi(cu)\psi_A(cv)\psi_A(cu)$$

$$= \sum_{\bar{u},\bar{v}} \Pi(v^{-1}u)\psi_A(c(v+u)) = \sum_{\bar{w}} \Pi(w) \sum_{\bar{v}} \psi_A(cv(1-w)) \ .$$

where we have put $w = v^{-1}u$. Write S_w for the internal sum.

Again we split this sum putting $v = y(1+x)$ where this time $\bar{y}$ runs over U_0/U_1 and $\bar{x}$ over J/J^{t+1} (note that $\psi_A(cv(1+w)) = \psi_A(v(1+w)c)$.)

$$S_w = \sum_{\bar{y},\bar{x}} \psi_A((1+x)y(w+1)c) = \sum_{\bar{y}} \psi(y(w+1)c) \sum_{\bar{x}} \psi_A(xy(w+1)c) \ .$$

Now the inner sum vanishes unless $y(w+1)c \in J^* = cJ^t$ and therefore unless $w \equiv -1 \bmod J^t$. (And in that case the inner sum is $|J/J^{t+1}| = N(J)^t$.) So restricting the original sum to $w \equiv -1 \bmod J^t$ and putting $w = -1+z$

$$\tau(\check{\Pi})\tau(\Pi)1_W = N(J)^t \sum_{\bar{z}\in J^t/J^{t+1}} \Pi(-1+z) \sum_{\bar{y}\in U_0/U_1} \psi_A(yzc)$$

$$= N(J)^t \sum_{\bar{y},\bar{z}} \omega_\Pi(-1)\psi_A(-zc)\psi_A(zcy) \qquad \text{(by 4.4, with } b = c)$$

$$= N(J)^t \ \omega_\Pi(-1) \sum_{\bar{y}} \sum_{\bar{z}} \psi_A(zc(y-1)) \ .$$

Yet again the inner sum is zero unless each term is 1 . For this we need $c(y-1) \in (J^t)^* = cJ$ so y must lie in U_1 . Therefore

$$\tau(\check{\Pi})\tau(\Pi) = N(J)^t \ \omega_\Pi(-1) \sum_{\bar{z}\in J^t/J^{t+1}} 1$$

$$= \omega_\Pi(-1)N(J)^{t+1} = \omega_\Pi(-1)N(\mathfrak{f}(\Pi)) \ .$$

We want to round off these results by relating $\overline{\tau(\Pi)}$, $\tau(\bar{\Pi})$ and $\tau(\check{\Pi})$. One can then calculate $|\tau(\Pi)|$ from $\tau(\Pi)\tau(\check{\Pi})$.

4.6 Theorem. Let $\varphi \in \mathrm{Aut}(\mathbb{C})$ and let $u(\varphi)$ be the element of $\mathbb{Z}_p^{\times}$ such that $\eta^{\varphi} = \eta^{u(\varphi)}$ for any p-power root of unity, η. Then

$$\tau(\Pi) = \tau(\Pi)^{\varphi}\omega_{\Pi}(u(\varphi))^{\varphi} .$$

In particular $\tau(\bar{\Pi}) = \overline{\tau(\Pi)}\,\omega_{\Pi}(-1)$.

Proof. This simply involves rearranging the sum using the relation $\psi_A(x)^{\varphi} = \psi_A(xu(\varphi))$.

Recalling the unramified characters of 4.2 we define

$$\mathrm{Mag}_{\Pi} = \varphi_{\lambda} \quad \text{where} \quad \lambda = |\omega_{\Pi}(\pi)|^{1/e} .$$

4.7 Theorem. **(i)** If $\mu^e = \omega_{\Pi}(\pi)$ and $\Pi_o = \Pi \otimes \varphi_{\mu}^{-1}$ then $|\Pi_o(G)| < \infty$.
(ii) Π_o and $\Pi \otimes (\mathrm{Mag}_{\Pi})^{-1}$ are unitary (given the right form on W).

Proof. **(i)** With $h \in G$ as in § 3, $h^e \in \pi U_o$. Now $\Pi_o(\pi) = 1$, by construction, so $\Pi_o(h)^e \in \Pi_o(U_o)$ and

$$|\Pi_o(G)| = |\Pi_o(G)/\Pi_o(U_o)|\cdot|\Pi_o(U_o)| \le e|U_o/U_j| .$$

(ii) Hence Π_o is unitary. But $\Pi \otimes (\mathrm{Mag}_{\Pi})^{-1} = \Pi_o \otimes \varphi_{\eta}$ where $\eta = \mu/|\mu|$ and this is unitary since φ_{η} , taking values in S^1 , is unitary.

4.8 Theorem. If φ is a linear non-ramified representation of G then

$$\tau(\Pi \otimes \varphi) = \tau(\Pi)\varphi(\mathfrak{f}(\Pi)^*) .$$

Proof. This follows easily from the definition of $\tau(\Pi)$.

4.9 Theorem. **(i)** $\tau(\bar{\Pi}) = \tau(\check{\Pi})\cdot(\mathrm{Mag}_{\Pi}(\mathfrak{f}(\Pi)^*))^2$
(ii) $|\tau(\Pi)| = \sqrt{|\tau(\Pi)\tau(\check{\Pi})|\ \mathrm{Mag}_{\Pi}(\mathfrak{f}(\Pi)^*)}$.

Proof. **(i)** Put $\Sigma = \Pi \otimes \mathrm{Mag}_{\Pi}^{-1}$ (So $\mathfrak{f}(\Sigma) = \mathfrak{f}(\Pi)$.) Then $\check{\Sigma} = \check{\Pi} \otimes \mathrm{Mag}_{\Pi}$ and $\bar{\Sigma} = \check{\Sigma}$ by 4.7 (ii). By 4.9

$$\begin{aligned}\tau(\bar{\Pi}) &= \tau(\bar{\Sigma})\ \mathrm{Mag}_{\Pi}(\mathfrak{f}(\Pi)) \\ &= \tau(\check{\Sigma})\ \mathrm{Mag}_{\Pi}(\mathfrak{f}(\Pi)) \\ &= \tau(\check{\Pi})\ \mathrm{Mag}_{\Pi}(\mathfrak{f}(\Pi))^2 .\end{aligned}$$

(ii) Clear.

Contemporary Mathematics
Volume **86**, 1989

THE FUNCTIONAL EQUATION ε-FACTORS

by J. Queyrut, Bordeaux

Introduction.

The Langlands conjecture in its vaguest form predicts a "nice" correspondence between some representations of the absolute Galois group G_F of a local field F and some representations of a simple central F-algebra. Here the word "nice" means that we want to translate the arithmetic description of G_F. The recent development of number theory shows that most of the arithmetic informations are contained or used in the description of the Artin L-function and the constant of its functional equation. Therefore one wants to define L-functions with functional equation associated to representations of simple central algebras.

Notation.

Let F be a non-archimedean local field of characteristic 0 and residue characteristic p. Thus F is a finite extension of the p-adic rational number field $\mathbb{Q}_p$.

We consider a simple central algebra A over F, with finite dimension over F. It is therefore the algebra of m by m matrices with entries in some division algebra D of center F and rank d^2 over the field F. The rank of A over F is then n^2, where $n = md$ and the multiplicative group A^* of A is identified to $GL_m(D)$. We identify the center of $GL_m(D)$ with F^*.

1. Fourier Transform

The field F has a canonical continuous additive character ψ_F defined by

$$\psi_F = \psi_{\mathbb{Q}_p} \circ \mathrm{Tr}_{F/\mathbb{Q}_p}$$

where $\psi_{\mathbb{Q}_p}$ is the composition of the following maps:

$$\mathbb{Q}_p \longrightarrow \mathbb{Q}_p/\mathbb{Z}_p \longrightarrow \mathbb{Q}/\mathbb{Z} \longrightarrow \mathbb{C}^*$$

the last one being $x \longmapsto e^{2i\pi x}$.

0271-4132/89 $1.00 + $.25 per page

The pairing $(x,y) \longmapsto \psi_F(xy)$, $x, y \in F$ is non-degenerate and may be used to identify the locally compact abelian group F with its Pontrjagin dual $\hat{F}$.

We write $\mathrm{Tred}_A: A \to F$ for the reduced trace. The definition $\psi_A = \psi_F \circ \mathrm{Tred}_A$ provides a canonical continuous additive character of A and the pairing $A \times A \to \mathbb{C}^*$ given by $(x,y) \longmapsto \psi_A(xy)$ is non-degenerate (because A is a separable algebra over F and the reduced trace does not degenerate).

Let dy be a Haar measure on A.

Definition. Call $S(A)$ the space of compactly supported locally constant functions from A to $\mathbb{C}$. The Fourier transform $\hat{\Phi}(x)$ of $\Phi \in S(A)$ is defined by:

$$\hat{\Phi}(x) = \int_A \Phi(y)\psi_A(xy)dy .$$

The Haar measure dy is chosen in such a way that for all Φ with $\Phi, \hat{\Phi} \in S(A)$,

$$\hat{\hat{\Phi}}(x) = \Phi(-x) .$$

2. ζ-Functions

Let Π be an admissible irreducible representation of A^* on a complex vector space V, so Π is a homomorphism $A^* \to \mathrm{Aut}_{\mathbb{C}}(V)$ such that

1) V does not contain any non-trivial A^*-submodule,

2) for every open compact subgroup H of G, the subvector space V^H of V stabilized by H is finite dimensional.

We write $v_F: F^* \to \mathbb{Z}$ for the canonical valuation of F and

$$\| x \|_F = q_F^{-v_F(x)} , \quad x \in F^* ,$$

where q_F is the cardinality of the residue class field of F.

We write $\mathrm{Nred}_A: A^* \to \mathbb{Q}^*$ for the reduced norm. So we get a multiplicative continuous homomorphism from A^* to $\mathbb{C}^*$:

$$\| x \|_A = \|\mathrm{Nred}_A(x)\|_F , \quad \forall\, x \in A^* .$$

Let d^*x be a Haar measure on A^*.

If V^* is the dual vector space of V, the function

$$\Pi_{v^*,v}: A^* \to \mathbb{C}$$

efined by $\Pi_{v^*,v}(g) = \langle v^*, \Pi(g)v \rangle = v^*(\Pi(g)(v))$ is called a coefficient of Π t (v^*,v), it is a locally constant function (which has compact support if and nly if Π is cuspidal with finite length).

In fact, we must restrict ourselves to the subspace $\tilde{V}$ of V^* defined by $= \bigcup_K V^{*K}$ where K runs through the set of compact open subgroups of A^*.

efinition. For $\Phi \in S(A)$, $v \in V$, $v^* \in \tilde{V}$, $s \in \mathbb{C}$, the ζ-function of A s defined by:

$$\zeta(\Phi, s, \Pi_{v^*,v}) = \int_A \Phi(x)\Pi_{v^*,v}(s) \, \| x \|_A^s \, d^*x \quad .$$

roposition. <u>There exists a real number</u> σ_o <u>such that for all</u> s <u>such that</u> $e(s) > \sigma_o$, <u>the integrals</u> $\zeta(\Phi, s, \Pi_{v^*,v})$ <u>converge absolutely for all coefficients</u> $_{v^*,v}$ <u>of</u> Π <u>at</u> (v^*,v) <u>and for all</u> $\Phi \in S(A)$.

. L-functions

We keep the notation of the previous section.

heorem. (R. Godement and H. Jacquet, [39]theorem 3.3). <u>Let</u> Π <u>be an admissible</u> <u>rreducible representation of</u> A^* ; <u>then there exists a unique function</u>

$$L_A(s,\Pi): \mathbb{C} \to \mathbb{C}$$

<u>uch that:</u>

) <u>There is a polynomial</u> $l(X) \in \mathbb{C}[X]$ <u>such that</u>

$$l(0) = 1$$

$$L_A(s,\Pi) = l(q^{-s})^{-1} \quad .$$

) <u>For any</u> $\Phi \in S(A)$, <u>and any coefficient</u> $\Pi_{v^*,v}$ <u>of</u> Π <u>at</u> (v^*,v) <u>there</u> <u>exists a polynomial</u> $P = P_{\Phi,\Pi,v^*,v} \in \mathbb{C}[X,X^{-1}]$ <u>such that</u>

$$\frac{\zeta(\Phi,\ s,\ \Pi_{v^*,v})}{L_A(s,\ \Pi)} = P_{\Phi,\Pi,v^*,v}(q^s) .$$

3. <u>The ideal of</u> $\mathbb{C}[X,X^{-1}]$ <u>generated by the polynomials</u> $P_{\Phi,\Pi,v^*,v}$ <u>for</u> $\Phi \in S(A)$, $v \in V$, $v^* \in \tilde{V}$ <u>is equal to</u> $\mathbb{C}[X,X^{-1}]$.

This last condition means that we can choose some coefficients f_i of Π and some Φ_i in $S(A)$ such that $L_A(s,\Pi) = \Sigma_i\ \zeta(\Phi_i,\ s,\ f_i)$. In fact in many cases, there exists Φ in $S(A)$ and coefficient f of Π such that $L_A(s,\Pi) = \zeta(\Phi,\ s,\ f)$.

Explicit formulae for $L_A(s,\Pi)$

1) Case $A = F$

The irreducible admissible representation Π of F^* is a continuous homomorphism from F^* to $\mathbb{C}^*$ (quasi-character); so there exists $s_o \in \mathbb{C}$ and a character χ, (quasi-character of absolute value 1) such that

$$\Pi(a) = \chi(a)\|a\|_F^{s_o} .$$

If $s_o = 0$, we say that Π is finite.

If Π is unramified, the conductor $f(\Pi)$ of Π is equal to the ring of integers $\mathcal{O}_F$ of F, so the restriction of Π to the group U_F of units of $\mathcal{O}_F$ is trivial. Therefore for all c in the prime ideal $\mathcal{P}_F$ of $\mathcal{O}_F$, $\Pi(c) = \|c\|_F^{s_o} = q^{s_o}$. In this case we write $f(\Pi)^* = U_F$.

Let π be any generator of $\mathcal{P}_F$, then

$$L_F(s,\Pi) = \frac{1}{1 - \Pi(\pi)\|\pi\|_F^s} .$$

If Π is ramified, the conductor $f(\Pi)$ of Π is equal to $\mathcal{P}_F^m$, where m is a positive integer. We write $f(\Pi)^* = 1 + \mathcal{P}_F^m$.

Then

$$L_F(s,\Pi) = 1 .$$

In these cases the map $\Phi = \mu(f(\Pi)^*)\chi_{f(\Pi)^*}$ is in $S(F)$ (where $\mu(f(\Pi)^*)$ is the measure of $f(\Pi)^*$ and $\chi_{f(\Pi)^*}$ is the characteristic function of $f(\Pi)^*$ if $m > 0$ and the characteristic function of $\mathcal{O}_F$ if $m = 0$) and satisfies the relation

$$L_F(s,\Pi) = \zeta(\Phi, s, \Pi) .$$

2) **Case** $A = D$

We have results which are analogous to the previous ones.

Let Π be an irreducible admissible representation of D^* on a vector space V. Then Π is finite dimensional (i. e. the dimension of V over F is finite). There exists a quasi-character ω_Π of F^* such that

$$\forall x \in F^* , \quad \Pi(x) = \omega_\Pi(x) Id_V .$$

The representation Π is called finite if $D^*/\ker \Pi$ is a finite group, so that Π is a homomorphism with open kernel and finite image.

Let Π be an irreducible admissible representation of D^*; then there exists a real number s_o and a finite irreducible admissible representation Π' of D^* such that

$$\forall a \in D^* \ \Pi(a) = \Pi'(a) \| a \|_D^{s_o} .$$

Moreover, Π is finite if and only if ω_Π is finite.

We denote by $\mathcal{O}_D$ the maximal order of D, $\mathcal{P}_D$ the maximal ideal of $\mathcal{O}_D$ and for each non-zero integer i by U_D^i the subgroup of the group U_D^o of units of $\mathcal{O}_D$ equal to $1 + \mathcal{P}_D^i$.

By the definition of an admissible representation, there exists a smallest integer m such that V is fixed by U_D^m. The conductor $f(\Pi)$ of Π is defined to be $\mathcal{O}_D$ if $m = 0$ and $\mathcal{P}_D^m$ if $m > 0$.

Let π_D be any generator of $\mathcal{P}_D$, then

$$L_D(s,\Pi) = \frac{1}{1 - \Pi(\pi_D) \|\pi_D\|_D^s} \qquad \text{if } f(\Pi) = \mathcal{O}_D$$

$$L_D(s,\Pi) = 1 \qquad \text{if } f(\Pi) \neq \mathcal{O}_D .$$

As before we put $f(\Pi)^* = U_D^m$. The map $\Phi = \mu(f(\Pi)^*)\chi_{f(\Pi)^*}$ is in $S(F)$ (where $\mu(f(\Pi)^*)$ is the measure of $f(\Pi)^*$ and $\chi_{f(\Pi)^*}$ is the characteristic function of

$f(\Pi)^*$ if $m > 0$ and the characteristic function of $\mathcal{O}_D$ if $m = 0$) and satisfies $L_D(s,\Pi) = \zeta(\Phi, s, \Pi)$.

3) Case $A = M_m(D)$

The two main results are the following ones:

- if Π is supercuspidal then $L_A(s,\Pi) = 1$
- if Π is parabolic, there exists an induction procedure in order to prove the existence of $L_A(s,\Pi)$.

4. Functional equation

Put $\Xi(\Phi, \Pi_{v^*,v}, s - \frac{1}{2}(n-1)) = \dfrac{\zeta(\Phi, s, \Pi_{v^*,v})}{L_A(s,\Pi)}$.

The contragredient representation $\hat{\Pi}$ of Π is defined by

$$\forall\ v \in V\ ,\quad \forall\ v^* \in \tilde{V}\ ,\quad < \hat{\Pi}(g)v^*\ ,\ v > = < v^*,\ \Pi(g^{-1})v > \ .$$

Then $\hat{\Pi}$ is also an irreducible representation of $\tilde{V}$, so Godement-Jacquet's theorem holds for $\hat{\Pi}$.

We denote by $\tilde{\Pi}_{v^*,v}$ the map:

$$g \longmapsto < v^*,\ \Pi(g^{-1})v > = \Pi_{v^*,v}(g^{-1})\ .$$

Theorem. <u>There exists a function</u> $\varepsilon(s, \Pi, \psi_A)$ <u>of</u> s , <u>which is a constant times a power of</u> q^{-s} <u>such that for all coefficients</u> $\Pi_{v^*,v}$ <u>all</u> Φ <u>in</u> $S(A)$ <u>and all</u> s <u>in</u> $\mathbb{C}$:

$$\Xi(\hat{\Phi},\ 1 - s,\ \hat{\Pi}_{v^*,v}) = (-1)^{m(d-1)}\varepsilon(s,\ \Pi,\ \psi_A)\Xi(\Phi,\ s,\ \Pi_{v^*,v})\ .$$

Definition. The Artin root number is the complex number

$$W(\Pi,\ \psi_A) = \varepsilon(\tfrac{1}{2},\ \Pi,\ \psi_A)\ .$$

Properties.

1) $\varepsilon(s, \Pi, \psi_A) = W(\Pi, \psi_A) \times (\text{power of } p^{\frac{1}{2}-s})$

2) Let ω_Π be the quasi-character of F^* such that

$$\forall\ a \in F^*\ \Pi(a) = \omega_\Pi(a)1_V\ ;$$

then $\varepsilon(s, \Pi, \psi_A)\varepsilon(1-s, \hat{\Pi}, \psi_A) = \omega_\Pi(-1)$.

3) If we change the additive character ψ_A , all the previous results are true. More precisely if ψ' is an other additive character, there exists b in A such that

$$\forall\ x \in A\ \psi'(x) = \psi_A(xb)\ ;$$

then

$$\varepsilon(s, \Pi, \psi') = \varepsilon(s, \Pi, \psi_A)\omega_\Pi(b)\|b\|_A^{n(s-\frac{1}{2})}\ .$$

5. ε-Factors for Galois Representations

Definition of ε-factors for Galois representations

Class field theory gives a homomorphism a from F^* to the quotient G_F^{ab} of the absolute Galois group G_F of F by its group of commutators.

If α is a 1-dimensional representation of G_F , α factorizes through G_F^{ab} , and by composition with a gives a character on F^* which we denote by $a(\alpha)$.

We define:

$$L(\alpha, \psi_F) = L_F(a(\alpha), \psi_F) \qquad f(\alpha) = f(a(\alpha))$$

$$\varepsilon(s, \alpha, \psi_F) = \varepsilon(s, a(\alpha), \psi_F) \qquad W(\alpha, \psi_F) = W(a(\alpha), \psi_F)\ .$$

The abelian Gauss sum $\tau(\alpha, \psi_F)$ is defined by the equality

$$W(\alpha, \psi_F) = \frac{\tau(\bar{\alpha}, \psi_F)}{\sqrt{N(f(\alpha))}}\ .$$

Then $W(\alpha, \psi_F)$ is a complex number of modulus 1 ; $\tau(\alpha, \psi_F)$ is an algebraic number; the Galois action is given by

$$\forall g \in G : g(\tau(g^{-1}(\alpha), \psi_F)) = \tau(\alpha, \psi_F)\ a(\alpha)(u_p(g))$$

where $u_p(g)$ is the unique p-adic unit such that $\eta^{\omega^{-1}} = \eta^{u_p(\omega)}$, for every p^n-th root of unity η in $\bar{\mathbb{Q}}$.

The first thing to do is to extend these definitions to all irreducible representations of G_F. For the L-function and the conductor, this was done by Artin who defined the so called Artin L-function and Artin conductor and gave explicite formulae. For the moment, no easy explicit formulae are known for W, ε and τ. The general method in order to extend functions defined on 1-dimensional representations of finite groups to all representations is:

1) extend by linearity these functions to the group of abelian representations

2) use the Brauer induction theorem.

In this case the non-uniqueness of the decomposition of a representation in a sum of induced representations gives troubles. However we have the following theorem:

Theorem. There exists a unique map τ from $\coprod_K R(G_K)$ to $\bar{\mathbb{Q}}^*$ where K runs through the set of finite extension of $\mathbb{Q}_p$ contained in a given algebraic closure $\bar{\mathbb{Q}}_p$ and $R(G_K)$ denotes the group of virtual characters of G_K, such that

1) If $\dim(\Sigma) = 1$, $\tau_K(\Sigma)$ is equal to the abelian Gauss sum $\tau(a(\Sigma), \phi_K)$.

2) For all fields $K \subset \bar{\mathbb{Q}}_p$ with finite degree over $\mathbb{Q}_p$, the map $\Sigma \longmapsto \tau_K(\Sigma)$ is an homomorphism.

3) If G_F is an open subgroup of G_K of finite index and $\Sigma \in R(G_F)$ with $\dim(\Sigma) = 0$, then

$$\tau_K(\mathrm{Ind}_{F/K}(\Sigma)) = \tau_F(\Sigma) .$$

The uniqueness is proved by using the Brauer induction theorem.

The first proof of the existence was given by Langlands who computed explicitly the Gaussian sums for representations in the kernel of the Brauer induction map. A second proof making use of global methods was then given by Deligne.

Explicite formula for the Gauss sum in the case $A = F$ and $\dim(\Sigma) = 1$

$$\tau(\Sigma) = \Sigma((-\,\mathrm{Frob})^{\delta})\, a(\Sigma)(\gamma d)^{-1} \sum_{x \in U_F^{o}/U_F^{n}} a(\Sigma)(x)\, \psi_F(x/\gamma d)$$

where

Frob is the Frobenius automorphism
$\delta = 0$ if Σ is ramified and $\delta = 1$ if Σ is unramified
d generates the different of F over $\mathbb{Q}_p$
γ generates $\mathcal{P}_F^{j+1}$ where j is the valuation of the Swan conductor.

Each factor in this decomposition has an interesting meaning:

1) $\Sigma((-\mathrm{Frob})^{\delta})$ can be extended to all representations Σ on a vector space V by

$$\Sigma \longmapsto \mathrm{Det}(-\,\mathrm{Frob},\ \Sigma^{G_F^{o}})$$

where $\Sigma^{G_F^{o}}$ is the representation of G_F on the subvector space invariant under the action of the first ramification group G_F^{o} . This function is called the unramified characteristic.

2) The extension of $\Sigma \longmapsto a(\Sigma(\gamma d))$ was proved by Deligne and Henniart for a suitable choice of γd .

3) The third factor was used by Taylor in order to prove Fröhlich conjecture on the structure of ring integers.

4) If $n = 1$, then $\tau(\Sigma)$ reduces to a Gauss sum on a finite field.

5) If n is even, then $W(\Sigma) = \Sigma(\gamma d)\psi_F(\gamma d)$ for Deligne-Henniart's choice of γd .

6) If n is odd, then $W(\Sigma)$ is the product of $\Sigma(\gamma d)\psi_F(\gamma d)$ for Deligne-Henniart's choice of γd and of a Gauss sum on a finite field for a quadratic character.

7) Fröhlich and Taylor gave a direct proof of the existence of the Galois Gaus sum for tame representations.

8) For a general direct proof, one must prove that the quadratic Gauss sums introduced in 6) can be extended in a function on the whole group of representations of G_F . It was done modulo roots of unity by Henniart.

References

[39] R. Godement, H. Jacquet: Zeta functions of simple algebras; Lecture Notes in Mathematics 260, Springer Verlag, Berlin-Heidelber-New York 1972.

Contemporary Mathematics
Volume 86, 1989

ROOT NUMBERS AND THE LOCAL LANGLANDS CONJECTURE

by M. Taylor, Cambridge

Introduction. In the first section we briefly recall a number of well-known definitions and results concerning Weil groups and Deligne representations. For a more complete account of such results the reader is referred to Deligne's article [25]. In Section 2 we describe the local functional equation for admissible, irreducible representations of GL_n . We are then able to state the local Langlands conjectures. (See also G. Henniart's paper [42] for a complete and up to date account of these conjectures). The remainder of this article is devoted to giving an account of I. G. Macdonald's paper [80], which establishes a Langlands like bijection between certain tame representations of Weil groups, and irreducible representations of $GL_n(k)$, k a finite field. This analogue will then be more fully developed in A. Fröhlich's article [F] in this volume.

1. **Weil groups and Deligne representations.** Let F denote a finite extension of the p-adic field Q_p . We let $\mathcal{O}_F$ (resp. $\mathcal{P}_F$) denote the valuation ring (resp. valuation ideal) of F . We write k for the residue class field $\mathcal{O}_F/\mathcal{P}_F$, and put $q = |k|$. We let $|\ | : F^* \to q^{\mathbb{Z}}$ denote the valuation with the property that $|x| = (\mathcal{O}_F : x\mathcal{O}_F)^{-1}$ for $x \in \mathcal{O}_F \smallsetminus \{0\}$.

Let $\bar{F}$ denote a fixed algebraic closure of F , and let $G_F = \mathrm{Gal}(\bar{F}/F)$ endowed with the Krull topology. Local class field theory yields a continuous dense injection $A:F^* \to G_F^{ab}$. The Weil group W_F is defined by the diagram:

$$\begin{array}{ccccccccc} 1 & \to & I_F & \to & G_F & \to & \hat{\mathbb{Z}} & \to & 1 \\ & & \| & & \uparrow & & \cup & & \\ 1 & \to & I_F & \to & W_F & \to & \mathbb{Z} & \to & 1 . \end{array}$$

Here I_F denotes the inertia group and $\mathbb{Z}$ denotes the subgroup of G_F/I_F generated by σ , the Frobenius automorphism. W_F is topologized by giving I_F the usual profinite topology and by giving W_F/I_F the discrete topology. In the sequel we write I_F^1 for the wild inertia subgroup, and we put $I = I_F/I_F^1$. The Artin map A induces an isomorphism $F^* \xrightarrow{\sim} W_F^{ab}$. We write a for the inverse of this isomorphism. We refer to a as the co-Artin map.

In order to obtain sufficient representations of W_F we refine our usual notion of representation as follows (though we note that the original motivation for this refinement was to introduce a notion of complex representation which corresponded to certain ℓ-adic representations): a Deligne representation of W_F is a pair (Π, N) where (a) Π is a finite dimensional, semi-simple, continuous, complex representation of W_F on V, (b) $N \in \mathrm{End}(V)$ is nilpotent, with the property that

$$\Pi(w)\, N \Pi(w)^{-1} = |a(w)| N \qquad w \in W_F .$$

We say that two Deligne representations (Π, N), (Π', N') are equivalent (resp. I_F-equivalent) if there is an isomorphism of W_F-modules (resp. I_F-modules) taking $\Pi \to \Pi'$, $N \to N'$.

Example. Let $V = \langle e_1, \ldots, e_n \rangle_{\mathbb{C}}$, let W_F act on V by $w\, e_i = |a(w)|^i e_i$, and put

$$N\, e_i = \begin{cases} e_{i+1} & i < n \\ 0 & i = n \end{cases} .$$

This defines a Deligne representation which we call $\mathrm{Sp}(n)$.

It can be shown that any indecomposable Deligne representation is of the form $\Sigma \otimes \mathrm{Sp}(n)$, where Σ is an irreducible representation of W_F. Clearly such a Deligne representation is irreducible in this category if, and only if, $n = 1$.

2. **Root numbers.** We fix $n \geq 1$ and we let Π denote an admissible, irreducible representation of $GL_n(F)$. In [Q] we have studied the functional equation

$$\frac{\zeta(\check{\chi}, 1 - s + \frac{n-1}{2}, \hat{\Phi})}{L(\check{\Pi}, 1 - s)} = \epsilon(\Pi, s) \frac{\zeta(\chi, s + \frac{n-1}{2}, \Phi)}{L(\Pi, s)}$$

where Φ is a Schwarz-Bruhat function of $M_n(F)$, χ is a coefficient function of Π, and $L(\Pi, s)$ is the L-function of Π in the complex variable s. In the sequel we write $\epsilon(\Pi)$ for the so-called local constant $\epsilon(\Pi, \frac{1}{2})$.

Tate's thesis. Tate obtained the above result in the case $n = 1$. Moreover his local functional equations can be pieced together to give a global equation; he thereby obtained a local factorisation of the abelian Artin root numbers.

More precisely let φ denote a character of a Galois group G of an extension of number fields. We write $\Lambda(s, \varphi)$ for the "extended" Artin L-function associated to φ ; then $\Lambda(s,\varphi)$ satisfies a functional equation $\Lambda(s,\varphi) = W(\varphi)\Lambda(1-s,\bar{\varphi})$. Then, if φ is abelian, we have the factorisation $W(\varphi) = \Pi W(\varphi_p)$, with

$$(*) \qquad W(\varphi_p) = \epsilon(\varphi_{|G_p} \circ A) .$$

More generally Landlands has shown that there is a unique "extendible" factorisation for all Artin root numbers $W(\varphi)$, regardless of whether φ is abelian. (See [106] for details). It is therefore natural to ask whether we can generalise Tate's result (*).

In the sequel we shall put

$$\tau(\chi) = W(\bar{\chi})\, N\mathfrak{f}(\chi)^{1/2}$$

for χ a local Galois character. Here $W(\bar{\chi})$ is the local Langlands root number of $\bar{\chi}$, and $N\mathfrak{f}(\chi)^{1/2}$ denotes the positive square root of the absolute norm of the Artin conductor of χ . We next introduce a root of unity twist of τ which was defined in [108]. This modified Gauss Sum τ^* turns out to be exactly the right invariant for the work in 3. We let $\mathcal{D}$ denote the absolute different of F/Q_p and we choose elements π , $c \in F^*$ such that $\pi\mathcal{O}_F = \mathcal{P}_F$, $c\mathcal{O}_F = \mathcal{P}_F\mathcal{D}$. For χ as above we then set

$$\tau^*(\chi) = \tau(\chi)\, \det_{\chi}{}_{I_F}(-A(\pi))^{-1}\, \det_{\chi}(A(c)) .$$

<u>**The Local Langlands Conjectures.**</u> If (Π,N) is a Deligne representation of W_F , then we define the local L-function

$$L((\Pi,N),\, s) = \det_{\Pi}{}_{I_F}(1-q^{-s}\sigma)^{-1}$$

where σ denotes the Frobenius automorphism over F . We let $\mathcal{W}_n$ (resp. $\mathcal{W}_n^{0}$) denote the set of equivalence classes of Deligne representations of W_F (resp. of irreducible Deligne representations of W_F) with dimension n , and we set $\mathcal{W} = \bigcup_n \mathcal{W}_n$. We let $\mathcal{A}_n$ (resp. $\mathcal{A}_n^{0}$) denote the set of equivalence classes of irreducible, admissible (resp. and supercuspidal) representations of $GL_n(F)$, and let $\mathcal{A} = \bigcup_n \mathcal{A}_n$.

Conjecture. There exists a unique bijection

$\beta : \mathcal{W} \to A$, taking $\mathcal{W}_n \longleftrightarrow A_n$, $\mathcal{W}_n^o \longleftrightarrow A_n^o$, with the property that for $\Sigma \in \mathcal{W}$

(1) $\beta(\check{\Sigma}) = \beta^{\vee}(\Sigma)$;

(2) $\beta(\Sigma) = \det(\Sigma) \circ a^{-1}$;

(3) If Σ_1 has degree 1, then

$$L(\Sigma_1 \otimes \Sigma_2, s) = L(\beta(\Sigma_1) \otimes \beta(\Sigma_2), s)$$

$$\epsilon(\Sigma_1 \otimes \Sigma_2, s) = \epsilon(\beta(\Sigma_1) \otimes \beta(\Sigma_2), s) .$$

3. The finite case.

Our aim now is to establish a "Langlands like" bijection for I-equivalence classes of tame representations. There are three essential components in play here:

I. We let $\hat{GL}_n(k)$ denote the isomorphism classes of irreducible $\mathbb{C}$-representations of $GL_n(k)$.

II. We denote the I-equivalence classes of tame Deligne representations of W_F, of dimension n, by $\Delta(n,F)$.

III. Let e denote a finite extension of k. We let $\hat{e}^*$ denote the group of abelian $\mathbb{C}$-characters of e^*, and we put $\hat{I} = \varinjlim_{e/k} \hat{e}^*$, where the direct limit is taken over all finite extensions e/k (within a given separable closure), with respect to the co-norm. For any $\chi \in \hat{I}$ there is a unique minimal field through which χ factors: we call this field the level of χ.

The Frobenius automorphism σ over k acts on $\hat{I}$ by the rule

$$\chi^{\sigma}(x) = \chi(x^{\sigma}) = \chi^{q}(x) .$$

We let $\underline{P}$ denote the set of all finite partitions, and we let $L(k) = \mathrm{Map}_{\sigma}(\hat{I}, \underline{P})$ denote the set of maps $f : \hat{I} \to \underline{P}$, with $f(\chi^{\sigma}) = f(\chi)$ for all $\chi \in \hat{I}$, and which are zero on almost all χ. We then define the degree map $\deg : L(k) \to \mathbb{Z}$ by $\deg(\ell) = \sum_{\chi \in \hat{I}} |\ell(\chi)|$ for $\ell \in L(k)$, where $|\ell(\chi)|$ denotes the sum of parts. In the sequel we set $L_d(k) = \deg^{-1}(d)$.

The main goal of this section is to describe bijections

$$\begin{array}{ccc} \hat{GL}_n(k) & \xleftarrow{\ \beta\ } & \Delta(n,F) \\ & {\scriptstyle\alpha}\searrow \quad \swarrow{\scriptstyle\gamma} & \\ & L_n(k) & \end{array}$$

which respect Gauss sums, contragredients, and where β takes irreducible representations to cuspidal representations.

The bijection α. Following Green, Macdonald has associated to an σ-orbit of $\hat{I}$ of t elements, $\omega = (\chi, \chi^{\sigma}, \ldots, \chi^{\sigma^{t-1}})$, a cuspidal irreducible representation $g_\omega \in GL_t(k)$.

Let $\omega_1, \ldots, \omega_m$ be m such orbits with $|\omega_i| = r_i$, $\Sigma r_i = r$. We thereby obtain a representation on $M = GL_{r_1} \times \ldots \times GL_{r_m} \to GL_r$; hence we get a representation η on the parabolic subgroup P of GL_r corresponding to M. We then induce η up to GL_r and so obtain a representation which we call $\omega_1 \circ \ldots \circ \omega_m$.

Case $\omega = \omega_i$, for all i. Then $\mathrm{Ind}_P^{GL_r} \eta$ breaks up into irreducible characters according as the conjugacy classes of S_r (the symmetric group on r letters); needless to say, $\hat{S}_r$ is naturally parametrised by $\underline{P}_r$, the partitions of r. We may therefore write the irreducible components of $\mathrm{Ind}_P^{GL_r} \eta$ as $\{\omega^\lambda\}$ for $\lambda \in \underline{P}_r$.

More generally, if $\omega_1, \ldots, \omega_s$ are distinct F-orbits, then $\omega_1^{\lambda_1} \circ \ldots \circ \omega_s^{\lambda_s} \in \hat{GL}_n$, where $n = \Sigma|\omega_i||\lambda_i|$. Furthermore every element of $\hat{GL}_n(k)$ arises in this way. The bijection α is now described by associating to $\omega_1^{\lambda_1} \circ \ldots \circ \omega_s^{\lambda_s}$ the map $\ell \in L_n(k)$ given by

$$\ell(\omega) = \begin{cases} \lambda_i & \text{if } \omega = \omega_i \text{ some } i \\ 0 & \text{otherwise} . \end{cases}$$

We next define the Kondo Gauss Sum for an irreducible representation $\alpha^{-1}(\ell) = \omega_1^{\lambda_1} \circ \ldots \circ \omega_s^{\lambda_s}$ of dimension d say, by

$$1_d \, G(\alpha^{-1}(\ell)) = \sum_{g \in GL_n(k)} \Psi(g) \, \alpha^{-1}(\ell)(g) .$$

Here Ψ is the canonical additive character

$$\begin{array}{ccccc} GL_n(k) & \xrightarrow{\text{trace}} & k & \xrightarrow{\text{trace}} & F_p \\ & & & \searrow^{\Psi_K} & \downarrow \; e^{\frac{2\Pi i -}{p}} \\ & & & & C^* \end{array} .$$

Definition. **(1)** Let e denote a finite extension of k, and let $\chi \in \hat{e}^*$. We then define the standard finite field Gauss sum $g(e, \chi)$ by

$$g(e, \chi) = - \sum_{x \in e^*} \chi(x)\, \Psi_e(x) .$$

(2) For $\ell \in L(k)$ we set t

$$G(k, \ell) = \prod_{y \in \hat{I}/\sigma} g(k_y, y)^{|\ell(y)|} .$$

Here k_y denotes the level of y, and the product extends over the σ-orbits of $\hat{I}$. We note that almost all elements in the product are 1.

Reformulating the work of Kondo in [66], Macdonald has shown that the above two Gauss sums are related by:

(3) $$G(\alpha^{-1}(\ell)) = G(k,\ell)\,(-1)^n\, q^{\frac{n(n-1)}{2}} .$$

The bijection γ. Let Π denote a tame, irreducible complex representation of G_F. By standard Clifford theory we know that $\Pi_{|I}$ is a single σ-orbit of characters in $\hat{I}$. Now let (Π, N) denote a Deligne representation of dimension n. By the work in §1 we may write

$$(\Pi, N) = \sum_i (\Pi_i \otimes \mathrm{Sp}(n_i)) .$$

We then define $\gamma(\Pi, N) \in L_n(k)$ by

$$\gamma(\Pi, N)(\chi) = \sum n_i\, (\chi, \Pi_i|_I)$$

where $\chi \in \hat{I}$ and $(\, , \,)$ is the standard inner product of representation theory.

The bijection β is now obtained by putting $\beta = \alpha^{-1} o \gamma$.

We conclude by considering the behaviour of the bijection γ with respect to Gauss sums. We recall that τ^* is the modified Gauss sum of 2. By using the induction formula for local Galois Gauss sums, it is easily shown that for $\ell \in L_n(k)$

(4) $$\tau^*(\gamma^{-1}(\ell)) = G(k,\ell)\,(-1)^{\deg(\gamma^{-1}(\ell)) + |\ell(\epsilon)|} .$$

Comparing (3) and (4) then shows that, up to a predictable factor, β does indeed preserve Gauss sums.

References

[25] P. Deligne: Formes modulaires et réprésentations de GL(2) , in: Modular Functions of one variable II (ed. P. Deligne, W. Kutzko), Lecture Notes in Mathematics No. 349, Springer Verlag, Berlin-Heidelberg-New York 1973, 55 - 105.

[42] G. Henniart: Les conjectures de Langlands locales pour GL(n) ; Seminar of Theory of Numbers, Paris 1982-83, Editor M.J. Bertin and C. Goldstein, Birkhäuser, Boston-Basel-Stuttgart, 1985.

[66] T. Kondo: On Gaussian sums attached to general linear groups over finite fields; J. Math. Soc. Japan, 15 (1963), 244 - 255.

[80] I.G. Macdonald: Zeta functions attached to finite general linear groups; Math. Annalen 249 (1980), 1 - 15.

[106] J.T. Tate: Local constants, in: Algebraic Number Fields (A. Fröhlich ed.), Academic Press, London 1977, 89 - 132.

[108] M.J. Taylor: On Fröhlich's conjecture for rings of integers of tame extensions; Invent. Math. 63 (1981), 41 - 79.

[F] A. Fröhlich, see this volume.

[Q] J. Queyrut, see this volume.

Contemporary Mathematics
Volume 86, 1989

ON THE EXCEPTIONAL REPRESENTATIONS OF GL_N

by Phil Kutzko, Iowa

1. The purpose of this lecture is to give an exposition of a rather technical result [69] concerning the representation theory of $GL_2(F)$, F a p-adic field, with an eye towards placing it in the context of more recent developments. To this end, I will fix a p-adic field F, a separable quadratic extension E/F, and a regular character θ of $E^\times$; that is a character θ with the property that $\theta \neq \theta^\tau$ where τ is the non-trivial galois automorphism of E/F. Then, by local abelian class field theory, we may think of θ as a one-dimensional representation of the absolute Weil group, $W(E)$, of E and so we obtain, by induction, a two-dimensional representation, $\Sigma(\theta) = \mathrm{Ind}_{W(E)}^{W(F)}\theta$, of $W(F)$, $\Sigma(\theta)$ being irreducible since θ is regular.

Now, the Langlands conjecture predicts that $\Sigma(\theta)$ should, in a natural way, parametrize an irreducible, admissible supercuspidal representation, $\Pi(\theta)$, of $GL_2(F)$. Thus, it is reasonable to ask the following questions.

1.1 What properties does $\Pi(\theta)$ have? In particular, given $\Pi(\theta)$, how can we recover E and θ?

1.2 What is the range of the map $\theta \longmapsto \Pi(\theta)$ when θ ranges over all regular characters of $E^\times$?

1.3 Given $\Pi(\theta)$, we know [67] that there exists a maximal open compact-mod-center subgroup K of $G = GL_2(F)$ and a representation $\varkappa$ of K such that $\Pi(\theta) \simeq \mathrm{Ind}_K^G \varkappa$. How do the data $(K,\varkappa)$ relate to the data (E,θ)?

In what follows, we will answer these questions for $GL_2(F)$ and discuss the significance of these answers to the representation theory of $GL_N(F)$, $N \geq 3$. We begin with an explicit description of $\Pi(\theta)$.

2. Weil representations

That $\Pi(\theta)$ can be explicity constructed in case $G = GL_2(F)$ is due to the happy coincidence that $SL_2(F)$ and $Sp_2(F)$ (or $Sp_1(F)$ depending on one's notation!) are the same group. As a consequence of this, the representation theory of the dual reductive pair (Sp_1, O_2) may be applied and, with a little work, one obtains a representation Π_E of G on the space $C_c^\infty(F^\times \times E)$ of locally constant, compactly supported functions on $F^\times \times E$ given explicitly by the following formulae: (Here, $\omega_{E/F}$ is the non-trivial character of $F^\times$ which factors through $F^\times/N_{E/F}E^\times$, ψ is a fixed non-trivial character of F, $\gamma_{E/F}$ is a complex number whose value may be found in Lemma 1.2 of [55] and $\hat{f}$ is the Fourier transform of f in the second variable with respect to $\psi \circ Tr_{E/F}$.)

$$\Pi_E\left(\begin{bmatrix} x & 0 \\ 0 & x^{-1}\end{bmatrix}\right) f(z,\beta) = \omega_{E/F}(x)|x|_E^{1/2} f(z, x\beta) \tag{2.1.1}$$

$$\Pi_E\left(\begin{bmatrix} 1 & y \\ 0 & 1\end{bmatrix}\right) f(z,\beta) = \psi(yzN_{E/F}\beta) f(z,\beta) \tag{2.1.2}$$

$$\Pi_E\left(\begin{bmatrix} 0 & 1 \\ -1 & 0\end{bmatrix}\right) f(z,\beta) = \gamma_{E/F}\omega_{E/F}(z)|z|_E^{1/2} \cdot \hat{f}(z, z\beta^\tau) \tag{2.1.3}$$

$$\Pi_E\left(\begin{bmatrix} w & 0 \\ 0 & 1\end{bmatrix}\right) f(z,\beta) = f(zw,\beta) \tag{2.1.4}$$

Now, $E^\times$ acts on $C_c^\infty(F^\times \times E)$ via the action $(\alpha \circ f)(z,\beta) = f(zN_{E/F}\alpha, \beta\alpha^{-1})$ and this action commutes with the action of G via Π_E. It follows that the subspace C_θ of functions f in $C_c^\infty(F^\times \times E)$ satisfying $\alpha \circ f = \theta(\alpha)|\alpha|^{1/2} f$ is G-stable. In fact, more is true.

2.2 Proposition. *Let θ be regular. Then the restriction, $\Pi(\theta)$, of Π_E to C_θ is an irreducible supercuspidal representation of G which is, up to equivalence, independent of ψ. Further, $\Pi(\theta_1) \simeq \Pi(\theta_2)$ if and only if either $\theta_2 = \theta_1$ or $\theta_2 = \theta_1^\tau$.*

2.3 Remarks. 1. For proofs of the above statements, see e. g., ([69] , §1). That $\Pi(\theta)$ corresponds, in the sense of Langlands, to $\Sigma(\theta)$ is proved in [55].

2. The answers to questions 1.1 and 1.2 may be obtained by direct (though arduous) calculation from the explicit form of $\Pi(\theta)$ given above. These calculations appear at the end of Chapter 7 of [75] and yield the following information.

2.3.2.1 Let E/F be separable quadratic and let $G_{E/F}$ be the subgroup of G consisting of elements g such that $\det g$ lies in $N_{E/F}E^\times$. Then an irreducible representation Π of G is equivalent to a representation $\Pi(\theta)$, θ a character of $E^\times$, if and only if the restriction of Π to $G_{E/F}$ is reducible.

2.3.2.2 Write $\Pi(\theta) \simeq_{G_{E/F}} \Pi^1(\theta) \oplus \Pi^2(\theta)$, let $\chi^i_{\Pi(\theta)}$, $i = 1,2$ be the character (see [40] for definitions) of $\Pi^i(\theta)$ and set $\chi_\theta = \chi^1_{\Pi(\theta)} - \chi^2_{\Pi(\theta)}$. Let x be a regular element of $G_{E/F}$; that is, an element with distinct eigenvalues. Then $\chi_\theta(x) = 0$ unless the eigenvalues, β_1 and β_2, of x lie in E. If β_1 and β_2 lie in E, then $\chi_\theta(x) = \pm(\theta(\beta_1) + \theta(\beta_2))/\delta$ where $\delta^2 = |\beta_1-\beta_2|^2/|\beta_1\beta_2|$.

3. With respect to Question 1.2, it is important to note that a given representation Π of G may have the form $\Pi(\theta)$ for characters θ of (the multiplicative groups of) different extensions E/F. Indeed, we have that $\Pi(\theta_1) \simeq \Pi(\theta_2)$ for characters θ_i of $E_i^\times$, $i = 1,2$ if and only if $\mathrm{Ind}_{W(E_1)}^{W(F)}\theta_1 \simeq \mathrm{Ind}_{W(E_2)}^{W(F)}\theta_2$. Having said this, one should note further that if Π has the form $\Pi(\theta)$, then [69] it is always possible to pick (θ,E) such that $\Pi \simeq \Pi(\theta)$ with θ a character of $E^\times$ and $f(\theta) \geq 2d(E/F) - 1$ where $f(\theta)$ is the exponent of conductor of θ and $d(E/F)$ is the exponent of the different of E/F. Also, E/F is unique with this property unless $f(\theta) = 2d(E/F) - 1$. We will need these facts below (see §3).

3. Construction of supercuspidal representations by induction

We now consider Question 1.3 and, to this end, we must describe the construction of supercuspidal representations by induction. Since the representations we will be constructing are as easily described for the groups $GL_N(F)$ as they are in the case $N = 2$, we will consider the more general case $G = GL_N(F)$.

Let A be the ring of $N \times N$ matrices over F (so that $G = A^\times$) and let $\mathcal{A}$ be a principal order in A; that is, $\mathcal{A}$ is a subring of A which, as a module over the ring of integers, $\mathcal{O}_F$, of F is free of rank N^2 and whose radical, $\mathcal{P}_{\mathcal{A}}$, is principal as a left (hence as a right) ideal. Then the normalizer, $K(\mathcal{A})$, in G of the group of units, $U(\mathcal{A})$, of $\mathcal{A}$ is a maximal compact-mod-center subgroup of G and any such subgroup has this form for some principal order $\mathcal{A}$ (see [12] for details).

Let $U^n(A) = 1 + [P(A)]^n$, $n \geq 1$, and let ψ be a character of F with conductor P_F . If b is an element of $[P(A)]^{1-n}$, then we may define a character, $\psi_b = \psi_{b,r,n}$ of $U^r(A)$, $\frac{n}{2} \leq r \leq n$, by the formula $\psi_b(x) = \psi(\text{tr } b(x-1))$, x in $U^r(A)$. We obtain

3.1 **Lemma.** The map $b \longmapsto \psi_b$ induces an isomorphism of $[P(A)]^{1-n}/[P(A)]^{1-r}$ with the topological dual, $(U^r(A)/U^n(A))^\wedge$, of $U^r(A)/U^n(A)$.

I will now single out certain elements, b , of $[P(A)]^{1-n}$ which are relevant to the study of supercuspidal representations.

3.2 **Lemma.** ([74]). Let E/F be a finite separable extension of fields, let O_E be the ring of integers in E , and pick α in E so that $E = F[\alpha]$. Then the following are equivalent.

i. $(v_E(\alpha), e(E/F)) = 1$ and $O_{E_{nr}} = O_F[N_{E/E_{nr}}\alpha]$ where, here, $v_E(\alpha)$ is the value of α in E , e(E/F) is the ramification degree of E/F and E_{nr}/F is the unramified part of E/F .

ii. Let f be the irreducible polynomial of α over F . Then $v_E(f'(\alpha)/\alpha^{[E:F]-1}) = d(E/F) - e(E/F) + 1$.

3.3 **Definition.** Call α as above E/F-minimal if it satisfies the conditions of 3.2.

3.4 **Definition.** Let A be a principal order in A . Call an element α of A A-alfalfa of degree R if

i. $F[\alpha]$ is a subfield of A , $[F[\alpha]:F] = R$ and α is $F[\alpha]/F$-minimal.

ii. Let A_α be the centralizer of α in A . Then $A_\alpha = A_\alpha \cap A$ is a principal $F[\alpha]$-order in A_α .

3.5. **Remarks:** 1. Clearly the degree of a A-alfalfa element must divide N . A-alfalfa elements of degree 1 are just scalars in A . On the other hand, A-alfalfa elements of degree N are precisely the S-cuspidal elements of Carayol [14]

2. Let e(A/F) be the ramification index of A/F ; that is, $[P(A)]^{e(A/F)} = P_F A$. Then if α is A-alfalfa, we have $[P(A_\alpha)]^r = [P(A)]^r \cap A_\alpha$, $r = 1,2,\ldots$; in particular, $e(A/F) = e(A_\alpha/F[\alpha])\cdot e(F[\alpha]/F)$.

3. Let the A-value, v_A, be defined in the usual way. Then $v_A(\alpha) = v_{A_\alpha}(\alpha) = e(A_\alpha/F[\alpha])\cdot v_{F[\alpha]}(\alpha)$. We conclude from 3.2.i that if α is A-alfalfa, then $(v_A(\alpha), e(A/F)) = e(A_\alpha/F[\alpha])$.

I may now state the main result of this section - a result which is a consequence of recent work of Bushnell [9]

3.6 Proposition. , : Let Π be an irreducible supercuspidal representation of G. Then there is a principal order A of A such that **either**

i. $e(A/F) = 1$ and $\Pi|U(A)$ contains a representation Σ with $U^1(A) < \ker \Sigma$ such that Σ, viewed as a representation of $GL_N(\mathcal{O}_F/P_F)$ is cuspidal **or**

ii. there is an integer $n \geq 2$ and an A-alfalfa element α with $v_A(\alpha) = 1 - n$ such that $\Pi|_{U^{n-1}(A)}$ contains ψ_α.

We now consider the situation where we have an irreducible supercuspidal representation Π of G which contains a $U^{n-1}(A)$ subrepresentation ψ_α for which $E = F[\alpha]$ has degree N over F. Then it is not hard to see that Π in fact contains a $U^{[(n+1)/2]}(A)$ subrepresentation ψ_α, α as above. Now the normalizer in $K(A)$ of ψ_α is just $E^\times\cdot U^{[n/2]}(A)$. If n is even, then it follows easily from Clifford theory that Π contains a representation of the form $\theta\psi_\alpha$ on $E^\times U^{n/2}(A)$ where θ is a character of $E^\times$ which extends the restriction of ψ_α to $U^{n/2}(A) \cap E^\times = U_E^{n/2}$ and also that, in fact, $\Pi \simeq \operatorname{Ind} \theta\psi_\alpha$. (If n is odd one may still associate to the data (θ,α) a representation, Φ, of $E^\times U^{(n-1)/2}(A)$ such that $\Pi = \operatorname{Ind} \Phi$; we will not need this fact here, however).

We see, therefore, that, at least for a large number of supercuspidal representations Π of G, there is a natural way to associate to Π an extension E/F and a character θ of $E^\times$. Question 1.3 may now be restated as

3.7 Question. Suppose that $\bar{E}/F$ is a quadratic separable extension and that $\bar{\theta}$ is a character of $\bar{E}^\times$ as in § 1. Suppose also that $\Pi(\bar{\theta}) \simeq \operatorname{Ind} \theta\psi_\alpha$ as above. How can we determine the data (θ,α) from the data $(\bar{\theta},\bar{E})$ and vice versa?

In answering this question, we begin with the case that $\bar{E}/F$ is unramified. Here, we have the following result due to Gerardin [34]

3.8 Proposition. Suppose that $\Pi = \mathrm{Ind}\,\theta\psi_\alpha$ is a supercuspidal representation of $GL_N(F)$ as above and that $E = F[\alpha]$ is unramified over F. Then Π corresponds, in the sense of Langlands, to $\mathrm{Ind}_{E/F}\theta\tilde{\omega}_{E/F}$ where $\tilde{\omega}_{E/F}$ is a character of $E^\times$ which is independent of θ.

We can consider the case that $\bar{E}/F$ is quadratic ramified and we make that following assumptions

1. We assume that $f(\bar{\theta}) \geq 2d(\bar{E}/F) - 1$ (see remark 2.3.3 above).
2. We assume that $f(\bar{\theta}) = \inf f(\bar{\theta}\cdot\chi \circ N_{\bar{E}/F})$ where χ runs through $\hat{F}^\times$; this is equivalent to the assumption that $f(\bar{\theta}) - d(\bar{E}/F)$ is odd.
3. We assume that, if $p = 2$, then ψ has the additional property that $\psi(x+x^2) = 1$ for all x in $\mathcal{O}_F$.

As a consequence of our first two assumptions we have that there is an element $\bar{\alpha}$ in $\bar{E}$ with $v_{\bar{E}}(\bar{\alpha}) = 2 - f(\bar{\theta}) - d(\bar{E}/F)$ such that $\theta(x) = \psi(\mathrm{Tr}_{\bar{E}/F}\bar{\alpha}(x-1))$ for all x in $U_{\bar{E}}^{[(d(\bar{E}/F)+1)/2]}$. Also, by the definition of $d(\bar{E}/F)$, there is an element $c(\bar{E}/F)$ in F with the property that $\omega_{\bar{E}/F}(x) = \psi(c(\bar{E}/F)(x-1))$ for all x in $U_F^{[(d(\bar{E}/F)+1/2]}$ and, clearly, we may (and will) take $c(\bar{E}/F) = 0$ if $\bar{E}/F$ is tamely ramified. Having said all this we may now state our main result.

3.9 Proposition. With the above assumptions, $\Pi(\bar{\theta}) \simeq \mathrm{Ind}\,\theta\psi_\alpha$ where α has the properties

1. $\mathrm{tr}\,\alpha = \mathrm{Tr}_{\bar{E}/F}\bar{\alpha} + c(\bar{E}/F)$
2. $\det\alpha = N_{\bar{E}/F}\bar{\alpha}$.

3.10 Remarks.

3.10.1 In the proof ([69] ; Proposition 2.3) of the above Proposition, α is specified as is the principal order implicit in the definition of ψ_α; indeed, a vector in $C_{\bar{\theta}}$ is produced which transformes under ψ_α. One can easily give an explicit formula for θ as well; it is not clear, however, what information it reveals.

3.10.2 Set $E = F[\alpha]$. Then if $d(\bar{E}/F)$ is small compared to $f(\bar{\theta})$, an application of Hensel's Lemma shows that $E \simeq_F \bar{E}$. However, as $f(\bar{\theta})$ approaches $2d(\bar{E}/F) - 1$, this need not be the case. In particular, if $f(\bar{\theta}) = 2d(\bar{E}/F) - 1$, it is possible that $d(E/F) \neq d(\bar{E}/F)$.

In light of the above remarks, it is clear that the question of whether a supercuspidal representation of G has the form $\Pi(\bar{\theta})$ for some pair $(\bar{\theta},\bar{E})$ must have a subtle answer. Call a supercuspidal representation which does not have this form exceptional. Then one has

3.11 **Proposition.** Let $\Pi = \mathrm{Ind}\,\theta\psi_\alpha$ be an irreducible representation of G. Then

1. If $2(2n+1) > 3d(E/F)$ then Π is not exceptional. (Recall that $E = F[\alpha]$ and that $n = 1 - v_E(\alpha)$.)
2. If $2(2n+1) \leq 3d(E/F)$ then Π is **not** exceptional if and only if the polynomial $X^3 - (\mathrm{tr}\,\alpha)X^2 + \det\alpha$ has a root over F. If $\bar{s}$ is such a root, then $\Pi = \Pi(\bar{\theta})$ where the extension $\bar{E}/F$ may be taken to be a splitting field over F for the polynomial $X^2 - \bar{s}X + \det\alpha$.

4. Conclusions.

We have now answered questions 1.1 - 1.3 for GL_2. All of these answers, however, hinged on the explicit construction of $\Pi(\theta)$ which was given in §2. Since this construction is not available for GL_N, $N > 2$, it is of interest to see what can be said in its absence.

To this end, we consider a cyclic extension E/F of degree N and a regular character, θ, of $E^\times$. Then Kazhdan [57] showed that there is an irreducible supercuspidal representation $\Pi(\theta)$ of $G = GL_N(F)$ with the following properties:

4.1 The restriction of $\Pi(\theta)$ to the subgroup $G_{E/F}$ (see 2.3.2.1 above) decomposes into a direct sum of N distinct irreducible $G_{E/F}$-subrepresentations.

4.2 Write $\Pi(\theta) \simeq_{G_{E/F}} \bigoplus_{j=1}^{N} \Pi_j(\theta)$ and set $\chi_\theta = \sum_{j=1}^{N} \xi^j \chi_{\Pi_j(\theta)}$ where ξ is a primitive N^{th}-root of unity. Then χ_θ is 0 at all regular elements except for those elements whose eigenvalues lie in E; if x is a regular element having an eigenvalue β in E then

$$\chi_\theta(x) = (c/\delta) \sum_\tau \theta(\beta^\tau)$$

where τ runs through $\mathrm{Gal}(E/F)$, δ is analogous to the function δ defined in 2.3.2.2 and where c is an Nth-root of unity.

Further, Henniart [44] showed that

4.3 $\Pi(\theta)$ corresponds in the sense of Langlands to $\mathrm{Ind}_{W(E)}^{W(F)}\theta$.

We see from the above result, in particular, from 4.1, that we have at least the possibility of using Mackey theory to generalize Proposition 3.11 above to GL_N . In particular, one should start with $\Pi = \mathrm{Ind}_{E^\times U^{(n-1)/2}(A)}^{G}\Phi$ (see §3 above) and ask about the behavior of Π when restricted to $SL_N(F)$. In case $N=2$, I have verified that Proposition 3.11 can be proved in this way, thus avoiding the Weil construction. In case N is prime, one would expect to obtain a polynomial as in 3.11.2 whose Galois group is either cyclic of degree $N + 1$ or dihedral of degree $2(N+1)$. In any case, it would seem to be a program well worth pursuing!

References

[9] C.J. Bushnell: Hereditary orders, Gauss sums and supercuspidal representations of GL_N ; J. für Reine und Angew. Math. .

[12] C.J. Bushnell, A. Fröhlich: Non-abelian congruence Gauss sums and p-adic simple algebras; Proc. London Math. Soc. (3), 50 (1985), 207 - 264.

[14] H. Carayol: Representations cuspidales du groupe lineaires; Ann. Sci. ENS 17 (1984), 191 - 225.

[34] P. Gérardin: Cuspidal unramified series for central simple algebras; Proc. Symp. Pure Math. 33 Amer. Math. Soc., Providence, R. I., 1979.

[40] Harish-Chandra: Admissible distributions on reductive p-adic groups, in: Lie Theories and Their Applications, Queens Papers (Queen's University, 1978), 281 - 347.

[44] G. Henniart: On the local Langlands conjecture for GL(n): the cyclic case; Ann. Math. 123 (1986), 143 - 203.

[55] H. Jacquet, R.P. Langlands: Automorphic Forms on GL(2) : Lecture Notes in Mathematics no. 114, Springer Verlag, Berlin-Heidelberg-New York, 1970.

[57] D. Kazhdan: On lifting, in: Lie Groups Representations II, Lecture Notes in Math. 1041 (Springer, 1983).

[67] P. Kutzko: On the supercuspidal representations of GL_2 ; Amer. J. Math. 100 (1978), 43 - 60.

[69] P. Kutzko: The exceptional representations of GL_2; Composio Math. 51 (1984), 3 - 14.

[74] P. Kutzko, D. Manderscheid: On intertwining operators for $GL_N(F)$, F a non-archimedean local field; pre-print.

[75] R.P. Langlands: Base Change for GL(2) ; Ann. of Math. Studies 96, Princeton University Press, 1980.

Contemporary Mathematics
Volume **86**, 1989

CHARACTERS OF REPRESENTATIONS OF D^*_n (TAMELY RAMIFIED CASE)

by L. Corwin, Rutgers University

As usual, let F be a local field of residual characteristic p, and let D_n be a division algebra whose center is F and whose index is n (thus $[D_n:F] = n^2$), Π an irreducible unitary representation of $D_n^\times$. The problem I will be concerned with here is:

Compute χ_Π, the character of Π.

This problem arises naturally in questions of harmonic analysis. One reason is the following: as is discussed in [83], there is a 1 - 1 correspondence between $(D_n^\times)^\wedge$, the unitary dual of D_n, and $GL(n,F)^\wedge_d$, the set of discrete series representations of $GL(n,F)$ under which characters match up to a possible factor of ± 1. (This means that if $x' \in D_n^\times$ and $x \in GL(n,F)$ satisfy the same minimal equation, and if that equation is of degree n, then the correspondence $\Pi' \longleftrightarrow \Pi$ gives $\chi_{\Pi'}(x') = \pm\chi_\Pi(x)$.) However, we do not have any explicit description of the pairing. In the tame case, where $(p,n) = 1$, we have parametrizations of both $(D_n^\times)^\wedge$ and $GL(n,F)^\wedge_d$ by the same set, and it is interesting to compare the correspondence given by the matching theorem with the one given by the parametrization.

Since character computations are easier on $D_n^\times$ than on $GL(n,F)$, it seems like a good idea to begin by computing characters on the division algebra side. For one thing, we can sometimes show that a character for $D_n^\times$ is determined by its values on a relatively small number of conjugacy classes; this reduces the work needed for $GL(n,F)$. For another, there are $\varphi(n)$ non-isomorphic D_n for each n (where φ is Euler's phi function), and we should also compare the characters of corresponding representations for different D_n.

In the tame case, $(p,n) = 1$, the paper [20] gives an inductive procedure for computing characters inductively. That paper had two defects, however, the inductive procedure wasn't carried out explicitly, and people found the arguments hard to follow. The results I give here are from joint work 22 with Paul Sally and Allen Moy, as well as with a forthcoming paper with Sally; we are able to remedy the first defect, at any rate, and we are giving a different version of the

* Supported by NSF Grant MCS 84-02104.

proofs in the hope of dealing with the second as well.

So assume from now on that we are in the tame case, and write D for D_n. From earlier lectures, we know that the elements of $(D^\times)^\wedge$ correspond to admissible characters ψ for extensions E of F with $[E:F]|n$. So let $\Pi \in (D_n^\times)^\wedge$ correspond to $(E/F,\psi)$. Then ψ has a factorization (the Howe factorization, also described in earlier lectures):

$$\psi = \psi_0 \cdot \psi_1 \cdot \ldots \cdot \psi_t \ , \quad \psi_j = \psi'_j \circ \mathrm{Tr}_{E/E_j} \ ,$$

corresponding to a tower of fields

$$F = E_0 \underset{\neq}{\subset} E_1 \underset{\neq}{\subset} \ldots\ldots \underset{\neq}{\subset} E_t = E \ .$$

I'll assume for simplicity that $\psi_0 \equiv 1$. (In general, ψ_0 twists Π by the 1-dimensional representation $\psi_0 \circ N_{D/F}$ of D; thus this assumption is harmless.)

We shall need a lot of information about the structure of D; this, too, can be found in earlier talks. Let P be the maximal ideal in the integers of D. Then one can write

$$D^* = (1 + P)\cdot R_n \cdot < \pi > \ ,$$

where R_n is a cyclic group of $(q^n - 1)^{th}$ roots of unity and π is a prime element of D with $\pi^n \in F$. In fact, one can even require that π normalize R_n. (The group $C_D = R_n \cdot < \pi >$ is a "complementary subgroup" in the terminology of Bushnell-Fröhlich [11].) There is an embedding of E in D such that

$$E^\times = (E \cap (1 + P))(E \cap C_D) \ ;$$

we shall assume that E is so embedded in D. By the Skolem-Noether Theorem, all embeddings of E are conjugate.

There is a natural valuation v on D, with $v(\pi) = 1$; give E (and the F_j) the valuation obtained by restricting v. (On F, for instance, this is n times the standard valuation.) Now define m_j, $1 \leq j \leq t$, by saying that ψ_j is non-trivial on $E_j \cap (1 + P^{m_j})$ and trivial on $E_j \cap (1 + P^{m_j+1})$; if ψ_j is trivial on $1 + P$, then $m_j = 0$. Note that $m_1 > m_2 > \ldots\ldots > m_t$. (If $m_j = 0$, then $j = t$ and E_t/E_{t-1} is unramified, by the definition of admissibility.) Let D_j be the division algebra of elements commuting with E_j, $0 \leq j \leq t$. Note that ψ_j defines a character $\psi_j \circ N_{D_j/E_j}$ on D_j. Recall the construction of Π from ψ: let $m'_j = m''_j = \frac{m_j + 1}{2}$ if m_j is odd, and let $m'_j = m''_j - 1 = \frac{m_j}{2}$ if m_j is even and >0. One extends $\psi_j \circ N_{D_j/E_j}$ to a representation Σ_j of $H_j = (i + p^{m_j+1}) D_j^{\times}(D_{j-1} \cap (1 + p^{m'_j}))$ trivial on $(1 + p^{m_j+1})$ such that Σ_j is a multiple of a character $\psi_j^{\#}$ on

$H_j^{\#} = (1+p^{m_j+1}) D_j^{\times}(D_{j-1} \cap (1 + p^{m''_j}))$. On $M = \bigcap_{j=1}^{t} H_j$, the Σ_j can be tensored together to give a representation Σ. Then $\mathrm{Ind}_H^{D^\times} \Sigma = \Pi$. It will be useful to denote $\mathrm{Ind}_H^{D_t^\times(1+P)}$ by Γ, and to let $\psi^{\#}$ be the product of the $\psi_j^{\#}$ on the intersection of the $H_j^{\#}$.

The first result is useful for describing the character formula and is indispensable for proofs. Recall that every element $\chi \in D^\times$ has a unique expression in the form

$$\chi = \sum_{j=j_0}^{\infty} \alpha_j \pi^j , \quad \alpha_j \in R_n \cup \{0\} , \quad \alpha_{j_0} \in R_n , \quad j_0 = v(\chi) .$$

Say that χ is <u>normal</u> if the monomials $\alpha_j \pi^j$ all commute. Note that normality is preserved under conjugation by any $\beta\pi^i$, $\beta \in R_n$.

Proposition 1. *For any $y \in D^\times$, there is a unique normal $x \in D^\times$ such that x is conjugate to y by an element of $1 + P$.*

Thus it suffices to compute χ_Π on normal elements. We may arrange matters so that in the embedding of E in D, every element is normal. If $x = \beta_0 \pi^{j_0}(1 + \sum_{j=1}^{\infty} \beta_j \pi^j)$ is normal, write $x_n = \beta_0 \pi^{j_0}(1 + \sum_{j=1}^{\infty} \beta_j \pi^j)$, $n \geq 0$.

The following result gives a nearly complete description of χ_Π. In it, all notation is as above.

Theorem 1. *Let $x \in D^\times$ be normal. Then:*

a) *If $\chi_\Pi(x) \neq 0$, then some conjugate $x' = \beta'_0 \pi^{j_0}(1 + \sum_{j=1}^{\infty} \beta'_j \pi^j)$ of x under C_D satisfies*

(*) $\quad x'_0 \in D_t$; $x'_n \in D_j$ *whenever* $n < m_j \quad (1 \leq j \leq t)$.

(As remarked above, x' is normal.)

b) *Let $C_D(x)$ be the set of distinct conjugates of x under C_D that satisfy (*), modulo the normal subgroup $1 + P^{m_j+1}$. (Thus two conjugates in the same*

$(1 + P^{m_j+1})$-coset are not regarded as distinct.) Then

$$\chi_{\Pi}(x) = \sum_{x' \in C_D(\chi)} \phi^{\#}(x')A(x')B(x')G(x') ,$$

where $\phi^{\#}$ was defined earlier (necessarily $x' \in H_j^{\#}$ for all j) ;

$A(x')$ is prime to p ;

$B(x') = q^{\alpha(x')/2}$, where $\alpha(x')$ is an integer;

$G(x')$ is a 4th root of unity (at least for odd p), and is obtained from a product of Gauss sums.

Both $A(x')$ and $B(x')$ depend only on x (and of course, Π) ; $G(x')$ is not always constant in $C_D(x)$.

There are explicit formulas for $A(y)$ and $B(y)$, but they are fairly complicated. For instance,

$$A(y) = \prod_{j=1}^{t} \frac{f(i,s(i))}{f(i-1,s(i))} \cdot \frac{q^{\frac{n}{e(i,s(i))}} - 1}{q^{\frac{n}{e(i-1,s(i))}} - 1} ,$$

where

$$E(i,s(i)) = E_i[y_{m_i-1}] , \quad E(i-1,s(i)) = E_{i-1}[y_{m_i-1}] ,$$

and, e.g., $e(i,s(i))$, $f(i,s(i))$ are respectively the ramification index and residue class degree of $E(i,s(i))$ over F . The formula for $B(y)$ is worse and requires much more notation; rather then given it here, I shall (unfairly) refer the reader to [22] . The $G(y)$ seem to be related to root numbers, but we have not yet investigated this matter.

One case of the character formula is particularly important: $x = 1$. Then one gets a dimension formula (which can also be obtained directly):

Theorem 2. For $0 \leq j \leq n$, let $e(E_j)$, $f(E_j)$, $n(E_j)$ be respectively the ramification index, residue class degree, and degree of E_j/F .
Then

$$\text{Dim } \Pi = f(E_t) \cdot \frac{q^n - 1}{q^{n/e(E_t)} - 1} \cdot q^{\alpha(\pi)/2}$$

where

$$\alpha(\Pi) = \frac{n}{e(E_t)} - n + \sum_{j=1}^{t} m_j\left(\frac{n}{n(E_{j-1})} - \frac{n}{n(E_j)}\right) .$$

Corollary: Dim Π <u>determines</u> $e(E_t)$ <u>and</u> $f(E_t)$ <u>(hence</u> $n(E_t)$) .

The proof goes as follows: the factor in Dim Π prime to p is $f \cdot \frac{q^n - 1}{q^{n/e} - 1}$ (where $e = e(E_t)$, etc.). As $ef|n$, it suffices to prove that if $a, b, c, d,$ are positive integers with $a|b$, $c|d$, and if $\frac{a}{q^b - 1} = \frac{c}{q^d - 1}$, then $a = c$ and $b = d$. This follows from some elementary number theory.

One consequence of the corollary, of course, is that Dim Π distinguishes those Π coming from fields E with $[E:F] = n$ from those Π coming with $[E:F] < n$. As Paul Sally's talk explains, this lets one show which Π match with supercuspidal representations of $GL_n(F)$ and which Π match with other discrete series representations.

While on the subject of matching representations, I should mention one other result.

Theorem 3. <u>Suppose</u> p <u>is odd. Let</u> D, D' <u>be two non-isomorphic division algebras over</u> F <u>of index</u> n . <u>Let</u> $(E/F,\psi)$ <u>be admissible, and let</u> $\Pi \in (D^\times)^\wedge$, $\Pi' \in (D'^\times)^\wedge$ <u>correspond to</u> $(E/F,\psi)$. <u>Then</u> $\chi_\Pi = \chi_{\Pi'}$, <u>in the following sense: if</u> $y \in D^\times$, $y' \in (D')^\times$ <u>satisfy the same minimal equation, then</u> $\chi_\Pi(\chi) = \chi_{\Pi'}(\chi')$.

This theorem was proved in [20] under the added assumption that n was odd. It probably holds for $p = 2$ as well; we have not considered that case because the proof involves binary forms over finite fields, and these behave differently when $p = 2$.

Now return to Theorem 1. It gives a great deal of interesting qualitative information about characters, as one can see from the following examples:

1. Let $t = 1$ (write m for m_1) , and suppose that $[E:F] = n$. Then $D_t = E$, and Theorem 1, (a) says that $\chi_\Pi(x) \neq 0 \Rightarrow x$ is conjugate to an element of $E^\times(1 + P^{m_1})$. (In fact, (b) says that supp $\chi_\Pi = E^\times(1 + P^{m_2})$.) Think of what this means for maximal tori. The maximal tori of D are the subfields E' of degree n over F . Unless E' is (conjugate to) E , χ_Π is 0 on E' except near the identity and on elements of E' that are also in some conjugate of E . One consequence is that $(E')^\times \not\subset$ supp χ_Π unless E' is conjugate to E . Let $E'_E = \{y \in E' : y$ is conjugate to an element of $E\}$. Then

$\operatorname{supp} \chi_\Pi \cap E'^\times = (E'_E \setminus \{0\})(E' \cap (1 + P^m))$. In particular, χ_Π determines E .

2. Again let $t = 1$, but suppose that $[E:F] < n$. Again, we look at a maximal torus E' . If E' contains (an isomorphic copy of) E , then χ_Π is nonzero everywhere on E' ; if not, $E' \not\subseteq \operatorname{supp} \chi_\Pi$. In fact, if one lets $E'_E = \{y \in E' :$ a conjugate of y commutes with $E\}$, then

$$\operatorname{supp} \chi_\Pi \cap E'^\times = (E'_E \setminus \{0\})(E' \cap (1 + P^m)) .$$

Here, too, supp χ_Π determines E ; now, however, there are a number of conjugacy classes of maximal tori in supp χ_Π .

When $t > 1$, it is still true that $(E')^\times \leq \operatorname{supp} \chi_\Pi \Leftrightarrow E'$ contains (a conjugate of) E ; the exact description of supp χ_Π becomes more complicated. Notice that supp χ_Π determines E . These results also indicate how characters of discrete series representations for $GL_n(F)$ behave on the regular elliptic elements.

The proof of Theorem 1 involves 5 main steps:

a) One reduces the problem to that of computing the characters χ_Γ (recall that $\Gamma = \operatorname{Ind}_H^{D_t^\times(1+P)}$).

b) One proves Proposition 1, to reduce the problem to normal elements.

c) One proves that $\chi_\Gamma(y) = 0$ unless $y \in D_1^\times (1 + P^{m_1})$.

d) Let Π_1 be the representation of $D_1^\times$ corresponding to $(\psi, E/E_1)$. One then computes $\chi_\Gamma(y)$ in terms of $\chi_{\Pi_1}(y)$ for $y \in D_1^\times$. (From there, one easily gets $\chi_\Gamma(y)$ on all of $D_1^\times(1 + P^{m_1})$.

e) Finally, one uses induction on t to compute χ_Π in terms of ψ .

Step (a) is routine (it's essentially the Frobenius formula for induced characters), and (e) is primarily a matter of bookkeeping. I'll illustrate the other steps by giving proofs of simple cases of (b) and (c). The calculation in (d) is done by a somewhat more sophisticated version of the argument in (c), and the general proofs in (b) and (c) differ from those in the special cases primarily in having more complicated notation. Thus the special cases below give a reasonable notion of the ideas behind the proof.

For Proposition 1, I need a few simple algebraic facts about D . Let $\bar{D}$ be the residue class field of D , and let k be that of F . Then $[\bar{D}:k] = n$. The elements of $R_n \cup \{0\}$ give representatives for $\bar{D}$, and in what follows I will sometimes regard them as elements of $\bar{D}$. The map $\alpha \longmapsto \alpha^\sigma = \pi\alpha\pi^{-1}$ of $R_n \cup \{0\}$ to itself generates $\operatorname{Gal}(\bar{D}/k)$. (These facts are routine; see [94].

Now consider one special case of Proposition 1:

$$x = \pi + \alpha_2\pi^2 + \alpha_3\pi^3 + \dots ,$$

where the α_j are in $R_n \cup \{0\}$. Conjugate with $1 + \gamma\pi$, $\gamma \in R_n \cup \{0\}$;

$$x_\gamma = (1 + \gamma\pi)x(1 + \gamma\pi)^{-1} = \pi + (\alpha_2 + \gamma - \gamma^\sigma)\pi^2 + \dots .$$

Now work mod P^3. Then we can ignore the ... terms and consider $\alpha_2 + \gamma - \gamma^\sigma$ mod P; that is, we can work with $\alpha_2 + \gamma - \gamma^\sigma$ as elements of $\bar{D}$. As γ varies, $\gamma - \gamma^\sigma$ runs through all elements β of $\bar{D}$ with $\mathrm{Tr}_{\bar{D}/\bar{F}}\beta = 0$. That is, we can change α_2 to any α_2' with $\mathrm{Tr}_{\bar{D}/k}\alpha_2 = \mathrm{Tr}_{\bar{D}/k}\alpha_2'$. But $\mathrm{Tr}_{\bar{D}/k}$ is faithful on k, because $(n,p) = 1$. Thus we can change α_2 to some $\alpha_2' \in F$. But now π and $\alpha_2'\pi^2$ commute. That is, we can make the first two terms of χ commute. Now conjugate with $1 + \gamma\pi^2$, and so on; induction and a limit argument give the proof.

For (c), I'll again give a special case; $t = 1$ and $E = F(\pi)$. Write m for m_1, $m > 0$. Then $\Gamma(1 + P^{m+1}) = 1$, and on elements of $1 + P^m$, Γ is a scalar. Thus

$$\Gamma(1 + \alpha\pi^m) = \psi_0(\alpha)I , \quad \alpha \in R_m \cup \{0\} \longleftrightarrow \bar{D}$$

where ψ_0 is an additive character of $\bar{D}$. We can therefore write

$$\psi_o(\alpha) = \psi' \mathrm{o}\, \mathrm{Tr}_{\bar{D}/\bar{F}}(\alpha\beta) , \quad \beta \in \bar{D} ,$$

where ψ' is some fixed nontrivial character of k; β is uniquely determined by ψ_0. Because of our assumption on E, it is not too hard to show that $(m,n) = 1$ and $\beta \in F$, $\beta \neq 0$.

Now suppose that $\chi_\Pi(y) \neq 0$, where $y = 1 + \gamma_j\pi^j + \gamma_{j+1}\pi^{j+1} + \dots$ is normal, $1 \leq j < m$, and $\gamma_j \neq 0$. I'll show that $\gamma_j \in k$. This doesn't quite prove that $y \in E(1 + P^m)$, but it's a major step in that direction; if $j = 1$ (or, more generally, if $(j,n) = 1$), it would show that $y \in E$ because of the normality of y. The idea is this: let

$$y_\delta = (1 + \delta\pi^{m+j})y(1 + \delta\pi^{m-j})^{-1} .$$

Then, of course, $\chi_\Gamma(y_\delta) = \chi_\Gamma(y)$ because y_δ and y are conjugate. But a computation gives

$$y_\delta \equiv y(1 + (\delta\gamma_j^{\sigma^{m-j}} - \gamma_j\delta^{\sigma^j})\pi^m) \bmod 1 + P^{m+1} .$$

Hence

$$\Gamma(y_\delta) = \psi_0(\delta\gamma_j^{\sigma^{m-j}} - \gamma_j\delta^{\sigma^j})\Gamma(y) \ , \ \chi\ (y_\delta) = \psi_0(\delta\gamma_j^{\sigma^{m-j}} - \gamma_j\delta^{\sigma^j})\chi_\Gamma(y) \ .$$

Thus, since $\chi_\Gamma(y) \nequiv 0$, we must have

$$\psi_0(\delta\gamma_j^{\sigma^{m-j}} - \gamma_j\delta^{\sigma^j}) = 1 \ , \quad \text{all} \quad \delta \in \bar{D} \ ,$$

or

$$1 = \psi' \circ \mathrm{Tr}_{\bar{D}/k}\beta(\delta\gamma_j^{\sigma^{m-j}} - \gamma_j\delta^{\sigma^j}) = \psi' \circ \mathrm{Tr}_{\bar{D}/k}\delta^{\sigma^j}(\beta^{\sigma^j}\gamma_j^{\sigma^m} - \beta\gamma_j) \ , \quad \forall\, \delta \in \bar{D} \ .$$

(In the last step, we used the fact that $\mathrm{Tr}_{\bar{D}/k}\alpha^\sigma = \mathrm{Tr}_{\bar{D}/k}\alpha$.) Since ψ' is non-trivial and δ^{σ^j} is arbitrary, this gives

$$\beta^{\sigma^j}\gamma_j^{\sigma^m} - \beta\gamma_j = 0 \ .$$

As $\beta \in \bar{F}$, $\beta^{\sigma^j} = \beta$; as $\beta \nequiv 0$, we get

$$\gamma_j^{\sigma^m} - \gamma_j = 0 \ .$$

But σ^m generates $\mathrm{Gal}(\bar{D}/k)$ (because $(m,n) = 1$). Hence γ_j is fixed by $\mathrm{Gal}(\bar{D}/k)$, and this implies that $\gamma_j \in k$, as desired.

References

[11] C.J. Bushnell, A. Fröhlich: Gauss sums and p-adic division algebras, Lecture Notes in Mathematics 987, Springer Verlag, Berlin-Heidelberg-New York, 1983.

[20] L. Corwin, R. Howe: Computing characters of Tamely Ramified Division Algebras; Pac. J. Math. 73 (1977), 461 - 477.

[22] L. Corwin, A. Moy and P. Sally: Degrees and Formal Degrees for Division Algebras and GL_n over a p-adic Field, submitted.

[83] C. Moreno: Matching theorems for division algebras and GL_n ; Not published.

[94] I. Reiner: Maximal Orders; Academic Press, 1975.

Contemporary Mathematics
Volume **86**, 1989

MATCHING AND FORMAL DEGREES FOR DIVISION ALGEBRAS AND GL_n OVER A p-ADIC FIELD

by P.J. Sally, Jr., Chicago *

§ 1. **Introduction.** Let F be a p-adic field of characteristic zero, and let $G = GL_n(F)$. Throughout this note, we assume that $(n,p) = 1$ (the tame case). The discrete series of G consists of (equivalence classes of) irreducible unitary representations of G whose matrix coefficients are square integrable (mod Z), where Z is the center of G . The discrete series splits into two distinct classes:

(1) Supercuspidal representations: irreducible unitary representations whose matrix coefficients are compactly supported (mod Z);

(2) Generalized special representations: irreducible unitary representations whose matrix coefficients are square integrable (mod Z), and which are subrepresentations of representations induced from a proper parabolic subgroup of G .

The supercuspidal representations of G were constructed by Howe [49], and the generalized special representations of G were characterized by Bernstein-Zelevinsky ([4], [113]). We note that the Bernstein-Zelevinsky construction uses the supercuspidal representations of $GL_m(F)$ where $m|n$ $(m < n)$. Since $(m,p) = 1$ in this case, the requisite supercuspidal representations can be obtained from Howe's constructions.

The key to the study of the supercuspidal representations of G is the notion due to Howe [49], of an admissible character of an extension of degree n over F (see § 2). In fact, the supercuspidal represention of G are parameterized by (conjugacy classes of) admissible characters of extensions of degree n over F , and generalized special representations are parameterized by (conjugacy classes of) admissible characters of extensions of degree m over F where $m|n$ $(m < n)$.

Now, let D_n be a division algebra of dimension n^2 over F , and let $G' = D_n^{\times}$, the multiplicative group of D_n . The irreducible representations of G'

* Research supported in part by the National Science Foundation.

were constructed by Corwin [18] and Howe [46].
In these constructions, the irreducible representations of G' are obtained from (conjugacy classes of) admissible characters of extensions of degree m over F where $m|n$ (including $m = n$). Thus, pending a proof that all supercuspidal representations of G are included in Howe's construction, we see that admissible characters provide a matching of irreducible representations of G' and the discrete series of G. The proof that Howe's representations exhaust the supercuspidal representations of G was provided by Moy [89] using the abstract matching theorem.

The abstract matching theorem was proved by Rogawski [98] and Deligne-Kazhdan-Vigneras [3]. Recall that, if E/F is an extension of degree n, then $E^\times$ can be embedded in both G and G'. In fact, any compact (mod center) Cartan subgroup of G (and G') is isomorphic to $E^\times$ for some extension of degree n.

<u>Abstract Matching Theorem.</u> There is a bijection $\Pi' \longleftrightarrow \Pi$ between irreducible representations of G' and the discrete series of representations of G with the following properties:

(1) if $\theta_{\Pi'}$ and θ_{Π} are the characters of Π' and Π respectively, and x is a regular element in a compact (mod center) Cartan subgroup $E^\times$, then

$$\theta_{\Pi'}(x) = (-1)^{n-1}\theta_{\Pi}(x) ;$$

(2) if the formal degree of the Steinberg representation [6] is normalized to be equal to one, then

$$d(\Pi') = d(\Pi) ,$$

where $d(\Pi')$ is the ordinary degree of the finite-dimensional representation Π' and $d(\Pi)$ is the formal degree of the infinite-dimensional representation Π ;

(3) if $\varepsilon(\Pi',\psi)$, $\varepsilon(\Pi,\psi)$ are the ε-factors of Π' and Π respectively [89] then $\varepsilon(\Pi',\psi) = (-1)^{n-1} (\varepsilon,\psi)$.
Here, ψ is a suitably chosen additive character on F.

<u>**Remarks.**</u> (1) Moy's proof [89] that the supercuspidal representations constructed by Howe and the generalized special representations constructed by Bernstein-Zelevinsky exhaust the discrete series of $GL_n(F)$ uses the abstract

matching theorem in an essential way. Thus, it is only after we use the abstract matching theorem that we can assert that the concrete matching by admissible characters is actually a bijection.

(2) In general, the abstract matching is not the same as the concrete matching via admissible characters. It would be of some interest to determine the exact relation between these two matchings (see [89] for additional remarks).

(3) Recently, using the theory of minimal K-types, Moy [M] has made progress towards an intrinsic proof of the exhaustion of supercuspidal representations of G. This would eliminate the need for the matching theorem in the exhaustion proof.

(4) The abstract matching theorem gives no indication as to which representations of G' correspond to the two distinct types of discrete series representations of G.

To sharpen our focus, we introduce the following distinction. If E/F is an extension of degree m, $m|n$ $(m < n)$, and θ is an admissible character of $E^\times$, we say that θ is <u>subadmissible</u> (for n). Thus, the term <u>admissible character</u> will be used only for extensions E/F of degree n. The conjugacy classes of admissible and subadmissible characters parameterize the irreducible representations of G' (see §2). As indicated above, the supercuspidal representations of G correspond to admissible characters, and the generalized special representations of G correspond to subadmissible characters. Thus it is natural to conjecture that, if π'_θ is the irreducible representation of G' corresponding to an admissible (resp. subadmissible) character, then the discrete series representation π of G which correspond to π'_θ by the abstract matching theorem is supercuspidal (resp. generalized special).

This last assertion is indeed the case, and it is the purpose of this note to indicate a proof using only the degrees of the representations. To this end, we consider the following sets:

$$\begin{aligned} A'_1 &= \{\Pi'_\theta \in (G')^\wedge \mid \theta \text{ is admissible}\} ; \\ A'_2 &= \{\Pi'_\theta \in (G')^\wedge \mid \theta \text{ is subadmissible}\} . \end{aligned} \tag{1.1}$$

Here Π'_θ is the representation of G' constructed from θ by Corwin and Howe, and $(G')^\wedge$ is the unitary dual of G'. In a similar fashion, we define

$$A_1 = \{\Pi_\theta \in \hat{G}_d \mid \theta \text{ is admissible}\} ;$$

$$A_2 = \{\Pi_\theta \in \hat{G}_d \mid \theta \text{ is subadmissible}\} . \tag{1.2}$$

In this case, we have the supercuspidal representations (resp. generalized special representations) constructed by Howe (resp. Bernstein-Zelevinsky), and $\hat{G}_d$ denotes the discrete series in the unitary dual of G .

Now, letting $d(\Pi)$ denote the ordinary or formal degree of a representation, we set

$$\Delta_1' = \{d(\Pi_\theta') \mid \Pi_\theta' \in A_1'\} ;$$

$$\Delta_2' = \{d(\Pi_\theta') \mid \Pi_\theta' \in A_2'\} ; \tag{1.3}$$

$$\Delta_1 = \{d(\Pi_\theta) \mid \Pi_\theta \in A_1\} ;$$

$$\Delta_2 = \{d(\Pi_\theta) \mid \Pi_\theta \in A_2\} . \tag{1.4}$$

If we assume that $d(\text{Steinberg}) = 1$, then (2) in the abstract matching theorem implies that $\Delta_1' \cup \Delta_2' = \Delta_1 \cup \Delta_2$. Here we show that

$$\Delta_1' = \Delta_1 \quad \text{and} \quad \Delta_2' = \Delta_2. \tag{1.5}$$

Since the trivial representation of G' is in A_2' , it follows that, under the abstract matching, representations in A_1' correspond to supercuspidal representations of G and representations in A_2' correspond to generalized special representations of G . It is interesting to note that the conductors of the representations Π_θ and Π_θ' appear naturally in the expressions for the formal degrees. This may be seen by comparing the formulas for the degrees in § 3 with the expressions for the conductor in [Mol].

One of the more important consequences of (1.5) is worth observing here. Using the standard Frobenius formula for induced characters, we are able to give explicit formulas for the characters of the representations $\Pi'_\theta \in (G')^\wedge$. It follows from (1) of the abstract matching theorem that these are (up to a sign) explicit formulas for the characters of the discrete series of G on the elliptic set. The distinction provided by the formal degrees tells us which of these are supercuspidal characters and which are generalized special characters. In turn, this allows us to analyze the differences between the two different classes of characters. This analysis is carried out in [23].

In the case $n = p$, Carayol [14] has determined the formal degrees of the supercuspidal representations of G and has observed the relationship between the formal degree and the conductor of the representation. Waldspurger [109] has computed the formal degrees of the discrete series of G with a normalization which differs from ours. His techniques for obtaining these formulas are also different.

The results presented in this note were obtained in collaboration with L. Corwin and A. Moy. Full details of the computation of the degrees along with some consequences will appear in [21].

§2. Admissible characters and representations.

The notion of admissible character, which is due to Howe [49], is the key to the construction of the representations of $G' = D_n^\times$ and the supercuspidal representations of $G = GL_n(F)$.

2.1. Definition. (Howe) Let E/F be an extension of degree n with $(n,p) = 1$. A character θ of $E^\times$ is admissible if

1) θ does not come via the norm from a subfield of E containing F,

2) if $\theta|_{1+P_E}$ comes via the norm from a subfield $E \supset L \supset F$, then E/L is unramified.

2.2 Remarks (1) In this note, the term character refers to unitary character, that is, a continuous homomorphism into the complex numbers of modulus one.

(2) If E/F is an extension of degree m, $m|n$ $(m < n)$ and θ is an admissible character on $E^\times$, we say that θ is a subadmissible character (for n).

2.3 Definition. Let θ_1 and θ_2 be admissible characters of E_1/F and E_2/F respectively. We say that θ_1 and θ_2 are conjugate if there is an F-isomorphism $\varphi: E_1 \to E_2$ such that $\theta_1 = \theta_2 \circ \varphi$.

We are now in a position to give a classification of the irreducible representations of $G' = D_n^\times$ and the discrete series representations of $G = GL_n(F)$.

2.4 Theorem. (Corwin [18], Howe [46])

The irreducible representations of G' may be parameterized by (conjugacy classes of) admissible and subadmissible characters of extensions of degree m over F $(m|n)$. More specifically, if θ is admissible or subadmissible, there is an open subgroup H'_θ of G' and an irreducible representation σ'_θ of H'_θ such that the induced representation

$$\Pi'_\theta = \mathrm{Ind}_{H'_\theta}^{G'} \sigma'_\theta$$

is an irreducible representation of G'. Moreover, Π'_{θ_1} is equivalent to Π'_{θ_2} if and only if θ_1 is conjugate to θ_2. The collection of equivalence classes $\{\Pi'_\theta\}$ is a complete set of irreducible representations for G', that is, $(G')^\wedge = \{\Pi'_\theta | \theta$ is admissible or subadmissible$\}$.

2.5 Theorem. (Howe [49])

If θ is an admissible character of $E^\times/F$ $([E:F] = n)$, then there is an open, compact subgroup H_θ of G and an irreducible representation σ_θ of $E^\times H_\theta$ such that the induced representation

$$\Pi_\theta = \mathrm{Ind}_{E^\times H_\theta}^{G} \sigma_\theta$$

is an irreducible supercuspidal representatin of G. Moreover, Π_{θ_1} is equivalent to Π_{θ_2} if and only if θ_1 is conjugate to θ_2.

Before we state a theorem about generalized special representations we need a little preparation. Let θ be a subadmissible character for $E^\times/F$. $[E:F] = m$, $m|n$, $m < n$. Then, by Theorem 2.5, there is an irreducible supercuspidal representation ρ_θ of $GL_m(F)$ corresponding to θ. Consider the parabolic subgroup P of G defined by

$$P = \begin{pmatrix} GL_m(F) & & & * \\ & GL_m(F) & & \\ & & \ddots & \\ 0 & & & GL_m(F) \end{pmatrix} .$$

We take the representation of P which is defined on the Levi component $M = \Pi\ GL_m(F)$ $(r = \frac{n}{m}$ copies) by $\bigotimes_{j=0}^{r-1} \rho_\theta\ |\cdot|^j$, where $|g| = |\det g|$, and extend trivially to the unipotent radical to obtain $(\bigotimes_{j=0}^{r-1} \rho_\theta\ |\cdot|^j) \otimes 1$.

2.6 Theorem. (Bernstein-Zelevinsky [4], [115])

The induced representation

$$\mathrm{Ind}_P^G[(\bigotimes_{j=0}^{r-1} \rho_\theta\ |\cdot|^j) \otimes 1]$$

contains a unique irreducible, square integrable (mod Z) *subquotient* Π_θ . *Moreover,* Π_{θ_1} *is equivalent to* Π_{θ_2} *if and only if* θ_1 *is conjugate to* θ_2 .

2.7 Remark. Theorem 2.6 is a slight rewriting of history. In fact, Bernstein-Zelevinsky show that *all* non-supercuspidal (i.e. generalized special) discrete series for G are obtained by choosing a supercuspidal representation of $GL_m(F)$ and proceeding as above. They make no reference to the explicit construction of the supercuspidal representations of $GL_m(F)$ nor do they impose any restrictions on the residual characteristic of F . Only after the exhaustion theorem of Moy [89] do we know (in the tame case) that all supercuspidal representations of $GL_m(F)$ can be constructed via subadmissible characters. This is a consequence of the following theorem.

2.8 Theorem. (Moy [89]). *If* $(n,p) = 1$, *then the irreducible supercuspidal representations of* $GL_n(F)$ *are parameterized by conjugacy classes of admissible characters of* $E^\times/F$, *where* E *is an extension of degree* n . *In particular, all the irreducible supercuspidal representations of* $GL_n(F)$ *are induced from open, compact* (mod Z) *subgroups.*

§ 3. Degrees and formal degrees.

Not surprisingly, the degrees and formal degrees of the representation Π'_θ and Π_θ depend on n, E and specific data related to the admissible or subadmissible character θ of $E^\times$. The specific data related to θ comes from the Howe factorization of θ [49,89]. For present purposes, we shall surpress the specific data and combine it into an expression $\alpha(\theta)$. The explicit form of $\alpha(\theta)$ is given in [21].

In order to compute the degrees and formal degrees so that (2) of the abstract matching theorem is satisfied, we must normalize measures in an appropriate fashion. First of all G'/Z is a compact group, and we set

$$\mathrm{vol}(G'/Z) = 1 . \tag{3.1}$$

Next, we normalize Haar measure on G so that the degree of the Steinberg representation is equal to one. To accomplish this, we let $K = GL_n(\mathcal{O}_F)$, a maximal compact subgroup of G. It is well-known that $d(\mathrm{Steinberg})\mathrm{vol}(KZ/Z) = \frac{1}{n}\prod_{i=1}^{n-1}(q^i - 1)$, where $q = |k|$ and $k = \mathcal{O}_F/\mathcal{P}_F$, the residue class field of F (see [98]). Thus we normalize

$$\mathrm{vol}(KZ/Z) = \frac{1}{n}\prod_{i=1}^{n-1}(q^i - 1) . \tag{3.2}$$

3.3 <u>Remark.</u> It should be observed that, in (3.1) and (3.2), we are simultaneously normalizing Haar measures on G' and Z, and G and Z respectively.

We have indicated in Theorem 2.4 and Theorem 2.5 that all the irreducible representations of G', and the supercuspidal representations of G can be constructed by inducing from compact (mod Z), open subgroups. It is a standard fact that, in such cases, the degree of the induced representation is equal to the degree of the inducing representation if the volume of the inducing subgroup (mod Z) is normalized to be equal to one. Of course, we have already fixed a normalization in (3.1) and (3.2). Hence, to compute the degree of the induced representations, it is necessary (and sufficient) to compute the degrees of the inducing representations and the indices of the inducing subgroups in some fixed subgroup whose measure is fixed. Here, we shall content ourselves with stating the final formulas.

3.4 <u>Theorem.</u> <u>Let</u> θ <u>be an admissible character of</u> $E^\times/F$ ($[E:F] = n$). <u>Let</u> $e = e(E/F)$ <u>and</u> $f = f(E/F)$. <u>Let</u> Π'_θ <u>be the irreducible representation of</u> G'

given by Theorem 2.4, and Π_θ the irreducible supercuspidal representation of G given by Theorem 2.5. Then, with the normalizations (3.1) and (3.2),

$$d(\Pi'_\theta) = d(\Pi_\theta) = f\,\frac{q^n - 1}{q^{n/e} - 1}\, q^{\frac{f}{2}(\alpha(\theta) + 2 - n - e)},$$ where $\alpha(\theta)$ is an expression depending only on data in the Howe factorization of θ.

If θ is a subadmissible character (for n), then the degree of the representation Π'_θ of G' given by Theorem 2.4 is computed in a fashion entirely similar to that for Π'_θ, θ admissible. On the other hand, the generalized special representations of G are not induced in the same manner as the supercuspidal representations, and very different techniques must be used to compute the formal degrees of these representations. These techniques are based on the theory of Hecke algebra isomorphisms developed in Howe-Moy [51] and further refined by the same authors.

3.5 Theorem. Let θ be a subadmissible character (for n) of $E^\times/F$ where $[E:F] = m$, $m|n$, $m < n$. Write $n = ma$. Let $e = e(E/F)$, $f = f(E/F)$ so that $ef = m$. Let Π'_θ be the irreducible representation of G' given by Theorem 2.4, and Π_θ the irreducible generalized special representation given by Theorem 2.6. Then, with the normalizations (3.1) and (3.2),

$$d(\Pi'_\theta) = d(\Pi_\theta) = f\,\frac{q^n - 1}{q^{n/e} - 1}\, q^{\frac{af}{2}[a\alpha(\theta) + a + 1 - an - e]},$$ where $\alpha(\theta)$ is the same expression for the present subadmissible θ as that for the admissible θ in Theorem 3.4.

We can now derive the results stated in (1.5). They are a consequence of the following Lemma.

3.6 Lemma. If a, b, c, d are positive integers such that $a|b$ and $c|d$, then $\dfrac{a}{q^b - 1} = \dfrac{c}{q^d - 1}$ if and only if $a = c$ and $b = d$.

3.7 Theorem. Let Δ'_1, Δ'_2, Δ_1, Δ_2 be the sets defined by (1.3) and (1.4). Then $\Delta'_1 \cap \Delta'_2 = \Delta_1 \cap \Delta_2 = \Delta'_1 \cap \Delta_2 = \Delta'_2 \cap \Delta_1 = \emptyset$. Thus $\Delta'_1 = \Delta_1$ and $\Delta'_2 = \Delta_2$.

This is an immediate consequence of Lemma 3.6 and the fact that $\Delta'_1 \cup \Delta'_2 = \Delta_1 \cup \Delta_2$.

3.8 Corollary. (1) If θ is an admissible character of $E^\times/F$ ($[E:F] = n$) and Π'_θ is the representation of G' given by Theorem 2.4, then the representation Π of G corresponding to Π'_θ under the abstract matching theorem is supercuspidal. Moreover, if $\chi_{\Pi'_\theta}$ is the character of Π'_θ, then $\chi_\Pi = (-1)^{n-1}\chi_{\Pi'_\theta}$ is a supercuspidal character on the elliptic set in G .

(2) If θ is a subadmissible character (for n) of $E^{\times}/F$, $[E:F] = m$, $m|n$, $m < n$, and Π'_{θ} is the representation of G' given by Theorem 2 , then the representation Π of G corresponding to Π'_{θ} under the abstract matching theorem is a generalized special representation of G . Moreover, if $\chi_{\Pi'_{\theta}}$ is the character of $\Pi_{\theta'}$, then $\chi_{\Pi} = (-1)^{n-1}\chi_{\Pi'_{\theta}}$ is a generalized special character on the elliptic set in G .

The Corollary follows from the fact that the trivial representation of G' is in A_2' and the Steinberg representation of G is in A_2 .

References

[3] J. Bernstein, P. Deligne, D. Kazhdan, M.-F. Vigneras: Représentations des algèbres centrales simples p-adiques, in: Représentations des Groupes Réductifs sur un Corps Local; Hermann, Paris, 1984 33 - 117

[4] I.N. Bernštein, A.V. Zelevinskii: Representations of the group GL(n,F) where F is a local non-archimedean field; Russian Math. Surveys 31 (1976), 1 - 68 (transl. from Uspekhi Mat. Nauk 31, no. 3 (1976), 5 - 70).

[6] A. Borel: Admissible representations of a semisimple group over a local field with vectors fixed under an Iwahori subgroup: Inv. Math. 35 (1976), 233 - 259.

[14] H. Carayol: Representations cuspidales du groupe lineaires; Ann.Sci. ENS 17 (1984), 191 - 225.

[18] L. Corwin: Representations of division algebras over local fields; Advance in Math. 13 (1974), 259 - 267.

[21] L. Corwin, A. Moy, P.J. Sally Jr.: Degrees and formal degrees for division algebras and GL_n over a local field. To appear in Pacific J. Mat

[23] L. Corwin, P.J. Sally Jr.: Discrete series characters on the elliptic set in GL_n . To appear.

[46] R. Howe: Representation theory for division algebras over local fields (tamely ramified case); Bull. AMS 77 (1971), 1063 - 1066.

[49] R. Howe: Tamely ramified supercuspidal representations of GL_n ; Pac.J. Math. 73 (1977), 437 - 460.

[51] R. Howe, A. Moy: Harish-Chandra Homomorphisms for p-adic Groups, CBMS Regic nal Conference, Series in Mathematics 59, Amer.Math.Soc. Providence, R.I., 1985.

[89] A. Moy: Local constants and the tame Langlands correspondence; Amer. J Math., 108 (1986), 863 - 930.

[98] J. Rogawski: Representations of GL(n) and division algebras over a p-adic field; Duke Math. J. 50 (1983), 161 - 196.

[109] J.-L. Waldspurger: Algèbres de Hecke et induites de représentations cuspidales pour GL(N) ; Crelle Journal 370 (1986), 127 - 191.

[115] A. Zelevinsky: Induced representations of reductive p-adic groups II: On irreducible representations of GL(n) ; Ann.Sci. ENS 13 (1980), 165 - 210.

[M] A. Moy, see this volume.

Contemporary Mathematics
Volume **86**, 1989

TAME REPRESENTATIONS AND BASE CHANGE

by A. Fröhlich, London

We are considering here firstly tame representations, in the sence of Deligne [25,26,T], of the absolute Weil group W_F of a local field F, a finite degree extension of Q_p, and secondly tame admissible irreducible representations of the chain group associated with a principal order in a finite dimensional central simple F-algebra. The term "tame" means here that in the first case the wild ramification group and in the second case the group of one-units of the principal order should lie in the kernel of the respective representation.

We shall call two Deligne representations (Π, N) and (Π', N') of W_F similar if firstly they are I-equivalent, cf.[T] , i.e. their restrictions to the inertia group I_F are equivalent, and secondly they have the same determinant, viewed here as a quasi-character of the multiplicative group $F^\times$ of F. This relation is stronger than I-equivalence, weaker than equivalence, but - and this is important - for irreducible representations similarity implies equivalence.

As shown in [T,80] the I-equivalence classes of tame representations are parametrized by the monoid $L(k)$, k the residue class field of F. The determinants are elements of the group of tame quasi-characters, which we shall denote by $\hat{F}^\times_t$ (the dual of $F^\times_t = F^\times$ mod one units). An obvious compatibility condition then leads us to a submonoid $\mathcal{L}(F)$ of $L(k) \times \hat{F}^\times_t$, which parametrizes similarity classes. We thus obtain a surjective homomorphism from the monoid $\mathcal{R}(F)$ of equivalence classes of tame Deligne representations of W_F onto $\mathcal{L}(F)$. This preserves the action of $\hat{F}^\times_t$ and of Galois groups which can be defined on both monoids, as well as the involution coming from the "contragredient".

We can moreover define restriction and induction maps on the monoids $\mathcal{L}$ of similarity classes compatible with the standard restriction and induction map on equivalence classes, i.e. on monoids $\mathcal{R}$. It is best to explain this separately for non-ramified and for totally ramified change of field.

Let first K be a non-ramified extension of F of finite degree. Let k_1 be

* This is a survey of some of the results to be published with proofs in a forthcoming long paper. [31]

0271-4132/89 $1.00 + $.25 per page

the residue class field of K. Clearly $L(k) \subset L(k_1)$. The restriction is then given by the commutativity of the diagram

$$\begin{array}{ccccc} \mathcal{L}(F) & \subset & L(k) & \times & \hat{F}_t^{\times} \\ \text{res} \downarrow & & \cap & & \downarrow \hat{N}_{K/F} \\ \mathcal{L}(K) & \subset & L(k_1) & \times & \hat{K}_t^{\times} \end{array} \qquad (1)$$

where $\hat{N}_{K/F}$ is the dual of the norm. Here, and in other diagrams in the sequel, one has of course to show that the compatibility condition defining $\mathcal{L}$ is preserved.

Next, the Galois group of k_1/k acts on the monoid $L(k_1)$ and we can thus define a trace map

$$t_{k_1/k} : L(k_1) \to L(k) ,$$

$$t_{k_1/k}(\ell) = \Sigma \ell^{\sigma} \quad \text{(sum over Galois group)} .$$

Now we define induction

$$\text{ind}: \mathcal{L}(K) \to \mathcal{L}(F)$$

by

$$\text{ind}(\ell, \varphi) = (t_{k_1/k}(\ell) , \varphi|_{F^\times} \cdot \Delta_{K/F}^{\deg(\ell)} , \quad \text{for} \quad (\ell,\varphi) \in \mathcal{L}(K) , \qquad (2)$$

where $\varphi|_{F^\times}$ is the restriction of φ to $F^\times$, $\Delta_{K/F}$ is the discriminantal character and $\deg(\ell)$ is the degree. Again one has to show that the right hand side actually lies in $\mathcal{L}(F)$.

Now let K/F be totally ramified of degree e; thus k is also identified with the residue class field of K. (Some care is needed here!)
We define an endomorphism

$$e_*: L(k) \to L(k)$$

by

$$(e_*\ell)(\chi) = \sum_{\substack{\theta \in \hat{I} \\ \theta^e = \chi}} \ell(\theta)$$

Then restriction is defined by the commutative diagram

$$\begin{array}{ccccc} \mathcal{L}(F) & \subset & L(k) & \times & \hat{F}_t^\times \\ \text{res} \downarrow & & \downarrow e_* & & \downarrow \hat{N}_{K/F} \\ \mathcal{L}(K) & \subset & L(k) & \times & \hat{K}_t^\times \end{array} \quad . \tag{3}$$

'inally define

$$e^*: \ L(k) \to L(k)$$

y

$$(e^*\ell)(\chi) = \ell(\chi^e) \ .$$

or induction we have to restrict K/F to be tame.
t is then defined by

$$\text{ind}(\ell, \varphi) = (e^*\ell, \varphi|_{F^\times} \ \Delta_{K/F}^{\deg(\ell)}) \ . \tag{4}$$

Next we turn to our other main object (See [12] for the background). Let be a principal order in a central simple F-algebra.

$$A \simeq M_m(D) \ , \quad \dim_F(D) = d^2 \ , \tag{5}$$

a division algebra, $M_m(R)$ the m by m matrizes over a ring R. Let G e the normalizer of A in the multiplicative group $A^\times$. (A is the endomorphism ing of an uniform chain, G the associated chain group). The one-units of A orm a normal subgroup of G, and we are interested in the set $\hat{G}_t$ of irreduc-ble, admissible representations of the quotient G_t of G modulo the one-units. nese are finite dimensional. G_t contains the image U_t of the group of units of as a normal subgroup and $G_t/U_t = \mathbb{Z}$ (with a canonical generator).

The first step is to parametrize the set $\hat{U}_t$ of irreducible representations of the finite group U_t . Here the notion of the Galois algebra M of the principal order A becomes important. As a k-algebra M is the center of the residue class-ring of A modulo its Jacobson radical. The Galois group is

$$C(A) = G_t/U_tF_t^{\times} \ (= G/UF^{\times})$$

the action being induced by the conjugation action of G on A . We then have

$$U_t = GL_s(M) \ , \quad \text{for some } s \ . \tag{6}$$

There is a natural extension of the functor $k \to L(k)$ to Galois algebras M , and this leads to a parametrization

$$L(M)_s \simeq \widehat{GL_s(M)} \ . \tag{7}$$

This extends the bijection of Macdonald (see [80])

$$L(k)_s \simeq \widehat{GL_s(k)} \ .$$

To go into some more detail, write $\bar{k}$ for the algebraic closure of k , and $\Omega = \mathrm{Gal}(\bar{k}/k)$ for the Galois group over k .
There is then a homomorphism

$$g: \Omega \to C(A) \ , \tag{8}$$

making $C(A)$ into an Ω-set by

$$c\omega = cg(\omega) \ ,$$

so that

$$\mathrm{Map}_{\Omega}(C(A),\bar{k}) \simeq M \tag{9}$$

(isomorphism of k-algebras and of $C(A)$-modules).
Here the algebra structure on the left is given by $(f_1 \dotplus f_2)(c) = f_1(c) \dotplus f_2(c)$, $c \in C(A)$, and the $C(A)$-module structure by

$$(fb)(c) = f(bc) \ , \quad b,c \in C(A) \ .$$

Viewing (9) as an identification we then have

$$L(M)_s = \mathrm{Map}_\Omega(C(A)\ ,\ L(\bar{k})_s) \tag{10}$$

where $L(\bar{k})_s$ is the set of all partition valued functions of degree s on the dual $\hat{I}$ of I - The set $L(k)_s$ in [T] is simply the set of functions on $L(\bar{k})_s$ fixed under the action of Ω .

Example: A is a maximal order precisely if the map g in (8) is surjective. In this case $M \simeq k_g$, the fixed subfield of $\mathrm{Ker}\, g$ in $\bar{k}$.

Now we define a notion of similarity on $\hat{G}_t$. The restriction to $F_t^\times$ of $\Pi \in \hat{G}_t$ is a scalar, defining the central character ω_Π of Π , an element of $\widehat{F_t^\times}$. We shall say that two irreducible representation Π , Π' are similar if firstly they are U-equivalent, i. e. their restrictions to U_t are equivalent, and secondly their central characters coincide. The U-equivalence classes are the orbits $\hat{U}_t/C(A)$, i. e. in view of (6) and (7) are parametrized by $L(M)_s/C(A)$, the orbit set best described via (10). Taking into account the obvious compatibility condition we conclude that the similarity classes of irreducible representations are parametrized by a subset $L(A)$ of $(L(M)_s/C(A)) \times \widehat{F_t^\times})$. The set $L(A)$, as well as the set $\hat{G}_t$, admits action by Galois groups, action by $\widehat{F_t^\times}$ and a "contragredient" involution $\vee$ and all these are respected by the map $\hat{G}_t \to L(A)$.

There are two special aspects to be considered. Firstly if A is the ring $M_s(O_F)$ of s by s matrizes over the ring O_F of integers in F then $G_t = F_t^\times U_t$ and hence similarity is the same as equivalence.

Secondly the notion of an irreducible representation of $GL_s(k)$ being cuspidal (cf [80]) can in a natural manner be extended to $GL_s(M)$. Such cuspidals will exist precisely if M is a field, i. e. A is a maximal order and they then form complete $C(A)$-orbits. It then makes sense to speak of cuspidals in $\hat{G}_t$, or in $L(A)$. For these similarity implies equivalence. More precisely if Π , Π' are, similar, and Π is cuspidal then Π' is equivalent to Π , hence cuspidal.

Remark. At this stage one might ask why we don't restrict ourselves to maximal orders and to cuspidals. In fact this would throw away an integral feature of the whole theory. Even if our attention were focused in the first place on cuspidals for maximal orders, any theory of base change would have to take it away from there: neither the maximality of an order, nor the cuspidal property of a representation is preserved under either extension or restriction of base.

The decisive step in the whole theory, which connects up tame Deligne representations of Weil groups and irreducible representations of chain groups is the definition of a map

$$t_A : \ L(A) \to L(F)_{md} \ , \tag{11}$$

where m, d are the numerical invariants of the underlying algebra A, as in (5) and where $L(F)_{md}$ is the set of similarity classes of degree md. This map exists for any principal order, although it has nicer properties for some, and it is defined in a perfectly simple and natural manner. In fact let $f \in L(M)_s$ (as given in (10)) and let $(f \bmod C(A), \psi)$ be an element of $(L(M)_s/C(A), \widehat{F_t^\times})$ lying in $L(A)$. Then

$$t_A(f \bmod C(A), \psi) = (\sum_c f(c), \psi) \tag{12}$$

$(c \in C(A))$ and the right hand side indeed belongs to $L(F)_{md}$. We outline the basic properties of t_A.

(a) t_A respects the action of Galois groups, the action of $F_t^\times$ and the contragredient involution.

(b) t_A converts central characters into determinants.

(c) t_A preserves root numbers up to a dimension factor $(-1)^{m(d-1)}$ (see (5)). Here the root numbers are the projections of the appropriate Gauss sums (Galois Gauss sums for Weil groups, congruence Gauss sums for principal orders) onto the unit circle. Equivalently: t_A preserves Gauss sums up to positive constants and up to a factor $(-1)^{m(d-1)}$.

(d) t_A commutes with base change (see below).

(e) If $A = M_s(O_F)$ then t_A is a bijection (of equivalence classes).

(f) If A is a maximal order, then t_A yields a bijection of cuspidals onto the irreducibles in $L(F)_{md}$. (Recall: In both cases similarity is equivalence). A consequence: If A, B are principal orders over O_F of the same rank, then there is a natural bijection of cuspidal equivalence classes from $\hat{G}_t(A)$ to $\hat{G}_t(B)$. Example: $A = M_s(O_F)$, B the ring of integers in a central F-division algebra of dimension s^2.

We finally briefly indicate the base change procedure for chain groups. Here the detailed analysis of embedding the multiplicative group of an extension field of F in a chain group G is fundamental (cf [30]). For going up, we may consider any extension field K of F of finite degree. We then associate with our given principal order $\mathfrak{A}$, over $\mathcal{O}_F$, in A a new principal order $\mathfrak{A}'$, over $\mathcal{O}_K$, in $A \otimes_F K$, and we get a natural extension map in the top row of the commutative diagram

$$\begin{array}{ccc}
\mathcal{L}(\mathfrak{A}) & \longrightarrow & \mathcal{L}(\mathfrak{A}') \\
\downarrow t_{\mathfrak{A}} & & \downarrow t_{\mathfrak{A}'} \\
\mathcal{L}(F)_{md} & \xrightarrow[\text{res}]{} & \mathcal{L}(K)_{md}
\end{array} .$$

$\mathfrak{A}'$ is unique mod conjugacy in $A \otimes_F K$. Under some restrictions we can actually make $\mathfrak{A}'$ completely unique: The chain group $G(\mathfrak{A}')$ is the unique chain group containing $G(\mathfrak{A})K^\times$. In particular if K/F is non-ramified then $\mathfrak{A}' = \mathfrak{A} \otimes_{\mathcal{O}_F} \mathcal{O}_K$ and if $\mathfrak{A} = M_s(\mathcal{O}_F)$ with K/F arbitrary then $\mathfrak{A}' = M_s(\mathcal{O}_K)$.

For <u>going down</u> let K be a tame extension of F, embedded in A. We assume that $K^\times \subset G(\mathfrak{A})$. There are rather pathological such embeddings which have to be excluded; this leads to the notion of a <u>sound</u> embedding. There are a number of criteria for these (cf [30]). In the present context what is relevant is that $K^\times$ is embedded in $G(\mathfrak{A})$ in such a way that the suborder $\mathfrak{B}$ of $\mathfrak{A}$, of elements commuting with K, is principal and that furthermore $G(\mathfrak{B}) \subset G(\mathfrak{A})$. Under these conditions we get a going down map in the top row of the commutative diagram

$$\begin{array}{ccc}
\mathcal{L}(\mathfrak{B}) & \longrightarrow & \mathcal{L}(\mathfrak{A}_1) \\
\downarrow t_{\mathfrak{B}} & & \downarrow t_{\mathfrak{A}_1} \\
\mathcal{L}(K)_{md} & \xrightarrow[\text{ind}]{} & \mathcal{L}(F)_{md}
\end{array} .$$

Here $\mathfrak{A}_1$ is a certain principal order over $\mathcal{O}_F$. If actually K/F is non-ramified then $\mathfrak{A}_1 = \mathfrak{A}$.

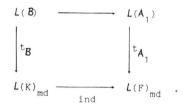

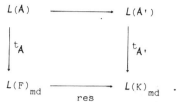

References

[12] C.J. Bushnell, A. Fröhlich: Non abelian congruence Gauss sums and p-adic simple algebras; Prod. London Math. Soc. (3), 50 (1985), 207 - 264.

[25] P. Deligne: formes modulaires et rèprésentations de GL(2), in: Modulár Functions of one variable II(ed. P. Deligne, W. Kutzko), Lecture Notes in Mathematics No. 349, Springer Verlag, Berlin-Heidelberg-New York 1973, 55 - 105.

[26] P. Deligne: Les constantes des équations fonctionnelles des fonctions L , in: Modular functions of one variable II(ed. P. Deligne, W. Kutzko), Lecture Notes in Math. 349, Springer Verlag, Berlin-Heidelberg-New York 1979, 501 - 597.

[30] A. Fröhlich: Principal orders and embedding of local fields in algebras; Proc. London Math. Soc. (3), 54 (1987), 247 - 266.

[31] A. Fröhlich: Tame representations of local Weil groups and of chain groups of local principal orders; Heidelberger Akad. d. Wiss. (1986), Springer Verlag, Berlin-Heidelberg-New York, 1 - 100.

[80] I.G. Macdonald: zeta functions attached to finite general linear groups; Math. Annalen 249 (1980), 1 - 15.

[T] M. Taylor, see this volume.

Contemporary Mathematics
Volume **86**, 1989

GAUSS SUMS AND SUPERCUSPIDAL REPRESENTATIONS OF GL_n

by C. J. Bushnell, London

Let F be a p-adic local field, and A a finite-dimensional central simple F-algebra. The Godement-Jacquet functional equation (see [39], [54] or Queyrut's talk [Q] in this volume) generalises, to irreducible admissible representations of $A^\times$, Tate's original local functional equation [105] for quasi-characters of $F^\times$. The generalisation is not, however, quite complete in that it does not include any analogue of Tate's explicit formula for the "local constant" or "ε-factor". We shall discuss this aspect of the matter, using the machinery of principal orders and non-abelian congruence Gauss sums developed in [12] and summarised in Wilson's talk [Wn]. Our approach is dependent on a certain hypothesis (see Theorem 1 below) concerning the structure of supercuspidal representations. It seems highly likely that this hypothesis holds, at least in the split case $A \simeq M_N(F)$. This is itself of wider interest, because it gives very useful-looking information in the direction of a complete classification of supercuspidal representations without recourse to global arguments.

We start by establishing our notation: this is consistent with that of Bushnell-Fröhlich [12], which contains a full account of the first part of this article. Throughout, F is a p-adic local field with discrete valuation ring $\mathfrak{o}_F$. We write $\mathfrak{p}_F$ for the prime ideal of $\mathfrak{o}_F$, and π_F for some prime element: $\mathfrak{p}_F = \pi_F\mathfrak{o}_F$. We also put $q = (\mathfrak{o}_F:\mathfrak{p}_F)$. Let ψ_F denote the "Iwasawa-Tate" character of the additive group of F. Thus ψ_F is, in particular, a continuous character of F, and the largest $\mathfrak{o}_F$-lattice contained in the kernel of ψ_F is the inverse different $\mathcal{D}^{-1}_{\mathfrak{o}_F}$ of F. (The whole theory can be developed relative to any non-trivial continuous additive character, but we find this canonical choice convenient, and it involves no loss of generality.)

Next, let A be a central simple F-algebra of finite dimension n^2, say $A \simeq M_m(D)$, where D is a central F-division algebra of dimension d^2, so that $n = md$. We put $\psi_A = \psi_F \circ \mathrm{Trd}_A$, where $\mathrm{Trd}_A: A \to F$ is the reduced trace map.

Let $\mathfrak{A}$ be a principal $\mathfrak{o}_F$-order in A, and let $\mathfrak{P}_{\mathfrak{A}}$ denote the Jacobson radical of $\mathfrak{A}$. Recall that $\mathfrak{A}$ is classified, up to $A^\times$-conjugacy, by its <u>central ramification index</u> $e = e(\mathfrak{A})$, where

$$A/P_A \simeq M_f(k_D)^e$$

as rings, where k_D is the residue class field of D and $ef = m$. We also have the group

$$G(A) = \{x \in A^\times : xAx^{-1} = A\}.$$

This is a <u>compact-mod-centre</u> subgroup of $A^\times$, in that it is a closed subgroup containing $F^\times$, and its image in $A^\times/F^\times$ is compact. Moreover, it is a <u>maximal</u> such subgroup. Indeed (see [12]):

Proposition 1. *The map $A \mapsto G(A)$ is a bijection, respecting $A^\times$-conjugacy, between the set of principal orders A in A and the set of maximal compact-mod-centre subgroups G of $A^\times$.*

Now let $G = G(A)$, and write $\hat{G}$ for the set of (equivalence classes of) irreducible admissible representations of G. Associated to a representation $\Sigma \in \hat{G}$, we have the <u>conductor</u> $\mathfrak{f}(\Sigma)$ of Σ, which is a non-negative power of P_A, and the <u>Gauss sum</u> $\tau(\Sigma)$ of Σ. This Gauss sum is a complex number which is zero for many Σ. However, it is easy to characterise the representations Σ for which $\tau(\Sigma) \neq 0$: let $\mathfrak{f}(\Sigma) = P_A^f$ and consider, for the moment, the case $f \geq 2$. Thus Σ is effectively a representation of $G/1 + P_A^f$, and we have

$$(1 + P_A^{f-1})/(1 + P_A^f) \simeq P_A^{f-1}/P_A^f,$$

$$1 + x \to x.$$

Thus $\Sigma | 1 + P_A^{f-1}$ splits as a direct sum of abelian characters α,

$$\Sigma | 1 + P_A^{f-1} = \oplus\, \alpha, \qquad (1)$$

each of which is of the form

$$\alpha(1 + x) = \psi_A(c_\alpha x), \quad x \in P_A^{f-1}, \qquad (2)$$

for some $c_\alpha \in A^\times$.

We say that a representation $\Sigma \in \hat{G}$ <u>is nondegenerate</u> if either

(i) $\mathfrak{f}(\Sigma)$ divides P_A, or

(ii) P_A^2 divides $\mathfrak{f}(\Sigma)$ and, for some α as in (1) there exists an element $c_\alpha \in G$ satisfying (2).

In case (ii) of this definition, if Σ is nondegenerate, then all c_α satisfying (2) lie in G, for all α occurring in $\Sigma|(1 + P_A^{f-1})$.
Now we have

Proposition 2. Let $\Sigma \in \hat{G}$. Then $\tau(\Sigma) \neq 0$ if and only if Σ is nondegenerate.

Let Π be an irreducible admissible representation of $A^\times$. With G as above, the restriction of Π to G splits as a discrete direct sum with finite multiplicities:

$$\Pi|_G = \bigoplus_{\Sigma \in \hat{G}} \Sigma^{n_\Pi(\Sigma)} . \tag{3}$$

The multiplicities $n_\Pi(\Sigma)$ are non-negative integers. Of course, we say Σ occurs in Π (or that $\Pi|_G$ contains Σ) if $n_\Pi(\Sigma) \geq 1$.

The existence of such a decomposition is a direct consequence of the definition of admissibility, and Π decomposes in exactly the same manner when G is replaced by any open compact-mod-centre subgroup of $A^\times$. For our present purposes, it is sufficient to consider only the maximal compact-mod-centre subgroups G given by the principal orders. Notice that conjugate subgroups G give equivalent decompositions (3), so there are only finitely many cases to be considered.

With Π as above, let $L(\Pi,s)$, $\varepsilon(\Pi,s)$ denote respectively the Godement-Jacquet Euler factor and local constant attached to Π. Here, s is a complex variable, and, in the notation of [39], our $\varepsilon(\Pi,s)$ is $\varepsilon(\Pi,s,\psi_A)$. See [39] or [Q] for these definitions. Let $\mathcal{A}$ be a principal order in A, and let $\mathcal{D}_\mathcal{A}$ denote its different. This is the invertible ideal of $\mathcal{A}$ whose inverse is the largest invertible fractional $\mathcal{A}$-ideal contained in $\mathrm{Ker}(\psi_A)$. (Explicitly, $\mathcal{D}_\mathcal{A} = P_\mathcal{A}^{e-1} \cdot \mathcal{D}_{o_F}$, where $e = e(\mathcal{A})$.) Finally, if $\mathfrak{a}$ is an ideal of $\mathcal{A}$, we put $N_\mathcal{A}\mathfrak{a} = (\mathcal{A}:\mathfrak{a})$.

With all this notation, we can state our first main result.

Theorem 1. (see [12]): Let Π be an irreducible admissible representation of $A^\times$, and $\mathcal{A}$ a principal order in A. Suppose

(a) $L(\Pi,s) = 1$ and

(b) there is a nondegenerate representation $\Sigma \in G(\mathfrak{A})^{\wedge}$ which occurs in $\Pi | G(\mathfrak{A})$.

Then

(i) Σ is ramified (i. e., $P_{\mathfrak{A}}$ divides $\mathfrak{f}(\Sigma)$) .

(ii) Suppose $\Sigma' \in G(\mathfrak{A})^{\wedge}$ also occurs in $\Pi | G(\mathfrak{A})$.

Then $\mathfrak{f}(\Sigma)$ divides $\mathfrak{f}(\Sigma')$, and moreover $\mathfrak{f}(\Sigma) = \mathfrak{f}(\Sigma')$ if and only if Σ' is also nondegenerate.

(iii) We have the formula

$$\varepsilon(\Pi,s) = (-1)^{m(d-1)} N_{\mathfrak{A}}(\mathcal{D}_{\mathfrak{A}}\, \mathfrak{f}(\Sigma))^{(\frac{1}{2}-s)/n} \frac{\tau(\check{\Sigma})}{\sqrt{N_{\mathfrak{A}}\, \mathfrak{f}(\Sigma)}} ;$$

where $\check{\Sigma}$ is the contragredient of the representation Σ .

Remarks.

(i) If $m = 1$, we have $A = D$, a division algebra. The only principal order in D is the unique maximal order $\mathcal{O}_D$, and $G(\mathcal{O}_D) = D^{\times}$. Further, all representations of G are nondegenerate, so hypothesis (b) is empty here. The hypothesis (a) is also quite unnecessary in this case: see [11] for a detailed discussion of division algebras. We henceforward assume that $m \geq 2$.

(ii) When seeking an explicit formula for $\varepsilon(\Pi,s)$ like (iii) above, it is only necessary to treat the case of Π supercuspidal (when $L(\Pi,s)$ is automatically 1) or one-dimensional (when (iii) holds without hypothesis). For, when Π is not supercuspidal and is therefore induced from a proper parabolic subgroup, its ε is given in terms of the ε's and L-factors of the factors of the inducing representation. These factors are supercuspidal or abelian. The critical point therefore is whether or not hypothesis (b) holds for supercuspidal representations. We shall return to this question below.

(iii) Part (i) of Theorem 1 is a rather different sort of result from the rest, and follows purely from the hypothesis $L(\Pi,s) = 1$. When $A \simeq M_N(F)$ and $\mathfrak{A}$ is a maximal order (i. e. $e(\mathfrak{A}) = 1$) , it is originally due to Jacquet. Macdonald's treatment [79] is valid for arbitrary A, with A maximal. The case of general A then follows from the functional equation, as in [12].

(iv) The ε-formula Theorem 1 (iii) reduces exactly to Tate's formula [105] in the case $A = F$. It also demonstrates that, modulo hypothesis (b), the local constant $\varepsilon(\Pi,s)$ has the same structure as the Langlands-Deligne local constant (see [26] or the expository account [106]). For simplicity, let Φ be an admissible complex representation of $\mathrm{Gal}(\bar{F}/F)$, where $\bar{F}$ is some algebraic closure of F. The Langlands-Deligne local constant $\varepsilon(\Phi,s)$ is then given by the formula

$$\varepsilon(\Phi,s) = N_{\mathcal{O}_F}(D_{\mathcal{O}_F}^{\dim(\Phi)}\mathfrak{f}(\Phi))^{\frac{1}{2}-s}\frac{\tau(\check{\Phi})}{\sqrt{N_{\mathcal{O}_F}\mathfrak{f}(\Phi)}}$$

where $\mathfrak{f}(\Phi)$ is the Artin conductor and $\tau(\Phi)$ the Galois Gauss sum. This Galois Gauss sum has proved very significant in several areas (see e. g. [29]), and seems a more fundamental invariant than ε.

There are some striking similarities between the properties of the two sorts of Gauss sums. For example, $\tau(\Phi)$ is a complex algebraic integer whose absolute value is $N_{\mathcal{O}_F}\mathfrak{f}(\Phi)^{\frac{1}{2}}$. On the other hand, if $\Sigma \in G(A)^\wedge$ is nondegenerate, if P_A^2 divides $\mathfrak{f}(\Sigma)$, and if the central quasi-character ω_Σ has finite order (these restrictions ar easily accounted for), then $\tau(\Sigma)$ is a complex algebraic integer of absolute value $N_A\mathfrak{f}(\Sigma)^{\frac{1}{2}}$. With the same restrictions, let $\omega \in \mathrm{Gal}(\bar{\mathbb{Q}}/\mathbb{Q})$, where $\bar{\mathbb{Q}}$ is an algebraic closure of $\mathbb{Q}$. Let $u_p\colon \mathrm{Gal}(\bar{\mathbb{Q}}/\mathbb{Q}) \to \mathbb{Z}_p^\times$ be the character given by

$$\zeta^{\omega u_p(\omega)} = \zeta,$$

for all p-power roots of unity $\zeta \in \bar{\mathbb{Q}}$. The representations Σ, Φ can both be realised over a subfield of $\bar{\mathbb{Q}}$, allowing us to form the Galois-twisted representations Φ^ω, Σ^ω. Then

$$\tau(\Sigma^{\omega^{-1}})^\omega/\tau(\Sigma) = \omega_\Sigma(u_p(\omega))$$

$$\tau(\Phi^{\omega^{-1}})^\omega/\tau(\Phi) = \det_\Phi(u_p(\omega))$$

for $\omega \in \mathrm{Gal}(\bar{\mathbb{Q}}/\mathbb{Q})$. Here, $\det_\Phi$ is the determinant of the representation Φ, viewed via class field theory as a character of $F^\times$. The local Langlands conjectures (which, in particular, relate the determinant with the central quasi-character) thus give a relation between the two types of Gauss sum. The congruence Gauss sums are given by explicit formulas and are correspondingly easy to manipulate: see [11] (and the related accounts in this volume) for a merciless exploitation of this. On the other hand, the Galois Gauss sum is defined very indirectly, using Brauer induction to reduce to the one-dimensional case which is then treated via class field theory.

Now we turn to a further discussion of hypothesis (b). The Godement-Jacquet functional equation implies a functional equation for the ε-factor

$$\varepsilon(\Pi,s)\,\varepsilon(\check{\Pi},\,1-s) = \omega_{\Pi}(-1),$$

where $\check{\Pi}$ is the contragredient of Π and ω_{Π} its central quasi-character. Since $\varepsilon(\Pi,s)$ is a Laurent polynomial in q^{-s}, it follows that

$$\varepsilon(\Pi,s) = C(\Pi)\,q^{-m(\Pi)s},$$

for some $C(\Pi) \in \mathbb{C}^{\times}$, $m(\Pi) \in \mathbb{Z}$.

Theorem 2. Let $A = M_n(F)$, and let Π be an irreducible supercuspidal representation of $A^{\times} = GL_n(F)$. Let $e(\Pi) = n/\gcd(n,m(\Pi))$, and let $\mathfrak{A}$ be a principal order in A with $e(\mathfrak{A}) = e(\Pi)$. Then $\Pi|G(\mathfrak{A})$ contains a nondegenerate representation Σ.

Caveat: At the time of writing, some details remain to be checked in certain cases (namely $f(\Pi) = 2,3$, in the notation below). In the remaining cases, the proof is complete and will appear elsewhere.

Some refinement of this result, in the direction of specifying more closely the nondegenerate component Σ, is certainly desirable.
However, in the case $e(\Pi) = n$, the discussion in Kutzko's talk [K] shows that the pair $(\mathfrak{A},\Sigma)$ is uniquely determined (up to the obvious equivalence) by Π, and the nondegenerate component Σ occurs in with multiplicity one. Exactly the same applies in the opposite extreme case $e(\Pi) = 1$ (we ignore minor details concerned with "twisting"). This effectively disposes of the case where n is prime. For the case $n = 4$, see Kutzko-Manderscheid [73].

Footnote added in proof: A complete proof of Theorem 2, different from the one one sketched here, will appear in J. reine angew. Math., under the title "Hereditary orders, Gauss sums and supercuspidal representation of GL_N".

We now sketch briefly the main steps of the proof. It is worth emphasizing that this uses only local methods, and indeed the only non-elementary contribution comes from the functional equation. We start by fixing a maximal order $\mathcal{M}$ in A. Thus $\mathcal{M} \simeq M_n(\mathcal{O}_F)$ and $\mathcal{P}_{\mathcal{M}} = \pi_F \mathcal{M}$, for a prime π_F of F. We write $U_0(\mathcal{M}) = \mathcal{M}^\times$ and, for $j \geq 1$ $U_j(\mathcal{M}) = 1 + \pi_F^j \mathcal{M}$. We define an integer $f = f(\Pi) \geq 0$ by

(i) Π has a non-zero $U_f(\mathcal{M})$-fixed vector;

(ii) f is minimal for condition (i).

It follows from Theorem 1 (i) that $f(\Pi) \geq 1$. If $f(\Pi) = 1$, the assertion is trivial, since any representation with conductor $\mathcal{P}_{\mathcal{M}}$ is nondegenerate. We therefore assume $f \geq 2$, and fix an integer g, $1 \leq g < f \leq 2g$. Let θ be the natural representation of $U_0(\mathcal{M})$ on the space of $U_f(\mathcal{M})$-fixed vectors. The restriction $\theta | U_g(\mathcal{M})$ splits as a direct sum of abelian characters χ, each of the form

$$\chi(1 + x) = \psi_A(\pi_F^{-d-f} b_\chi x), \quad x \in \pi_F^g \mathcal{M}, \tag{4}$$

for some $b_\chi \in \mathcal{M}$, $\notin \pi_F \mathcal{M}_F$, where $\mathcal{P}_F^d = \mathcal{D}_F$. We fix such a character χ occurring in $\theta | U_g(\mathcal{M})$. An element $t \in A^\times$ is called uniform if it lies in $G(\mathcal{B})$, for some principal order $\mathcal{B}$ in A. Equivalently, t is uniform if the roots of its characteristic polynomial (in some splitting field) all have the same absolute value.

Lemma. There exists $b_\chi = b \in A^\times$ satisfying (4) and such that

(i) $|\det b| = q^{m(\Pi)-n(d+f)}$,

(ii) b is uniform and regular.

Here, $|\ |$ is the normalized absolute value on F: $|\pi_F| = q^{-1}$.

To prove this lemma, we express the canonical projection on the $U_f(\mathcal{M})$-fixed vectors as a zeta integral and then apply the functional equation. Evaluating "the other side" appropriately gives an intertwining operator with nonzero trace, essentially $\Pi(U_g(\mathcal{M}) \cdot b \cdot U_g(\mathcal{M}))$, with $b = b_\chi$. The fact that the matrix coefficients of Π are compactly supported (mod centre) then forces the double coset $U_g b U_g$ to contain a uniform element.

The element b given by this lemma lies in some $G(\mathcal{A})$. If $f - g \geq 2$ (i. e., $f \geq 4$: small conductor cases need more care) and if also $\mathcal{A} \subset \mathcal{M}$, we would be through, since some component of $\theta|G(A)$ would be nondegenerate. The structure theory of principal orders tells us that b normalises some $A \subset M$ precisely in the following situation. Let L be an indecomposable left M-lattice (e. g. $L = \mathcal{O}_F^n$ if $M = M_n(\mathcal{O}_F)$); then the set of lattices $\{b^i \pi_F^j L : i,j \in \mathbb{Z}\}$ is linearly ordered. (These lattices then form a "chain" of which b is an automorphism, so b normalises the chain order of this chain.)

We indicate the first step (in the case $f \geq 4$) in getting to this picture. ...urring in $\theta|U_g(M)$, choose any b satisfying (4). This element b defines a k-endomorphism of the k-space $(k = \mathcal{O}_F/\mathfrak{p}_F)$ $L/\pi_F L$ whose rank we denote by $rk_L(b)$. This quantity depends only on $\chi|U_{f-1}$, so we sometimes denote it by $rk(\chi)$.

Theorem 3. Let χ be a character occurring in $\theta|U_{f-1}$ such that $rk(\chi)$ is minimal. Let $b = b_\chi$ be as in the Lemma (with $g = f - 1$). Then, for any indecomposable M-latice L, we have $L \supset bL \supset \pi_F L$.

The index $(L:bL)$ is $|\det b|^{-1}$, so we have

$$1 \leq q^{n(d+f)-m(\Pi)} < q^n ,$$

the strict inequality coming from the fact $b \notin \pi_F M$ (so $bL \not\subset \pi_F L$). This gives a new proof of the following Corollary, (due originally to Kazdan-Novodvorsky in the supercuspidal case: see [56] for a fuller discussion of results of this kind).

Corollary. Let Π be an irreducible admissible representation of $GL_n(F)$. Let $\mathcal{D}_F = \mathfrak{p}_F^d$. Then, if $\varepsilon(\Pi,s) = C(\Pi)q^{-m(\Pi)s}$, where $C(\Pi) \in \mathbb{C}$ and $m(\Pi) \in \mathbb{Z}$, we have $m(\Pi) \geq nd$. If Π is supercuspidal, with conductor $f = f(\Pi)$ as above, then $m(\Pi) > n(d + f - 1)$.

The basic idea of the proof of Theorem 3 (when $f(\Pi) \geq 4$) is as follows. Let L' be an $\mathcal{O}_F$-lattice, $L \supset L' \supset \pi_F L$, and let M' be the associated maximal order $M' = \mathrm{End}_{\mathcal{O}_F}(L')$. Then $U_{g+1}(M') \subset U_g(M)$. Suppose also that $b_\chi L' \subset L'$ (i. e., $b_\chi \in M'$). Then the character $\chi' = \chi|U_{g+1}(M')$ is null on $U_f(M')$, and the rank $rk(\chi')$ is just $rk_{L'}(b_\chi)$. The natural representation of $U_0(M')$ on the $U_f(M')$-fixed vectors is equivalent to θ (under a conjugacy taking M' to M) and Theorem 3 follows from an easy contradiction argument based on the following piece of algebra:

Lemma. *Let* $c \in \mathcal{M}$, $\notin \pi_F\mathcal{M}$, *be uniform and regular. Let* L *be an indecomposable left* $\mathcal{M}$*-lattice and suppose that* $cL \not\supset \pi_F L$. *Then there exists an* $\mathcal{O}_F$*-lattice* L' *satisfying*

(i) $L \supset L' \supset \pi_F L$;

(ii) $cL' \subset L'$;

(iii) $rk_{L'}(c) < rk_L(c)$.

The proof comes from the theory of lattices over commutative orders, and is related to the ideas in Singer's proof of the Dade-Taussky-Zassenhaus theorem (see Curtis-Reiner [24]).

The proof of Theorem 2 proceeds by showing that $\chi|U_{f-2}$ progressively defines the lattices $b^i\pi_F^jL$ sufficiently well to allow the use of a "rank reduction" argument like theorem 3 applied to elements $\pi_F^{-j}b^i$. The method goes through simply if $f(\Pi) \geq 6$. The small conductor cases need more care and more explicit computation. We say nothing about this here.

References

[11] C.J. Bushnell, A. Fröhlich: Gauss sums and p-adic division algebras; Lecture Notes in Mathematics 987, Springer Verlag, Berlin-Heidelberg-New York, 1983.

[12] C.J. Bushnell, A. Fröhlich: Non-abelian congruence Gauss sums and p-adic simple algebras; Proc. London Math. Soc. (3), 50 (1985), 207 - 264.

[24] C.W. Curtis, I. Reiner: Methods of representation theory I; Wiley, New York, 1981.

[26] P. Deligne: Les constantes des équations fonctionnelles des fonctions L , in: Modular functions of one variable II (ed. P. Deligne, W. Kutzko), Lecture Notes in Mathematics 349, Springer Verlag, Berlin-Heidelberg-New York, 1979, 501 - 597.

[29] A. Fröhlich: Galois module structure of algebraic integers; Springer Verlag, Berlin-Heidelberg-New York, 1983.

[39] R. Godement, H. Jacquet: Zeta functions of simple algebras; Lecture Notes in Mathematics 260, Springer Verlag, Berlin-Heidelberg-New York, 1972.

[54] H. Jacquet: Principal L-functions of the linear group, in: Automorphic forms, representations and L-functions (A.Borel & w. Casselman ed.), Proc. Symp. P. Math. 33 vol 2 (American Math. Soc. Providence, 1979), 63 - 86.

[56] H. Jacquet, I.I. Piatetski-Shapiro, J. Shalika: Conducteur des représentations du groupe linéaire; Math. Ann. 256 (1981), 199 - 214.

[73] P. Kutzko, D. Manderscheid: On the supercuspidal representations of GL_4, I; Duke Math. J. 52 (1985), 841 - 867.

[79] I.G. Macdonald: Symmetric functions and Hall polynomials; Oxford University Press, 1979.

[105] J.T. Tate: Fourier analysis in number fields and Hecke's zeta functions; Thesis, Princeton University 1950. Also: Algebraic Theory (J.W.S. Cassels & A. Fröhlich edd.) (Academic Press, London, 1967) 305 - 347.

[106] J.T. Tate: Local constants, in: Algebraic Number Fields (A. Fröhlich ed.), Academic Press, London 1977, 89 - 132.

[Q] J. Queyrut, see this volume.

[Wn] S. Wilson, see this volume.

[K] Ph. Kutzko, see this volume

Contemporary Mathematics
Volume 86, 1989

IDENTITIES ON DEGREE TWO GAMMA FACTORS

by P. Gérardin*, Paris and Wen-Ch'ing Winnie Li*, Pennsylvania

Introduction

The Euler gamma function $\Gamma(s)$ appears on the half plane $\mathrm{Re}(s) > 0$ as $\int_0^\infty e^{-t} t^{s-1} dt$. Using a suitable integration process, we obtain $2(2\pi)^{-s}\Gamma(s)e^{i\pi s/2}$ as the value of the integral $2\int_0^\infty e^{2\pi i t} t^s dt/t$, which can be expressed as a sum of two integrals

$$\int_{\mathbb{R}^\times} e^{2\pi i t}|t|^s dt/|t| \quad + \quad \int_{\mathbb{R}^\times} e^{2\pi i t}\eta_{\mathbb{C}/\mathbb{R}}(t)|t|^s dt/|t| ,$$

where $\eta_{\mathbb{C}/\mathbb{R}}$ is the sign function on $\mathbb{R}^\times$. In the latter two integrals, we integrate a multiplicative character of the field $\mathbb{R}$ against a unitary additive character of the field. This makes sense if $\mathbb{R}$ is replaced by any local field F : our basic gamma factors will be defined in this way, again with a suitable integration process. We introduce them in section 1, through the Fourier transforms of $F^\times$-homogeneous tempered distributions on F , following Tate's thesis. This method generalizes to any division algebra over a local field; this is shown in detail in section 2, where the general definition of gamma factors is also given, following Godement and Jacquet [39], for any separable finite dimensional semi-simple algebra over F . For such an algebra A of rank 2 , gamma factors have been introduced before by Jacquet and Langlands [55]; they appear as functions of a multiplicative character of F and a non-trivial (unitary) additive character of F , and can be seen essentially as a Fourier transform of the character of the representation viewed as a function on $F^\times \times F$ through reduced trace and reduced norm. Recall that A is isomorphic to $F \times F$ (type (I)), or a separable quadratic extension K of F (type (II)), or a quaternion algebra H (type (III)), or the matrix algebra $M_2(F)$ (type (IV)). In section 3, we give relations between the gamma factors of different types, called identities of the first kind. These identities reflect correspondences between irreducible representations of the group of units of these algebras, in particular, regular characters of $K^\times$ define irreducible representations

* Research supported in part by N.S.F. grant no. DMS-8404083

of $H^\times$ and discrete series of $GL_2(F)$, and the irreducible representations of $H^\times$ parametrize the discrete series of $GL_2(F)$; also, all constituents of a principal series representation have the same gamma factors. Our methods give a direct proof of the fact that, for a non-archimedean field F with odd residual characteristic, the discrete series of $GL_2(F)$ is described by the regular characters of the multiplicative groups of the different separable quadratic extensions of F (as is the case for $GL_2(\mathbb{R})$). The identities of the second kind involve two gamma factors of degree two; the main one is the multiplicative formula (4.3), which gives a characterization of the functions which are degree two gamma factors, others relate together two gamma functions of type (II) for two quadratic étale F-algebras.

1. The basic gamma factors

We follow essentially the presentation given by J. Tate in his thesis.

The base field F is a local field, that is, a locally compact non-discrete field; hence, either F is archimedean and then isomorphic (in a unique way) to $\mathbb{R}$ or isomorphic (in two different ways) to $\mathbb{C}$, or F is non-archimedean and then it has a residue field of order q, called the module of F.

The additive characters of F considered will always be unitary; they form a one-dimensional vector space $\hat{F}$ over F. In the archimedean case, we have a natural isomorphism between $\hat{F}$ and F:

for F real, by $(x,y) \longmapsto e^{2\pi ixy}$,

for F complex, by $(x,y) \longmapsto e^{2\pi iT(xy)}$, where T is the trace form from F to $\mathbb{R}$.

In any case, the field F has a canonical absolute value $|\ |$ (which can be seen from the action of the group $F^\times$ on any Haar measure on F: $d(yx) = |y|dx$), and so does the group $\hat{F}$; in the non-archimedean case, the absolute value on $\hat{F}$ is given by the smallest $|\psi| \in |F|$ such that

$$|\psi|\ |x| \leq 1 \Rightarrow \psi(x) = 1, \quad \text{for } \psi \text{ in } \hat{F},\ x \text{ in } F.$$

Let $\mathcal{S}(F)$ be the Fréchet space of Schwartz-Bruhat functions on F; its dual space $\mathcal{S}'(F)$ is the space of tempered distributions on F. The group $F^\times$ acts on these spaces. Let $A(F^\times)$ be the group of characters of $F^\times$, that is, the group of all continuous homomorphisms from $F^\times$ to $\mathbb{C}^\times$, endowed with the compact-open topology. As the group F_u of elements of $F^\times$ of absolute value 1 is the maximal compact subgroup of $F^\times$, the identity component of $A(F^\times)$ is the group of unramified characters of $F^\times$, that is, the group of characters of the image $|F^\times|$ of the absolute value; hence, it is the image of $\mathbb{C}$ by $s \longmapsto (x \longmapsto |x|^s)$; in the non-archimedean case, we identify it with $\mathbb{C}^\times$ by $Z \longmapsto (x \longmapsto Z^{\operatorname{ord} x})$, where $\operatorname{ord} x$ is given by $|x| = q^{-\operatorname{ord} x}$. The positive characters of $F^\times$ are the positive characters of $|F^\times|$; to each χ in $A(F^\times)$, we associate a positive character $|\chi|$ of $F^\times$ given by $|\chi|(x) = |\chi(x)|$, x in $F^\times$. Note that in case F is non-archimedean and χ is given by Z in $\mathbb{C}^\times$, $|\chi|$ is the usual absolute value of Z .

Let dx be the standard Haar measure on F ; it is also the self-dual Haar measure associated to any element of $\hat{F}$ with absolute value 1 .
Define a measure d^* on F by

$$d^*x = |x|^{-1/2}\, dx .$$

(This measure is a "geometric mean" of an additive and a multiplicative Haar measure.) A character $\chi \in A(F^\times)$ is said to be singular if the measure χd^* on $F^\times$ does not extend as a tempered distribution on F . The singular characters are the following:

for F real : $x^{-n}|x|^{-1/2}$, $n \in \mathbb{N}$,

for F complex : $x^{-p}\bar{x}^{-q}$, $p,q \in \mathbb{N}$,

for F non-archimedean : $|x|^{-1/2}$.

If a character χ satisfies $|\chi(x)| < |x|^{-1/2}$ for $|x| < 1$, then it is non-singular, and the corresponding tempered distribution is given by the integral $\int f(x)\chi(x)d^*x$, $f \in S(F)$; in general, for a non-singular χ , it is the so-called finite part of the integral which gives the tempered distribution extending χd^* : for each f in $S(F)$, the integral $\int f(x)\chi(x)d^*x$ is analytic in χ in the above domain, and represents a meromorphic function on $A(F^\times)$, the poles being at the singular characters; these poles are simple, and their residues give tempered distributions on F , which are, up to a non-zero constant, the following:

for F real : $\delta_0^{(n)}$, the n-th derivative of the Dirac measure at 0 , $n \in \mathbb{N}$,

for F complex : $\delta_0^{(p,q)}$, the (p,q) mixed derivative of the Dirac measure at 0 , $p,q \in \mathbb{N}$,

for F non-archimedean : δ_0 , the Dirac measure at 0 .

So, for each χ in $A(F^\times)$, we have defined a non-zero tempered distribution on F ; for χ non-singular, it is the finite part of χd^* , and for χ singular, it is its residue. Each of these distributions is transformed under the action of $y \in F^\times$

by the factor $\chi(y)|y|^{1/2}$, and, up to constant multiples, these are exactly the eigenvectors of the action of $F^\times$ on $\mathcal{S}'(F)$.

Now, each non-trivial ψ in $\hat{F}$ defines an isomorphism of F with $\hat{F}$ via $(x,y) \longmapsto \psi(xy)$; this defines a self-dual Haar measure $d_\psi x$ on F, equal to $|\psi|^{1/2}dx$, and a Fourier transform on $\mathcal{S}(F)$:

$$\hat{f}(x) = \int_F f(y)\psi(xy)d_\psi y .$$

By duality, we have also a Fourier transform $T \longmapsto \hat{T}$ on $\mathcal{S}'(F)$. This Fourier transform on tempered distributions permutes the eigenspaces of the action of $F^\times$. As they are one-dimensional, we arrive at the definition of the basic gamma factor attached to F : for χ and χ^{-1} both non-singular, the gamma factor $\gamma^F(\chi,\psi)$ associated to χ in $A(F^\times)$ and to a non-trivial ψ in $\hat{F}$ is the non-zero complex number given by

$$\widehat{\chi d^*} = \gamma^F(\chi,\psi)\chi^{-1}d^* . \tag{1.1}$$

It is an analytic function of χ, meromorphic on $A(F^\times)$, the poles being at the singular characters and the zeros at the inverses of the singular characters.

We list the first properties of the basic gamma factors $\gamma^F(\chi,\psi)$:

$\gamma^F(1,\psi) = 1$, which expresses that d^* is its own Fourier transform with respect to any non-trivial ψ ;

$$\gamma^F(\chi,\psi') = \chi(t)^{-1}\gamma^F(\chi,\psi) \quad \text{if} \quad \psi'(x) = \psi(tx) , \quad t \in F^\times ; \tag{1.2}$$

$$\gamma^F(\chi,\psi)\gamma^F(\chi^{-1}\,\bar{\psi}) = 1 \quad \text{(complement formula), which comes} \tag{1.3}$$

from the fact that Fourier inversion with respect to ψ is the Fourier transform with respect to $\bar{\psi}$.

The eigenspaces in $\mathcal{S}'(F)$ of $F^\times$ which are preserved by Fourier transform correspond to characters with trivial square; for such a character η, the complement formula reads $\gamma^F(\eta,\psi)^2 = \eta(-1)$.

By class field theory, the closed subgroups of index two in $F^\times$ correspond, by the image of the norm map, to the isomorphism classes of separable quadratic extensions of F. We write $\eta_{K/F}$ for the order two character associated to K : $\eta_{K/F}(x) = 1$ if and only if x is a norm of $K^\times$; then, we write

$$\lambda_{K/F}(\psi) = \gamma^F(\eta_{K/F}, \psi) , \tag{1.4}$$

a fourth root of 1 with square $\eta_{K/F}(-1)$. In particular, $\lambda_{\mathbb{C}/\mathbb{R}}(\psi) = i$ for $\psi(x) = e^{2\pi ix}$.

The basic gamma factors for archimedean fields have simple expressions in terms of the Euler gamma function $\Gamma(s)$; for the standard ψ above,

$$\gamma^{\mathbb{R}}(|\ |^s, \psi) = \pi^{-s}\Gamma(\frac{1/2 + s}{2})/\Gamma(\frac{1/2 - s}{2}) \tag{1.5}$$

$$= L_{\mathbb{C}}(S)\cos\frac{\pi S}{2} , \quad \text{with } S = s + \frac{1}{2} \text{ and } L_{\mathbb{C}}(S) = 2(2\pi)^{-S}\Gamma(S) ;$$

$$\gamma^{\mathbb{R}}(|\ |^s\eta_{\mathbb{C}/\mathbb{R}}, \psi) = i\pi^{-s}\Gamma(\frac{3/2 + s}{2})/\Gamma(\frac{3/2 - s}{2}) \tag{1.6}$$

$$= L_{\mathbb{C}}(S) i \sin\frac{\pi S}{2}, \quad \text{with } S = s + \frac{1}{2} ;$$

$$\gamma^{\mathbb{C}}(x^a\bar{x}^b, \psi \circ T) = i^{a-b}(2\pi)^{-(a+b)}\Gamma(\frac{1}{2} + a)/\Gamma(\frac{1}{2} - b) \tag{1.7}$$

where a,b are two complex numbers with $a-b$ an integer, and $T(x) = x+\bar{x}$.

This formulae are obtained by testing the distributions in (1.1) on suitable functions of $\mathcal{S}(F)$: for (1.5), we take $f(x) = e^{-\pi x^2}$ which is its own Fourier transform for the character ψ , and for (1.6) we take its derivative $f'(x)$, the Fourier transform of which is $if'(x)$; for (1.7), we observe first that

$$\gamma^{\mathbb{C}}(\bar{\chi}, \psi \circ T) = \gamma^{\mathbb{C}}(\chi, \psi \circ T) , \quad \text{where } \bar{\chi}(x) = \chi(\bar{x}) , \tag{1.8}$$

so we assume $a - b = n \geq 0$, and we take for f the partial derivative $f^{(o,n)}(x) = (\partial^n/\partial\bar{x}^n)e^{-\pi T(Nx)}$, $Nx = x\bar{x}$, the Fourier transform of which being $i^n f^{(n,o)}(x) = i^n f^{(o,n)}(\bar{x})$.

The basic gamma factor for $\mathbb{C}$ is related to the basic gamma factor for $\mathbb{R}$ by the formula

$$\lambda_{\mathbb{C}/\mathbb{R}}(\psi)\gamma^{\mathbb{C}}(x^a\bar{x}^b, \psi \circ T) = \gamma^{\mathbb{R}}(|\ |^a, \psi)\gamma^{\mathbb{R}}(|\ |^b\eta_{\mathbb{C}/R}^{a-b+1}, \psi) \tag{1.9}$$

$$= \gamma^{\mathbb{R}}(|\ |^a\eta_{\mathbb{C}/\mathbb{R}}, \psi)\gamma^{\mathbb{R}}(|\ |^b\eta_{\mathbb{C}/\mathbb{R}}^{a-b}, \psi) .$$

All these formulae are obtained from the complement formula for the Euler gamma function $\Gamma(s)\Gamma(1-s) = \pi/\sin\pi s$ and the duplication formula $\Gamma(s)\Gamma(s+\frac{1}{2}) = 2^{1-2s}\pi^{1/2}\Gamma(2s)$.

For a non-archimedean field F of module q , we have

$$\gamma^F(|\ |^s,\psi) = (q^{1/2} - q^s)/(q^{1/2} - q^{-s}) \quad \text{for} \quad |\psi| = 1 . \tag{1.10}$$

When the character χ does not factor through the absolute value, the expression for $\gamma^F(\chi,\psi)$ involves gaussian sums. More precisely, denote by $\mathcal{O}$ the ring of integers of F , and by $\mathcal{P}$ its maximal ideal; if the conductor of χ is 1 , choose any ψ with $|\psi| = q$, then we have

$$\gamma^F(\chi,\psi) = q^{-1/2} \sum_{(\mathcal{O}/\mathcal{P})^\times} \chi(x)\psi(x) ;$$

if the conductor a of χ is ≥ 2 , let $a' = [a/2]$, $a'' = a - a'$, and choose any ψ satisfying $\chi(1+y)\psi(y) = 1$ for y in $\mathcal{P}^{a''}$, then we have

$$\gamma^F(\chi,\psi) = \psi(1) \hat{\int}_{\mathcal{P}^{a'}/\mathcal{P}^{a''}} \chi(1+y)\psi(y) ,$$

where the symbol $\hat{\int}_{\mathcal{P}^{a'}/\mathcal{P}^{a''}}$ means $[\mathcal{P}^{a'}/\mathcal{P}^{a''}]^{-1/2} \sum_{\mathcal{P}^{a'}/\mathcal{P}^{a''}}$.

2. Gamma factors for finite dimensional simple algebras

We first examine the case of a division algebra D of dimension d^2 over its center F , a locally compact non-discrete field. Let T and N be the reduced trace and reduced norm on D . Then, the map $(x,y) \longmapsto T(xy)$ on $D \times D$ is non-degenerate symmetric and bilinear; hence, given a non-trivial additive character ψ of F , the map $(x,y) \longmapsto \psi \circ T(xy)$ identifies the additive group of D with its Pontrjagin dual. We write $d_{\psi\circ T}$ for the corresponding self dual Haar measure on D. For the Hamilton quaternion algebra $\mathbb{H}$ over $\mathbb{R}$, this measure is $4|\psi|^2dx$ with the Lebesque measure dx ; over a non-archimedean field F , this measure is $q^{-d(d-1)/2}|\psi|^{d^2/2}dx$ where dx gives the volume 1 to the ring of integers $\mathcal{O}_D$. The kernel of the norm map $N: D^\times \to F^\times$ is the commutator subgroup of $D^\times$, and the image is $\mathbb{R}_+^\times$ for $D = \mathbb{H}$, and all $F^\times$ otherwise. The factor group $D^\times/F^\times$ is compact, so the irreducible representations of $D^\times$ are finite dimensional. We write $A(D^\times)$ for the set of isomorphism classes of irreducible representations of $D^\times$. By composition with N , we get a map from $A(F^\times)$ to $A(D^\times)$, injective when D is not isomorphic to $\mathbb{H}$, and an action of $A(F^\times)$ on $A(D^\times)$ by twist, that is, $\chi \in A(F^\times)$

acts on the representation Σ by giving the representation $x \longmapsto \chi(Nx)\Sigma(x)$; this gives on $A(D^\times)$ a structure of a one dimensional complex manifold. By restriction to the center $F^\times$, an irreducible representation of $D^\times$ defines a character of $F^\times$, called its central character, and this gives a map from $A(D^\times)$ to $A(F^\times)$; under the twist by $\chi \in A(F^\times)$, the central character is multiplied by χ^d . A positive character of $D^\times$ is a homomorphism from $D^\times$ to $\mathbb{R}_+^\times$; it is of the form $x \longmapsto |Nx|^a$ for some $a \in \mathbb{R}$ and its restriction to $F^\times$ yields a positive character of $F^\times$.

The absolute value on D is defined by $|x|_D = |Nx|$; for a Haar measure dx on the additive group of D , we have $d(yx) = |y|_D^d dx$. The kernel of the absolute value on $D^\times$ is the maximal compact subgroup D_u , and the image is the value group $|F|^\times$. For t in F , one has $|t|_D = |t|^d$. To each $c \in A(D^\times)$ we associate a positive character $|c|$ of $D^\times$ as follows: it is the unique positive character χ such that the twist of a representation of class c by χ^{-1} is unitary; one has $|\omega| = |c|^d$, where ω is the central character of c . Also, $|\mu \circ N| = |\mu|$ for $\mu \in A(F^\times)$.

To an irreducible representation Σ of $D^\times$, we associate its reduced character $c_\Sigma = \mathrm{Tr}\,\Sigma/\dim\Sigma$; the reduced characters parametrize the isomorphism classes of irreducible representations of $D^\times$, so it gives a description of $A(D^\times)$. They are class functions on $D^\times$, taking the value 1 at 1 , and they satisfy a functional equation

$$\iint c(\xi\eta)\,d\xi\,d\eta = c(x)c(y) , \quad x, y \in D^\times , \tag{2.1}$$

where ξ and η run through the conjugacy class of x and y respectively, and the measures on the conjugacy classes are the unique $D^\times$-invariant measures with total volume 1 . This relation (2.1) characterizes the irreducible reduced characters of $D^\times$ among the continuous non-zero functions on $D^\times$.

Let $S(D)$ be the space of Schwartz-Bruhat functions on the additive group of D. In the non-archimedean case, the elements of $S(D)$ are the compactly supported locally constant functions on D . The choice of a non-trivial additive character ψ of F defines a Fourier transformation on $S(D)$: $\hat{f}(x) = \int f(y)\psi\circ T(xy)\,d_{\psi\circ T}y$, which is an automorphism of $S(D)$ with inverse given by $f(y) = \int \hat{f}(x)\psi\circ T(-xy)\,d_{\psi\circ T}x$. The space $S(D)$ is naturally a Fréchet space, and its topological dual $S'(D)$ is called the space of tempered distributions on D . The function $x \longmapsto |x|_D^{-1/2} = |Nx|^{-d/2}$ is locally integrable on D for a Haar measure, and defines a tempered distribution $d^*x = |x|_D^{-1/2}dx$ on D . This distribution is its own Fourier transform. We write $d^*_{\psi\circ T}$ for $|x|_D^{-1/2}d_{\psi\circ T}x$. For each c in $A(D^\times)$, we have a measure cd^* on $D^\times$; if $|x|_D < 1 \Rightarrow |c|(x) < |x|_D^{-1/2}$, then this measure extends to a measure on D , giving the tempered distribution

$f \longmapsto \int f(x)c(x)d^*x$; taking the finite parts of these integrals, we get a meromorphic map from $A(D^\times)$ to $S'(D)$, still written as $c \longmapsto cd^*$; we call singular the elements of $A(D^\times)$ where a pole appears.

Let $\check{c}$ denote the contragredient of c . Then, if c and $\check{c}$ are both non-singular, there exists a non-zero complex number $\gamma^D_{c,\psi}$ such that

$$\widehat{cd^*} = \gamma^D_{c,\psi}\, \check{c}d^* . \tag{2.2}$$

We give a proof here. By analyticity of the maps $\chi \longmapsto \chi \circ N\, cd^*$ from $A(F^\times)$ to $S'(D)$, it is sufficent to prove it when both distributions cd^* and $\check{c}d^*$ are represented by integrals, that is, when the c's are such that $|x|_D^{1/2} < |c|(x) < |x|_D^{-1/2}$ for $|x|_D < 1$. We start from the formula

$$\int_D (x)g(x)dx = \int_D f(x)\hat{g}(x)dx , \qquad \hat{f}, g \in S(D) ,$$

which is due to the symmetry of the Fourier kernel $(x,y) \longmapsto \psi \circ T(xy)$. Let $y \in D^\times$; we replace in this formula the function $g(x)$ by the function $g(yx)$, so the Fourier transform $\hat{g}(x)$ is replaced by $|y|_D^{-1}\hat{g}(xy^{-1})$, and the formula becomes

$$|y|_D^{1/2} \int_D \hat{f}(x)g(yx)dx = |y|_D^{-1/2} \int_D f(x)\hat{g}(xy^{-1})dx .$$

Let Σ be a representation of class c ; we multiply both sides by $\Sigma(y)^{-1}|y|_D^{-1}$ and integrate along the measure d^*y ; changes of variables give

$$\int_{D\times D} \hat{f}(x)g(y)\Sigma(xy^{-1})d^*xd^*y = \int_{D\times D} f(x)\hat{g}(y)\Sigma(x^{-1}y)d^*xd^*y ,$$

or

$$\int_D \hat{f}(x)\Sigma(x)d^*x \int_D g(y)\Sigma(y^{-1})d^*y = \int_D f(x)\Sigma(x^{-1})d^*x \int_D \hat{g}(y)\Sigma(y)d^*y .$$

With $f = g$, we see that the operators $\int \hat{f}(x)\Sigma(x)d^*x$ and $\int f(x)\Sigma(x^{-1})d^*x$ commute; now, there is an f such that $\int f(x)\Sigma(x^{-1})d^*x$ is invertible: we take $f(tu) = f_1(t)\, c_\Sigma(u)$, $t \in F^\times$, $u \in D_u$, for which the operator is scalar, and non zero for a suitable f_1.This proves the existence of a complex number $\gamma^D_{c,\psi}$ such that

$$\int_D \hat{g}(y)\Sigma(y)d^*y = \gamma^D_{c,\psi} \int_D g(y)\Sigma(y^{-1})d^*y , \quad \text{for} \quad g \in S(D) . \tag{2.3}$$

Applying the formula again to the transposed operator of $\int g(y)\Sigma(y^{-1})d^*y$, we get

$$\gamma^D_{c,\psi}\gamma^D_{\check{c},\bar{\psi}} = 1, \quad \text{or} \quad \gamma^D_{c,\psi}\gamma^D_{\check{c},\psi} = \omega(-1),$$

where ω is the central character of c. These two formulae prove (2.2). We define the gamma factor of c (with respect to F) as the function of a couple (χ,ψ) with χ in $A(F^\times)$ and ψ a non-trivial additive character of F, analytic on $A(F^\times)$,

$$\gamma^F_c(\chi,\psi) = \lambda_{D/F}\,\gamma^D_{c\otimes\chi,\psi} \quad \text{where} \quad \lambda_{D/F} = (-1)^{d-1}. \tag{2.4}$$

The behavior under change of ψ is given by

$$\gamma^F_c(\chi\ \psi') = \omega(t)^{-1}\chi(t)^{-d}\gamma^F_c(\chi,\psi) \quad \text{if} \quad \psi'(x) = \psi(tx). \tag{2.5}$$

Formally, we view $\gamma^D_{c,\psi}$ as the integral of $\psi\circ T$ for $cd^*_{\psi\circ T}$.

In particular, for c factoring through the absolute value: we have the following explicit formula:

$$\gamma^{\mathbf{H}}_{|\ |^s,\psi} = (2\pi)^{-2s}\Gamma(1+s)/\Gamma(1-s) \quad \text{if} \quad |\psi| = 1,$$ in the archimedean case,

$$\gamma^D_{|\ |^s,\psi} = q^{(d-1)s}(q^{d/2}-q^s)/(q^{d/2}-q^{-s}) \quad \text{if} \quad |\psi| = 1,$$ in the non-archimedean case.

Assume now that A is a finite dimensional separable semi-simple algebra over the local field F. Let $A(A^\times)$ be the set of classes of irreducible representations of $A^\times$; there is again an action of $A(F^\times)$ on $A(A^\times)$ by twist: $c \longmapsto c\otimes\chi$, $\chi\in A(F^\times)$, coming from the twist by $\chi\circ N$ on representations, where N is the reduced norm on A. This gives to $A(A^\times)$ the structure of a one-dimensional complex manifold. Given a non-trivial additive character ψ of F, Godement and Jacquet [39] have defined a meromorphic map $c \longmapsto \gamma^A_{c,\psi}$ from $A(A^\times)$ to $S'(A)$, the space of tempered distributions on A: when c is represented by a cuspidal representation Π, the method is similar to the one above, using integrals $\int f(x)\Pi(x)d^*x$, $f\in S(A)$, $d^*x = |Nx|^{-1/2}dx$, with dx a Haar measure on A. In general, an irreducible representation of $A^\times$ occurs as a component of an induced representation from a parabolic subgroup of $A^\times$ of a cuspidal representation of the Levi quotient, which is the multiplicative group of a similar algebra $'A$, and $\gamma^A_{c,\psi} = \gamma^{'A}_{'c,\psi}$, with obvious notations. The algebra A is the direct product of its simple ideals, each one of them being isomorphic to an algebra $M_n(D)$ where D is a division algebra of finite dimension over F ; the space $A(A^\times)$ decomposes accordingly, and $\gamma^A_{c,\psi}$ is the product of the gamma factors of the components of c . For A isomorphic to $M_n(D)$ with D central over F , we write $\lambda_{A/F}(\psi) = \lambda_{D/F}(\psi)^n$ and $\gamma^F(\chi,\psi) = \lambda_{A/F}\gamma_{c \otimes \chi,\psi}$ for $\chi \in A(F^\times)$. In particular, for A isomorphic to $M_n(F)$, then $\lambda_{A/F}(\psi) = 1$ and $\gamma^F_c(\chi,\psi) = \gamma^A_{c \otimes \chi,\psi}$. When A is isomorphic to a product of matrix algebras over central division algebras over F , $\lambda_{A/F}(\psi)$ is defined as the product of the corresponding λ's , and γ^F_c , as the product of the corresponding γ's .

Assume now that D is a finite dimensional division algebra with center F , a non-archimedean field with module q ; let d be the square root of the dimension of D over F . Call $\mathcal{O}_D$ the ring of integers of D and $\mathcal{P}_D$ its maximal ideal; then, the subgroups $1 + \mathcal{P}^n_D$, $n \geq 1$, are normal in $D^\times$, and the subquotients $(1+\mathcal{P}^m_D)/(1+\mathcal{P}^n_D)$ are commutative for $2m \geq n$. The conductor $f(\Sigma)$ of an irreducible representation Σ of $D^\times$ is 0 if Σ factors through the absolute value, otherwise it is the smallest integer f such that $1 + \mathcal{P}^f_D$ lies in the kernel of Σ . For $f(\Sigma) \leq 1$, the representation Σ factors through $D^\times/(1+\mathcal{P}_D)$; choose an unramified field F_d of degree d over F in D , so $D^\times/(1+\mathcal{P}_D)$ appears as the semi-product of $F^\times_d/(1+\mathcal{P}_d)$ by the Galois group C_d of F_d over F , a cyclic group of order d coming from $|D^\times|/|F^\times_d|$; a character θ of $F^\times_d$ trivial on $1 + \mathcal{P}_d$ and regular under C_d gives by induction an irreducible representation Σ_θ of $D^\times$ of degree d , with character supported on $F^\times_d(1+\mathcal{P}_D)$ given here by $t(1+x) \longmapsto \theta(t)$, so $\gamma^D_{\Sigma_\theta,\psi} = \gamma^{F_d}(\theta,\psi \circ T)$. In general, for each divisor $'d$ of d , let $F_{'d}$ be the subextension of F_d with degree $'d$ over F ; the action of C_D on F_d given by D induces an action of $C_{'d}$ on $F_{'d}$ which defines a division algebra $'D$ central over F and split by $F_{'d}$ and we have a norm map from $D^\times/(1+\mathcal{P}_D)$ onto $'D^\times/(1+\mathcal{P}_{'D})$; a regular character $'\theta$ of $F^\times_{'d}$ trivial on $1 + \mathcal{P}_{'D}$ defines as above an irreducible representation $'\Sigma_{'\theta}$ of $'D^\times$ of degree $'d$, hence by composition with the norm, an irreducible representation $\Sigma_{'\theta}$ of $D^\times$, and its gamma factor $\gamma^D_{\Sigma_{'\theta}}$ is $\gamma^{F_{'d}}('\theta,\psi \circ {'T})$. This describes all irreducible representations of $D^\times$ with conductor at most 1 . If now the irreducible representation Σ of $D^\times$ has a conductor $f = f(\Sigma) > 1$, we set $f' = [f/2]$ and $f'' = f - f'$. The representation factors through $D^\times/(1+\mathcal{P}^f)$; the subgroup $(1+\mathcal{P}^{f''})/(1+\mathcal{P}^f)$ is normal and commutative; the characters of this subgroup occuring in the restriction of Σ form one orbit under conjugation of $D^\times$, due to irreducibility of Σ ; let ξ be one of them. Let $\mathrm{ord}_D \psi \circ T = d\, \mathrm{ord}\psi + d - 1$, so $d_{\psi \circ T} x = |\psi \circ T|^{d/2} dx$ where dx is the Haar measure on D giving to $\mathcal{O}_D$ the volume 1 . There is a ξ_ψ in $\mathcal{P}_D^{-\mathrm{ord}_D \psi \circ T - f}$, unique mod $\mathcal{P}_D^{-\mathrm{ord}_D \psi \circ T - f'}$, such that $\xi(1+y)\psi \circ T(\xi_\psi y) = 1$ for all y in $\mathcal{P}^{f''}_D$. We choose in (2.3) the function g to be the characteristic function of $1+P_D^f$, and assume that the reduced character c of Σ satisfies $|c|<q^{d/2}$, so (2.3) reads:

$$\gamma^D_{c,\psi}I_\Sigma = \int_{P_D^{-f-\mathrm{ord}_D\psi\circ T}} \Sigma(x)\psi\circ T(x)d^*_{\psi\circ T}x ,$$

where I_Σ is the identity operator on the space of Σ. Let $\Sigma(x)_\xi$ be the composition on both sides of $\Sigma(x)$ with the projection on the ξ-isotypic component and I_ξ be the identity operator of this subspace; then

$$\gamma^D_{c,\psi}I_\xi = \int_{P_D^{-f-\mathrm{ord}_D\psi\circ T}} \Sigma(x)_\xi\psi\circ T(x)d^*_{\psi\circ T}x .$$

For y in $P_D^{f''}$, one has $\Sigma(x(1+y))_\xi = \Sigma(x)_\xi\Sigma(1+y) = \Sigma(x)_\xi\psi\circ T(-\xi_\psi y)$, and

$$\gamma^D_{c,\psi}I_\xi = \int_{P_D^{-f-\mathrm{ord}_D\psi\circ T}} \Sigma(x)_\xi\psi\circ T((x-\xi_\psi)y)\, d^*_{\psi\circ T}x , \qquad y\in P_D^{f''} ;$$

by averaging over all y in the compact group $P_D^{f''}$, we see that it is enough to take x in $\xi_\psi + P_D^{-f''-\mathrm{ord}_D\psi\circ T} = \xi_\psi(1+P_D^{f'})$, and, using the fact that $\Sigma(x)_\xi\psi\circ T(x)$ is constant on $\xi_\psi(1+P_D^{f''})$, we arrive at

$$\gamma^D_{c,\psi}I_\xi = \hat{\int}_{\xi_\psi(1+P_D^{f'})/(1+P_D^{f''})} \Sigma(x)_\xi\psi\circ T(x) \quad \text{where the symbol}$$

$$\hat{\int}_{\xi_\psi(1+P_D^{f'})/(1+P_D^{f''})} \quad \text{means} \quad [P_D^{f'}:P_D^{f''}]^{-1/2}\sum_{\xi_\psi(1+P_D^{f'})/(1+P_D^{f''})} .$$

This shows that for f even

$$\gamma^D_{c,\psi}I_\xi = \psi\circ T(\xi_\psi)\Sigma(\xi_\psi)_\xi .$$

For f odd, we introduce the operators

$$Q_{\Sigma,\psi,\xi_\psi}(z) = \psi\circ T(\xi_\psi z)\Sigma(1+z)_\xi , \qquad z\in P_D^{f'} ;$$

they satisfy the quadratic identity

$$Q_{\Sigma,\psi,\xi_\psi}(z+z') = \psi\circ T(\xi_\psi zz')Q_{\Sigma,\psi,\xi_\psi}(z)Q_{\Sigma,\psi,\xi_\psi}(z') , \quad z,z'\in P_D^{f'} .$$

The scalar $\psi\circ T(\xi_\psi zz')$ depends only on z and z' mod $P_D^{f''}$, and gives a self duality on $P_D^{f'}/P_D^{f''}$, which is of dimension one over O_D/P_D for f odd. Finally, we get the formula

$$\gamma^D_{c,\psi} I_\xi = \psi\circ T(\xi_\psi)\Sigma(\xi_\psi)_\xi \int^{\wedge}_{P_D^{f'}/P_D^{f''}} Q_{\Sigma,\psi,\xi_\psi}(z) \ .$$

3. Degree two gamma factors, first set of identities and correspondences

Assume that A is a separable rank 2 semi-simple algebra over the local field F. On A, there is an involution $x \longmapsto \bar{x}$, a reduced norm N and a reduced trace T; moreover, A is isomorphic to one of the following algebras:

(I) $F \times F$, for $x = (u,v)$, $\bar{x} = (v,u)$, $Nx = uv$, $Tx = u+v$; its automorphism group is the two element group, generated by the involution;

(II) K, a separable quadratic extension of F, $\bar{x}$ being the Galois conjugate of x, and N, T the norm and trace from K to F; its automorphism group is the two element group, generated by the involution;

(III) H, a quaternion algebra with center F, N and T being the reduced norm and reduced trace; the automorphism group is $H^\times/F^\times$;

(IV) $M = M_2(F)$, the algebra of 2×2 matrices over F; the involution send $x = \begin{pmatrix} u & v \\ w & t\end{pmatrix}$ to $\bar{x} = \begin{pmatrix} t & -v \\ -w & u\end{pmatrix}$, $N = \det$, $T = \mathrm{tr}$; its automorphism group is $M^\times/F^\times$.

In each case, the map

$$(N,T)\colon A^\times \longmapsto F^\times \times F \tag{3.1}$$

is constant on the orbits by the group $\mathrm{Aut}_F A$ of F-automorphisms of A; this map is proper in types (II) and (III), when $A^\times/F^\times$ is compact; in types (I) - (III) it is injective from $A^\times/\mathrm{Aut}_F A$ onto a closed subset of $F^\times \times F$; in type (IV), it gives a surjective map from $M^\times/\mathrm{Aut}_F M$, which is the set of conjugacy classes of $M^\times$, onto $F^\times \times F$, and this map is injective except that the two conjugacy classes of elements having a given central element as semi-simple part have the same image.

Given a non-trivial ψ in $\hat{F}$, in each case we introduce a fourth root of 1 as follows. The function $\psi\circ T$ defines a tempered distribution on A, and its

Fourier transform with respect to $\psi \circ T$ is $\psi^{-1} \circ N$ up to the product by a fourth root of 1 denoted by $\lambda_{A/F}(\psi)$: it is 1 for A of type (I) or (IV), -1 for A of type (III), and is equal to $\gamma^F(\eta_{K/F},\psi)$ in case (II), where $\eta_{K/F}$ is the non-trivial character of $F^\times$ trivial on the image of $K^\times$ under N . If in case (I) we write $\eta_{A/F}$ for the trivial character of $F^\times = NA^\times$, then $\lambda_{A/F}(\psi)(=1)$ is also equal to $\gamma^F(\eta_{A/F},\psi)$.

The group $\mathrm{Aut}_F(A)$ acts on the space $\mathcal{A}(A^\times)$; this action commutes with the action of $\mathcal{A}(F^\times)$; it is trivial for the cases (III) and (IV). We denote by $\mathcal{A}_F(A^\times)$ the set of orbits of $\mathrm{Aut}_F(A)$ in $\mathcal{A}(A^\times)$. Hence, for type (I), $\mathcal{A}_F(A^\times)$ is the set of pairs $\{\mu,\nu\}$ of characters of $F^\times$; for type (II), $\mathcal{A}_F(K^\times)$ is either a regular character θ of $K^\times$ determined up to Galois action $(\bar\theta(x) = \theta(\bar x))$ or a character μ of $F^\times$ determined up to its product by $\eta_{K/F}$: μ and $\mu\eta_{K/F}$ are the same when composed with N ; for types (III) and (IV), since any F-automorphism is inner we have $\mathcal{A}_F(A^\times) = \mathcal{A}(A^\times)$.

For a character χ of $F^\times$ and a representation Π of $A^\times$, write $\chi \otimes \Pi$ for the representation $x \longmapsto \chi(Nx)\Pi(x)$; this gives an action of $\mathcal{A}(F^\times)$ on $\mathcal{A}(A^\times)$ commuting with the action of $\mathrm{Aut}_F A$, hence an action of $\mathcal{A}(F^\times)$ on $\mathcal{A}_F(A^\times)$, called the twist by characters of $F^\times$. By the gamma factor of an element $\Pi \in \mathcal{A}_F(A^\times)$, we mean the function γ_Π^F :

$$(\chi,\psi) \longmapsto \lambda_{A/F}(\psi)\gamma^A_{\chi \otimes \Pi,\psi} \tag{3.2}$$

where Π represents the given element of $\mathcal{A}_F(A^\times)$, χ is a character of $F^\times$, and ψ is a non-trivial unitary character of F . The central character ω_Π of Π is, for types (I), (III), (IV) the character of $F^\times$ by which $F^\times \subset A^\times$ acts; for type (II) it is the product of this character by $\eta_{K/F}$. By definition, a degree two local factor for F is a function $\gamma(\chi,\psi)$ equal to a $\gamma_\Pi(\chi,\psi)$ for some Π in $\mathcal{A}_F(A^\times)$ for some A as above. At least for type (I) - (III), the gamma factor of an irreducible representation of A appears, up to the factor $\lambda_{A/F}(\psi)$, as a Fourier transform on the commutative group $F^\times \times F$ of the product of the reduced character of the representation by the jacobian of the direct image of the measure $d^*_{\psi\circ T}$ under the map (N,T) ; we call it also the Fourier transform of the representation.

The degree two gamma factors for F have the following properties. The first one, called the additive condition, indicates the behavior with respect to the non-trivial additive character:

$$\gamma(\chi,\psi') = (\omega\chi^2)(t^{-1})\gamma(\chi,\psi) \quad \text{if } \psi' \text{ is } u \longmapsto \psi(tu) , \tag{3.3}$$

where ω is the central character. The second property states that in χ the function $\gamma(\chi,\psi)$ is meromorphic on $A(F^\times)$, with poles and zeros of order at most 2, and, if F is non-archimedean, $\gamma(\chi,\psi)$ is a rational function on $A(F^\times)$. The third property, called the complement formula, is

$$\gamma(\chi,\psi)\gamma(\chi^{-1}\omega^{-1},\psi^{-1}) = 1 ; \tag{3.4}$$

it comes from the functional equation (2.3), the relation $\lambda_{A/F}(\psi)\lambda_{A/F}(\psi^{-1}) = 1$, and the fact that the identity $\bar{x}^{-1} = x(Nx)^{-1}$ implies $\Pi(\bar{x}^{-1}) = (\omega^{-1}\otimes\Pi)(x)$ so $\omega^{-1}\otimes\Pi$ is a realisation of the contragredient representation of Π. The next property - which is similar to the Davenport-Hasse identity - expresses the gamma factor of a representation of $A^\times$ which factors through N in terms of a gamma factor attached to a degree one representation of $F^\times\times F^\times$. Observe first that $\mu\otimes\nu : (u,v) \longmapsto \mu(u)\nu(v)$ for $\mu,\nu \in A(F^\times)$ has gamma factor:

$$\gamma^F_{\mu\otimes\nu}(\chi,\psi) = \gamma^F(\chi\mu,\psi)\gamma^F(\chi\nu,\psi) ; \tag{3.5}$$

the relations are:

$$\lambda_{K/F}(\psi)\gamma^K((\mu\chi)N , \psi\circ T) = \gamma^F(\mu\chi\eta_{K/F},\psi)\gamma^F(\mu\chi,\psi) \quad \text{for type (II),} \tag{3.6}$$

and

$$\lambda_{A/F}(\psi)\gamma^A_{\chi\circ N,\psi} = \gamma^F(\chi|\ |^{1/2},\psi)\gamma^F(\chi|\ |^{-1/2},\psi) \quad \text{for types (III) - (IV).} \tag{3.7}$$

For each pair μ,ν of characters of $F^\times$, we introduce $\Pi(\mu,\nu) \in A(M^\times)$ as follows: let $\Sigma(\mu,\nu)$ be the representation of $M^\times$ induced from the character $\begin{pmatrix} u & w \\ o & v\end{pmatrix} \longmapsto \mu(u)\nu(v)$ of the upper triangular subgroup; it is realized in the space of smooth functions on $M^\times$ satisfying $F(\begin{pmatrix} u & w \\ o & v\end{pmatrix}g) = \mu(u)\nu(v)|u/v|^{1/2}F(g)$, the action of $M^\times$ being by right translations. When $\Sigma(\mu,\nu)$ is irreducible, call it $\Pi(\mu,\nu)$; otherwise, $\Pi(\mu,\nu)$ is the unique irreducible component which is finite dimensional: this defines $\Pi(\mu,\nu)$. Then, the gamma factor of $\Pi(\mu,\nu)$ is of type (I):

$$\gamma^F_{\Pi(\mu,\nu)}(\chi,\psi) = \gamma^F(\chi\mu,\psi)\gamma^F(\chi\nu,\psi) . \tag{3.8}$$

More precisely, the irreducible finite dimensional representations of $M^\times$ occur as follows:

a) $F = \mathbb{R}$: $\Pi(\mu,\nu)$ for $(\mu\nu^{-1})(x) = x^n\eta_{\mathbb{C}/\mathbb{R}}(x)$, n a non-zero integer;

b) $F = \mathbb{C}$: $\Pi(\mu,\nu)$ for $(\mu\nu^{-1})(x) = x^m\bar{x}^n$, m, n non-zero integers of same sign;

c) F non-archimedean: $\Pi(\mu,\nu)$ for $(\mu\nu^{-1})(x) = |x|^{\pm 1}$.

Assume $F = \mathbb{R}$. Let $n \geq 1$ be a positive integer and ω a character of $\mathbb{R}^\times$ satisfying the condition $\omega(-1) = -(-1)^n$. To (n,ω) , we associate:

a) a regular element $\theta_{n,\omega}$ in $A_{\mathbb{R}}(\mathbb{C}^\times)$ by taking the two characters $x \longmapsto x^a\bar{x}^b$ of $\mathbb{C}^\times$ defined by the relations

$$a - b = \pm n \ , \ \omega(x) = x^{a+b} \quad \text{for} \quad x > 0 \ ;$$

in this way, we get a parametrization of the set $A_{\mathbb{R}}(\mathbb{C}^\times)'$ of regular elements of $A_{\mathbb{R}}(\mathbb{C}^\times)$ by this set of (n,ω)'s ;

b) an element $\Sigma_{n,\omega}$ in $A_{\mathbb{R}}(\mathbb{H}^\times) = A(\mathbb{H}^\times)$: it is the class of irreducible degree n representations of the quaternion group $\mathbb{H}^\times$ with central character ω ; it is obtained from the standard degree n representation Σ_n in the (n-1)th symmetric tensor product of the natural degree two representation of $\mathbb{H}^\times$ by twisting it with any $\mu \in A(\mathbb{R}^\times)$ such that $x^{n-1}\mu(x)^2 = \omega(x)$; there are two such μ's , giving rise to equivalent representations; then $A(\mathbb{H}^\times)$ is parametrized by our (n,ω)'s ;

c) two distinct elements $\Pi^{\pm}_{n,\omega}$ in $A(M^\times)$, representing the two irreducible degree n representations of $M^\times$ with central character ω : they are obtained from the standard degree n representation Π_n of $\mathbb{M} = M_2(\mathbb{R})$ in the space of degree $n-1$ homogeneous polynomials in two variables by twisting it with the characters μ and $\mu\eta_{\mathbb{C}/\mathbb{R}}$ where $x^{n-1}\mu(x)^2 = \omega(x)$, $\mu(-1) = 1$; in particular, $\Pi^+_{1,1}$ is the trivial representation of $M^\times$; also, $\Pi_n = \Pi^+_{n,x^{n-1}} = \Pi(u^{n-1}|u|^{1/2} , |v|^{1/2})$;

d) an element $\Delta_{n,\omega}$ in $A(M^\times)$, representing the irreducible representation of the discrete series with formal degree $-n$ and central character ω ; then, the discrete series of $M^\times$ is parametrized by our (n,ω)'s ; further, $\Delta_{n,\omega}$ is the second component of the 2 principal series representations in which $\Pi^+_{n,\omega}$ and $\Pi^-_{n,\omega}$ occur;

e) two distinct elements $\alpha^{\pm}_{n,\omega}$ of $A_{\mathbb{R}}(\mathbb{R}^\times \times \mathbb{R}^\times)$ defined by the (distinct) pairs $\{\omega|\ |^{-a}, |\ |^a\}$ and $\{\omega|\ |^{-a}\eta_{\mathbb{C}/\mathbb{R}}, |\ |^a\eta_{\mathbb{C}/\mathbb{R}}\}$, where a is as above (in a)).

We extend also to $(0,\omega)$ for $\omega(-1) = -1$; let μ be any character such that $\mu(x)^2 = \omega(x)$ for $x > 0$; then, we define

a) $\alpha_{o,\omega}$ in $A_{\mathbb{R}}(\mathbb{R}^\times \times \mathbb{R}^\times)$ to be the pair $\{\mu, \mu\eta_{\mathbb{C}/\mathbb{R}}\}$,

b) $\theta_{o,\omega}$ in $A_{\mathbb{R}}(\mathbb{C}^\times)$ to be the non-regular character $\mu \circ N$ of $\mathbb{C}^\times$,

c) $\Pi_{o,\omega}$ in $A(M^\times)$ to be the irreducible principal series $\Pi(\mu, \mu\eta_{\mathbb{C}/\mathbb{R}})$.

Hence, we have defined for each couple (n,ω) , $n \in \mathbb{N}$, $\omega \in A(\mathbb{R}^\times)$ with $\omega(-1) = -(-1)^n$, a family of seven irreducible representations when $n \geq 1$, and three irreducible representations when $n = 0$. Then, all representations attached to the same (n,ω) have the same gamma factor. (Note that $n = 0$ is included in (3.5), (3.6), (3.8)). This statement, with (3.5), (3.8), gives rise to the identities between degree two gamma factors for $\mathbb{R}$, and these cases are the only ones: two different elements of $A_{\mathbb{R}}(A^\times)$, $A_{\mathbb{R}}(A'^\times)$ have the same gamma factor if and only if it is $\Pi(\mu,\nu)$ and $\mu \otimes \nu$ for some characters μ,ν of $\mathbb{R}$, or there exists (n,ω) as above and they are in the family associated to it.

When F is $\mathbb{C}$, the types (II) and (III) do not exist; also, the elements of $A(M^\times)$ are exactly the $\Pi(\mu,\nu)$ for μ,ν in $A(\mathbb{C}^\times)$. The finite dimensional irreducible representations of $M^\times = GL_2(\mathbb{C})$ are parametrized by their bidegree (m,n) , a couple of positive integers, and their central character ω as follows: ω satisfies $\omega(-1) = (-1)^{m+n}$, and $\Pi_{m,n;\omega}$ is the class of the representation $(\Pi_m \otimes \bar{\Pi}_n) \otimes \mu$ where Π_m is the standard degree m representation of $M_2(\mathbb{C})$ in the space of degree $m-1$ homogeneous polynomials in two variables, $\bar{\Pi}_n$ is the transform of Π_n by complex conjugation, and μ is the character of $\mathbb{C}^\times$ such that $x^{m-1}\bar{x}^{n-1}\mu(x)^2 = \omega(x)$. Then

$$\Pi_{m,n;\omega} = \Pi(u^m\bar{u}^n|u|^{-1/2}, |v|^{-1/2})$$

and the other component $\Pi'_{m,n;\omega}$ of the principal series representation $\Sigma(u^m\bar{u}^n|u|^{-1/2}, |v|^{-1/2}) \otimes \mu$ is $\Pi(u^m|u|^{-1/2}, \bar{v}^n|v|^{-1/2}) \otimes \mu$. The two elements $\Pi_{m,n;\omega}$ and $\Pi'_{m,n;\omega}$ have the same gamma factor, and the identity obtained is the only case of equality of gamma factors between elements of $A(M^\times)$, and also between elements of $A_{\mathbb{C}}(\mathbb{C}^\times \times \mathbb{C}^\times)$.

Assume now that F is a non-archimedean field. While the degree two gamma factors for archimedean fields are all of type (I), this is no more true now. We have the following results:

a) except for A of type (IV), elements of $A_F(A^\times)$ are characterized by their gamma factors; for A of type (IV), the only distinct elements of $A_F(A^\times)$ that give rise to the same gamma factor are the two components of the reducible principal series $\Sigma(|\ |^{1/2}, |\ |^{-1/2}) \otimes \mu$, $\mu \in A(F^\times)$, that is, the one-dimensional representation $\mu \circ N = \Pi(|\ |^{1/2}, |\ |^{-1/2}) \otimes \mu$ and the special representation $\sigma(\mu)$;

b) the identity (3.8) is the only case of equality between gamma factors of type (I) and type (II);

c) two gamma factors attached to two distinct separable quadratic extensions K and K' of F are equal if and only if, B being their composite extension $K \otimes K'$, they come from characters θ of $K^\times$ and θ' of $K'^\times$ satisfying $\bar{\theta} = \theta\eta_{B/K}$, $\bar{\theta}' = \theta'\eta_{B/K'}$, $\theta \circ N_{B/K} = \theta' \circ N_{B/K'}$ (where $^-$ indicates the Galois conjugation):

$$\lambda_{K/F}(\psi)\gamma^K(\theta\chi \circ N, \psi \circ T) = \lambda_{K'/F}(\psi)\gamma^{K'}(\theta'\chi \circ N', \psi \circ T') ; \qquad (3.9)$$

d) let $A_F(K^\times)'$ be the subset of elements in $A_F(K^\times)$ coming from regular characters of $K^\times$; then, for each θ in $A_F(K^\times)'$, there is a unique Σ_θ in $A(H^\times)$ with the same gamma factor:

$$\lambda_{K/F}(\psi)\gamma^K(\theta\chi \circ N, \psi \circ T) = -\gamma^H_{\Sigma_\theta \otimes \chi, \psi} , \qquad \theta \in A_F(K^\times)' ; \qquad (3.10)$$

if the residual characteristic of F is not 2 , and only in this case, the Σ_θ's together with the one-dimensional representations $\mu \circ N$ exhaust $A(H^\times)$, the only coincidences being given by c) above;

e) for each θ in $A_F(K^\times)$; there is a unique Π_θ in $A(M^\times)$ with the same factor; if θ is regular, then Π_θ is (super)cuspidal; if θ is not regular, it is a $\mu \circ N$ and the corresponding Π_θ is the irreducible principal series $\Pi(\mu, \mu\eta_{K/F})$; so,

$$\lambda_{K/F}(\psi)\gamma^K(\theta\chi \circ N, \psi \circ T) = \gamma^M_{\Pi_\theta \otimes \chi, \psi} , \qquad \theta \in A_F(K^\times) ; \qquad (3.11)$$

f) for each Σ in $A(H^\times)$, there exists a unique Π_Σ in $A(M^\times)$ with the same gamma factor and which is infinite dimensional:

$$-\gamma^H_{\Sigma \otimes \chi, \psi} = \gamma^M_{\Pi_\Sigma \otimes \chi, \psi} , \qquad \Sigma \in A(H^\times) , \qquad (3.12)$$

and this gives a bijection between $A(H^\times)$ and the discrete series of $M^\times$.

g) the degree two gamma factors $\gamma(\chi, \psi)$ fixed by some non-trivial element η of $A(F^\times)$ (acting by translation on the χ variable) are exactly those of type (II), and then η is a $\eta_{K/F}$.

The assertions d) - g) are true for $F = \mathbb{R}$.

4. Degree two gamma factors, second set of identities

From now on, we assume the local field to be non-archimedean (for the analogue of (4.3) in the archimedean case, see the note [91] by H. Prado). We write $\mathcal{O}$ for the ring of integers, $\mathcal{P}$ for the ideal of the valuation, $\mathcal{O}^{\times}(=F_u)$ for the group of units; the conductor $a(\chi)$ of a character $\chi \in \mathbf{A}(F^{\times})$ is the smallest integer for which the restriction of χ to $1+\mathcal{P}^n$ is trivial. We write $d\chi$ for the differential form on $\mathbf{A}(F^{\times})$ given on each connected component $\mathbb{C}^{\times}\chi$ consisting of all characters $x \longmapsto z^{\operatorname{ord} x}\chi(x)$, $z \in \mathbb{C}^{\times}$, by $(2\pi i)^{-1}\frac{dz}{z}$; the contour integrals are taken with respect to the measure defined by $d\chi$.

Each degree two gamma factor for F satisfies the following identity, called the deep twist property:

$$\gamma(\chi,\psi) = \gamma^F(\chi,\psi)\gamma^F(\chi\omega,\psi) \quad \text{for} \quad a(\chi) \quad \text{large enough} \tag{4.1}$$

where ω is the central character of γ. Such a relation determines ω uniquely. Assume now that γ and γ' are degree two local factors satisfying:

a) the poles of $\gamma(\chi,\psi)$ and the poles of $\gamma'(\chi^{-1},\psi)$ are disjoint,

b) the central characters ω and ω' verify $\omega\omega' \neq q$.

Here q means the character $x \longmapsto q^{\operatorname{ord} x} = |x|^{-1}$ of $F^{\times}$. Under these two conditions we attach to the couple (γ,γ') a complex number, called their inner product

$$<\gamma|\gamma'>_{\psi} = \oint_{\mathbf{A}(F^{\times})}\gamma(\chi,\psi)\gamma'(\chi^{-1},\bar{\psi})d\chi\,, \tag{4.2}$$

by taking the finite part of the integral on a simple loop around 0 containing the poles of $\chi \longmapsto \gamma'(\chi^{-1},\psi)$ but not the poles of $\chi \longmapsto \gamma(\chi,\psi)$: the condition a) assures the existence of the contour; the integral converges for $|\omega\omega'| > q$, and the condition b) assures that its finite part exists. To take ψ' instead of ψ modifies the inner product as follows: $<\gamma|\gamma'>_{\psi'} = (\omega\omega')(t)^{-1} <\gamma|\gamma'>_{\psi}$ for $\psi'(x) = \psi(tx)$. We have also $<\gamma|\gamma'>_{\psi} = <\gamma'|\gamma>_{\bar{\psi}} = (\omega\omega')(-1)<\gamma'|\gamma>_{\psi}$.

The inner product can be computed explicitely when one of the gamma factors is of type (I), that is, given by (3.5):

$$< \gamma|\gamma' >_\psi = \frac{(\mu\nu)(-1)}{\Gamma(\omega\mu\nu q^{-2},\psi)} \gamma(\mu q^{-1/2},\psi)\gamma(\nu q^{-1/2},\psi) \text{ for } \gamma = \gamma^F_{\mu\otimes\nu} ; \tag{4.3}$$

where

$$\Gamma(\chi,\psi) = (1-q^{-1})^{-1}(\psi)^{-1/2}\gamma^F(\chi q^{1/2},\psi) ,$$

this is called the multiplicative formula: it gives the product of two values of a degree two gamma factor as a (in general infinite) linear combination of values of this gamma factor, using the residue formula. For γ of type (I) or (II), the multiplicative formula expresses the functional equation of elements $\theta \in A(A^\times)$ amongst the continuous functions on $A^\times$: $\theta(xy) = \theta(x)\theta(y)$. For γ of type (III) it expresses the functional equation of irreducible reduced characters of $H^\times$

$$\int_{H^\times/F^\times} c(xzyz^{-1})d^\cdot z = c(x)c(y)$$

where $d^\cdot z$ is the normalized Haar measure on the compact group $H^\times/F^\times$. Finally, when γ is of type (IV), the multiplicative formula expresses the fact that the generators of $GL_2(F)$ given by the elements of the upper triangular subgroup and the element $w = \begin{pmatrix} 0 & 1 \\ -1 & 0 \end{pmatrix}$ satisfy the relation

$$[w\begin{pmatrix} 1 & 1 \\ 0 & 1 \end{pmatrix}]^3 = 1 .$$

Let now B be a separable biquadratic extension of F ; as its Galois group is isomorphic to $(\mathbb{Z}/2\mathbb{Z}) \times (\mathbb{Z}/2\mathbb{Z})$, the extension B contains exactly three quadratic subextensions. In each of them, say K , every element of F is a norm from B to K , so the number $\lambda_{B/K}(\psi o T_{B/K})$ is the independent of the non-trivial character ψ of F and is a sign; we write it $\lambda_{B/K}$; it satisfies

$$\lambda_{B/K}\lambda_{K/F}(\psi) = \lambda_{K'/F}(\psi)\lambda_{K''/F}(\psi) , \lambda_{B/K'}\lambda_{B/K''} = \lambda_{K/F}(\psi)^2 = \eta_{K/F}(-1) ,$$

where K' and K'' are the two other quadratic subextensions. We define a sign by the formula

$$\lambda_{B/F} = \lambda_{B/K}\lambda_{B/K'}\lambda_{B/K''} = \lambda_{K/F}(\psi)\lambda_{K'/F}(\psi)\lambda_{K''/F}(\psi) .$$

If F has characteristic 2, one has $\lambda_{B/F} = 1$. If the characteristic of F is not 2, then $\lambda_{B/K} = (K'|K'')$, the Hilbert symbol attached to K',K'' (that is, the cup product of the order two characters $\eta_{K'/F}$ and $\eta_{K''/F}$ of $F^\times$, seen as element of $H^2(\mathrm{Gal}\ B/F,\mu_2(F)) = \mathbb{Z}/2\mathbb{Z}$). Now for any character β of $B^\times$, we define

$$\gamma^F_\beta(\chi,\psi) = \lambda_{B/F}\gamma^B(\beta\chi o N_{B/F},\psi o T_{B/F}).$$

The second identity says that if K and K' are two non-isomorphic separable quadratic extensions of F , and B is their composite extension, then for θ in

$A(K^\times)$ and θ' in $A(K'^\times)$, we have under the conditions a), b) above:

$$< \gamma^F_\theta | \gamma^F_{\theta'} >_\psi \quad \frac{\gamma^F_\beta(q^{-1/2},\psi)}{\Gamma(\omega\omega' q^{-2},\overline{\psi})}\theta(-1) , \tag{4.4}$$

where $\beta = \theta \circ N_{B/K}\theta' \circ N_{B/K'}$, and ω, ω' are the central characters of θ, θ', that is, $\omega = \eta_{K/F}\theta|_{F^\times}, \omega' = \eta_{K'/F}\theta'|_{F^\times}$. In particular,

$$< \gamma^F_\theta | \gamma^F_{\theta'} >_\psi = \frac{\lambda_{B/K}}{\Gamma(\eta_{K''/F}q^{-1},\overline{\psi})} \quad \text{if} \quad \theta|_{F^\times}\theta'|_{F^\times} = q . \tag{4.5}$$

with the sign $\lambda_{B/K}$ given by the Hilbert symbol $(K'|K'')$ if the characteristic is not 2, and, in characteristic 2, the sign $\lambda_{B/K} = 1$.

These two formulas remain valid when K' is taken to be the split extension $F \times F$: in this case the composite extension of K and $F \times F$ is $K \times K$; one has $\lambda_{(K \times K)/K} = 1$, and $\lambda_{(K \times K)/F} = \lambda_{(K \times K)/K}\lambda_{K/F}(\psi)^2$. The formulas (4,4) and (4.5) are then special cases of (4.3). When K' is isomorphic to K, then B is $K \times K$ and (4.3) is still valid.

Call the contragredient $\check{\gamma}$ of a degree two gamma factor γ with central character ω the degree two gamma factor $\gamma(\chi\omega^{-1},\psi)$; for types (III) or (IV), $\check{\gamma}$ is the gamma factor attached to the contragredient representation (for $\bar{x}^{-1} = x(Nx)^{-1}$ in $A^\times$) with gamma factor γ. There is a characterization of the contragredient: if γ is a degree two gamma factor without zero, and γ' a degree two gamma factor the chentral character of which is the inverse of the central character of γ, then $< \gamma | \gamma' >_\psi = 0$ is equivalent to $\gamma' = \check{\gamma}$.

5. About the proofs

The multiplicative formula (4.3) which is the main identiy satisfied by the degree two gamma factors, is proved in [78], where it is also shown that any rational function in $\chi \in A(F^\times)$ depending also on a non-trivial additive character ψ of F, which satisfies the additive condition (3.3), and the complement formula (3.4), is a degree two gamma factor of type (IV). In [36] it was proved that the complement formula is a consequence of the other identities. For degree two gamma factors of type (III), the proofs are in [35]; the explicit inversion formula $c \longrightarrow \gamma^H_{c,\psi}$ is given, and so are the consequences on the corresponding between

$A(H^\times)$ and $A(M^\times)$, including character values on elliptic elements and (formal) degrees. In [37], we prove the identities (4.3) - (4.5) on degree two gamma factors of type (II), and give some consequences.
In each case, the proof of the multiplicative formula (4.3) uses the fact that the tempered distribution $f \longmapsto \int f(x)\psi \circ N(x)dx$ on A has Fourier transform with respect to ψ the distribution $f \longmapsto \lambda_{A/F}(\psi)\int f(x)\bar{\psi} \circ N(x)dx$; this is written as

$$\lambda_{A/F}(\psi) = \int \psi \circ N(x) d_{\psi \circ T} x .$$

In [38], we show that the degree two gamma factors of type (II) are characterized among those of type (IV) as the ones which are fixed by a non-trivial character of $F^\times$; this gives a direct proof of a key lemma needed by Langlands in his work on Base Change for GL(2) (Lemma 7.17, p. 128 in [75]), about it he wrote "it is a shame that our proof is so uninspired". In this paper, we proved also that unless F is a non-archimedean field of residual characteristic 2 , any degree two gamma factor is of type (I) or (II).

Gamma factors occur also with the Galois group $Gal(\bar{F}/F)$ of a separable algebraic closure $\bar{F}$ of F , or, more precisely, with the Weil group W_F([26]):

a) for F complex, $\bar{F} = F$ and $W_F = F^\times$;

b) for $F = \mathbb{R}$, $\bar{F} = \mathbb{C}$, $W_{\mathbb{R}}$ is the normalizer in the quaternion group of the diagonal subspace of the quaternion algebra $\mathbb{H}$, subspace isomorphic to $\mathbb{C}$ by $x \longmapsto \begin{pmatrix} x & \\ & \bar{x} \end{pmatrix}$; its commutator subgroup is $W_{\mathbb{C}} = \mathbb{C}^\times$, the factor group being the two element group acting on $\mathbb{C}^\times$ by complex conjugation;

c) for F non-archimedean, the reciprocity map $F^\times \to Gal(\bar{F}/F)^{ab}$ describes the abelianized group of $Gal(\bar{F}/F)$ as the profinite compactification of $F^\times$; the group W_F is the fiber product of $F^\times$ by $Gal(\bar{F}/F)$ according to these maps in $Gal(\bar{F}/F)^{ab}$; the group $F^\times$ appears as the abelianized group of W_F .

In all cases, we have a continuous homomorphism from W_F to $Gal(\bar{F}/F)$. For any finite extension E of F in $\bar{F}$, we take $\bar{E} = \bar{F}$; then, the Weil group W_E appears as a subgroup of W_F : it is the inverse image of the subgroup $Gal(\bar{F}/E)$ of $Gal(\bar{F}/F)$ by the preceding map. In this way, we get a parametrization of the closed subgroups of W_F with finite index by the finite extensions of F in $\bar{F}$.

For each integer $n \geq 1$, let $\mathcal{W}_F(n)$ be the set of classes of degree n continuous semi-simple representations of W_F over $\mathbb{C}$, and $\mathcal{W}_F$ be the union of all $\mathcal{W}_F(n)$, $n \geq 1$. In particular, $\mathcal{W}_F(1)$ is canonically identified with $A(F^\times)$. Taking direct sum (resp. tensor product) of representations, gives a map:

$\mathcal{W}_F \times \mathcal{W}_F \longrightarrow \mathcal{W}_F$ which add (resp. multiply) degrees. Let E be a finite extension of F in $\bar{F}$; by induction of representations from W_E to W_F we get a map $\mathrm{Ind}_E^F : \mathcal{W}_E \longrightarrow \mathcal{W}_F$, which multiply degrees by $[E:F]$. We write $\mathcal{W}_F^*(d) \subset \mathcal{W}_F(d)$ for the part coming from these representations which are direct sums of representations induced from one-dimensional characters, and $\mathcal{W}_F^*$ denote the union of all $\mathcal{W}_F^*(d)$, $d \geq 1$. By Brauer theorem, for any Σ in $\mathcal{W}_F$, there is a Φ in $\mathcal{W}_F^*$ such that $\Sigma + \Phi$ lies in $\mathcal{W}_F^*$. Taking the determinant of a representation gives a map $\Sigma \longmapsto \omega_\Sigma$ from $\mathcal{W}_F$ to $A(F^\times)$.

Given a non-trivial additive character ψ of F , Deligne and Langlands ([26]) show that there is a unique way of defining for each finite extension E of F in $\bar{F}$ a complex number $\lambda_{E/F}(\psi)$ such that:

a) $\lambda_{F/F}(\psi) = 1$;

b) for any θ in $A(E^\times)$, the number $\lambda_{E/F}(\psi)\gamma^E(\theta\ \psi \circ T_{E/F})$ depends only on the representation $\Sigma = \mathrm{Ind}_E^F \theta$ (not on the choice of E,θ) ; it is noted $\gamma^F_{\Sigma,\psi}$

c) $\Sigma \longmapsto \gamma^F_{\Sigma,\psi}$ extends by additivity to a function on $\mathcal{W}_F$; in particular, if $\Sigma + \Phi = \Phi'$ with Φ,Φ' in $\mathcal{W}_F^*$, then $\gamma^F_{\Sigma,\psi} = \gamma^F_{\Phi',\psi}/\gamma^F_{\Phi,\psi}$.

The number $\gamma^F_{\Sigma,\psi}$ is called the gamma factor of Σ ; we can also see it as a function on $\mathcal{W}_F$ by $\Phi \longmapsto \gamma^F_{\Sigma \otimes \Phi,\psi}$. We write $\gamma^F_\Sigma(\chi,\psi)$ for $\gamma^F_{\Sigma \otimes \chi,\psi}$ when χ is a character of $F^\times$. They are the gamma factors we consider.

P. Kutzko proved in [68] that the gamma factors $\gamma^F_\Sigma(\chi,\psi)$ for Σ in $\mathcal{W}_F(2)$ are exactly the degree two gamma factors introduced in section 2, with respect to the field F , and that an element of $\mathcal{W}_F(2)$ is determined by its gamma factor. These correspondences with $\mathcal{W}_F(2)$ associate

a) to the reducible representation of W_F sum of the two characters μ and ν the element of $A_F(F^\times \times F^\times)$ defined by $\mu \otimes \nu$ and the element $\Pi(\mu,\nu)$ of $A(M^\times)$;

b) to the induced representation $\mathrm{Ind}_K^F \theta$, K a quadratic extension in $\bar{F}$, the element of $A_F(K^\times)$ defined by θ and the element Π_θ of $A(M^\times)$; when θ is regular, it associates the element Σ_θ of $A(H^\times)$.

The identity (3.6) reflects then that the representation of W_F induced from a character $\mu \circ N$ of $K^\times$ is reducible, sum of μ and $\mu\eta_{K/F}$. The identity (3.8) reflects the fact that the involved representation can be obtained by inducing from different index two closed subgroups. The identity (4.3) gives a functional equation for gamma factors coming from $\mathcal{W}_F(2)$; we can rewrite it as

$$< \gamma_{\Sigma}^F | \gamma_{\Sigma'}^F >_{\psi} = \omega(-1) \frac{\gamma^F_{\Sigma \otimes \Sigma'}(q^{-1/2}, \psi)}{\Gamma(\omega\omega' q^{-2} \overline{\psi})} \qquad \textbf{(5.1)}$$

where Σ, Σ' are in $\mathcal{W}_F(2)$, $\omega = \det \circ \Sigma, \omega' = \det \circ \Sigma'$ and Σ' is reducible: in fact, if Σ' is a sum of two characters μ and ν , then $\Sigma \otimes \Sigma'$ is the sum of $\Sigma \otimes \mu$ and $\Sigma \otimes \nu$, and $\gamma^F_{\Sigma \otimes \Sigma'}(q^{-1/2}, \psi) = \gamma^F(\mu q^{-1/2}, \psi)\gamma^F(\nu q^{-1/2}, \psi)$. The formula (4.4) has also the form (5.1): in fact, with the notations introduced there, we have $\mathrm{Ind}_K^F \theta \otimes \mathrm{Ind}_{K'}^F \theta' = \mathrm{Ind}_B^F \beta$. It is likely that (5.1) holds in general. The analogous formula is proved for degree two gamma factors in [77].

References

[26] P. Deligne: Les constantes des équations fonctionnelles des fonctions L , Modular Functions in One Variable II. Lecture Notes in Mathematics 349, Springer Verlag, Berlin-Heidelberg-New York, 1973, 501 - 597.

[33] P. Gerardin: Représentations du groupe SL(2) d'un corps local (d'après Gel'fand, Graev et Tanaka), Séminaire Bourbaki 332, Nov. 1987, Benjamin, 1969.

[35] P. Gerardin and W.-C.W. Li: Fourier transforms of representations of quaternions; J. reine angew. Mathematik 359 (1985), 121 - 173.

[36] P. Gerardin, W.-C.W. Li: A functional equation for degree two local factors; Canadian Math. Soc. Bull. 28 (3) (1985), 355 - 371.

[37] G. Gerardin, W.-C.W. Li: Identities on quadratic Gauss sums, preprint.

[38] P. Gerardin, W.-C.W. Li: Degree two monomial representations of local Weil groups and correspondences (preprint).

[55] H. Jacquet, R.P. Langlands: Automorphic Forms on GL(2); Lecture Notes in Mathematics no. 114, Springer Verlag, Berlin-Heidelberg-New York, 1970.

[68] P. Kutzko: The Langlands conjecture for GL_2 of a local field; Ann. Math. 112 (1980), 381 - 412.

[75] R.P. Langlands: Base Change for GL(2) ; Ann of Math. Studies 96, Princeton University Press, 1980.

[77] W.-C.W. Li: On the representations of GL(2) I, ε-factors and n-closeness, J. reine und angew. Math. 313 (1980), 27 - 42; II, ε-factors of the representations of GL(2) x GL(2), J. reine angew. Math. 314 (1980), 3 - 20.

[78] W.-C.W. Li: Barnes identities and representations of GL(2) II, Non-archimedean local case; J. reine angew. Math. 345 (1983), 69 - 92.

[91] H. Prado: Représentations de $GL(2,\mathbb{R})$ et identités de type Barnes pour la fonction L ; C.R.Ac. Sc. Paris t. 300 Série I, no. 4 (1985), 97 - 100.

[105] J.T. Tate: Fourier analysis in number fields and Hecke's zeta functions; Thesis, Princeton University 1950, Also: Algebraic Number Theory (J.W.S. Cassels & A. Fröhlich edd.)(Academic Press, London 1967), 305 - 347.

ontemporary Mathematics
olume **86**, 1989

A CONJECTURE ON MINIMAL K-TYPES FOR GL_n OVER A p-ADIC FIELD

by A. Moy, Seattle

Let F be a p-adic field, and let $G = GL_n(F)$ be the F-rational points of the lgebraic group GL_n. Based on the work of [82,101], it has been conjectured that very supercuspidal representations (Π, V) of G can be induced from an open comact mod center group L, i. e. $\Pi = \mathrm{ind}_L^G \Omega$ for some irreducible representation Ω f L. This conjecture has been verified in several cases [13,50,67,72,73]. The air (L,Ω) should be thought of as a p-adic analogue of a minimal K-type as in 'ogan's theory for real groups. I would like to present here a conjecture which I elieve to be a partial analogue of Vogan's theory of minimal K-types for all irreducible admissible representations of G. More specifically, certain pairs (L,Ω) onsisting of an open compact subgroup L of G and an irreducible representation Ω are singled out and it is conjectured that every irreducible admissible represeentation (Π, V) of G contains a pair (L,Ω) i. e. the decomposition of the restriction $\Pi_{|L}$ contains a representation equivalent to Ω. The groups L are filtration subgroups J_i of a parahoric subgroup J of G. These subgroups were first introduced by Prasad-Raghunathan for general reductive groups in [92] and used by [45] in the construction of supercuspidal representations. The representation Ω will be a representation of $L = J_i$ trivial on J_{i+1} and satisfy either a cuspidality or semisimplicity condition.

In order to review the filtrations of Prasad-Raghunathan in the context of GL_n, let us first establish some notation. Let $\mathcal{O}$ denote the ring of integers in F, $\mathcal{P}$ the prime ideal of $\mathcal{O}$, and π a prime element. Let $\mathbb{F}_q = \mathcal{O}/\mathcal{P}$ be the residual field. Let A denote the diagronal matrices in G and let

$$d = \begin{bmatrix} a_1 & & & & \\ & a_2 & & & \\ & & \cdot & & \\ & & & \cdot & \\ & & & & \cdot \\ & & & & & a_n \end{bmatrix} \tag{1}$$

0271-4132/89 $1.00 + $.25 per page

denote a typical element of A. The normalizer $N = N_G(A)$ of A is the set of monomial matrices. Let $X = \mathrm{Hom}(A, GL_1)$ be the characters of A. The Lie algebra $\mathfrak{g}$ (resp. $\mathfrak{a}$) of G (resp. A) is of course the set of $n \times n$ (resp. diagonal) matrices. Let $E_{i,j}$ $(1 \leq i, j \leq n)$ be the $n \times n$ matrix all of whose entries are zero except the i,j-th one which is 1. Let

$$\Phi = \{\alpha_{i,j} \in X \mid \alpha_{i,j}(d) = a_i/a_j \ , \ i \neq j\} \tag{2}$$

denote the roots of $\mathfrak{g}$ with respect to A. For $\alpha \in \Phi$, let

$$\mathfrak{g}_\alpha = \{x \in \mathfrak{g} \mid dxd^{-1} = \alpha(d)x\} \tag{3}$$

be the α root space. If $\alpha = \alpha_{i,j}$, then $E_{i,j}$ is a basis for $\mathfrak{g}_\alpha$. The Lie algebra $\mathfrak{g}$ can of course be decomposed into

$$\mathfrak{g} = \mathfrak{a} \oplus \sum_{\alpha \in \Phi} \mathfrak{g}_\alpha \ . \tag{4}$$

For each root α, there is a root group $U_\alpha \subset G$. If $\alpha = \alpha_{i,j}$, then

$$U_\alpha = \{1 + xE_{i,j} \mid x \in F\} \ . \ \text{The map } i\colon \mathfrak{g}_\alpha \longrightarrow U_\alpha$$

$$i(xE_{i,j}) = 1 + xE_{i,j} \tag{5}$$

is an isomorphism. We can define filtrations of $\mathfrak{g}_\alpha$ and U_α by

$$\mathfrak{g}_{\alpha,t} = \{xE_{i,j} \mid x \in \mathcal{P}^t\} \tag{6}$$

$$U_{\alpha,t} = \{1 + xE_{i,j} \mid x \in \mathcal{P}^t\} \ ,$$

where t is an integer. We shall later view the pair (α,t) as an element in $X \times \mathbb{Z}$. We can also define a filtration on the group $A(\mathcal{O}) = \{d \mid a_i \text{ is } \in \mathcal{O}^\times\}$. For m a natural integer, set

$$U_{(0,0)} = A(\mathcal{O}) \tag{7}$$

$$U_{(0,m)} = \{d \in A(\mathcal{O}) \mid a_i \equiv 1 \bmod \mathcal{P}^m\} \ .$$

The Borel subgroup of upper triangular matrices in G determines a partial ordering $<$ on X, hence Φ. In this ordering, a root $\alpha = \alpha_{i,j}$ is positive precisely when $i < j$. An Iwahori subgroup B is defined to be the group

$$B = U_{(0,0)} \prod_{\alpha>0} U_{(\alpha,0)} \prod_{\alpha<0} U_{(\alpha,1)} \tag{8}$$

$$= \{g \in G(\mathcal{O}) \mid g \text{ is upper triangular modulo } \mathcal{P}\} .$$

A compact parahoric subgroup is a compact subgroup containing B. In order to give a description of the compact parahoric subgroups let

$$w_i = I - E_{i,i} - E_{i+1,i+1} + E_{i,i+1} + E_{i+1,i} \qquad (1 \le i \le n-1) \tag{9}$$

$$w_n = I - E_{n,n} - E_{1,1} + \pi^{-1}E_{1,n} + \pi E_{n,1} .$$

and set $S = \{w_1, w_2, \ldots . w_n\}$. The subgroup of N generated by S is a Coxeter group with Dynkin diagram

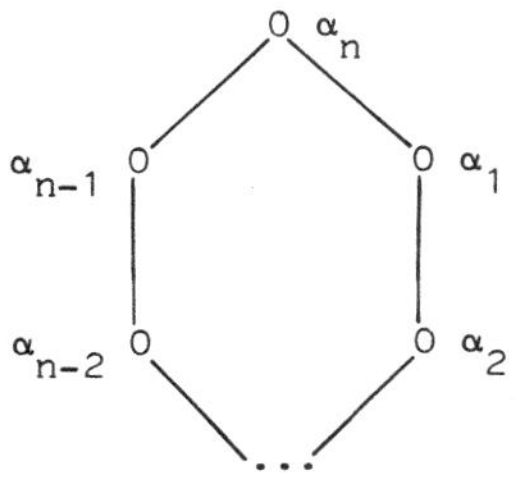

(10)

If $S' \subseteq S$, then $P_{S'} = \langle S', B \rangle$, the subgroup generated by S' and B, is a compact parahoric subgroup, and all compact parahoric subgroups are obtained in this fashion. Let us now recall the Prasad-Raghunathan filtrations [ ]. Each w_i in S has associated to it a root. For $i = 1,2,\ldots,n-1$ the root associated to w_i is the simple root

$$\alpha_i = \alpha_{i,i+1} . \tag{11}$$

To w_n associate the inverse of the longest root,

$$\alpha_n = \{\prod_{i=1}^{n-1} \alpha_i\}^{-1} . \tag{12}$$

Consider the group $X \times \mathbb{Z}$. Each affine root β (in $(\Phi \cup \{0\}) \times \mathbb{Z}$ has a unique decomposition

$$\beta = (\alpha_1,0)^{h_1(\beta)} (\alpha_2,0)^{h_2(\beta)} \cdots (\alpha_{n-1},0)^{h_{n-1}(\beta)} (\alpha_n,1)^{h_n(\beta)} \tag{13}$$

e. g. $(0,1) = (\alpha_1,0)(\alpha_2,0) \ldots (\alpha_{n-1},0)(\alpha_n,1)$. For $S' \subsetneq S$, define a height function $ht_{S'}$ on $(\Phi \cup \{0\}) \times \mathbb{Z}$ by

$$ht_{S'}(\beta) = \sum_{w \notin S'} h_i(\beta) . \tag{14}$$

The Prasad-Raghunathan filtration of $P = P_{S'}$ is then

$$P_0 = P \tag{15}$$

$$P_t = < U_\beta \mid ht_{S'}(\beta) \geq t > \text{, the group generated by the } U_\beta\text{'s} .$$

The groups P_t $(t \geq 1)$ are normal in P .

Example: If $S' = \{w_1,\ldots,w_{n-1}\}$, then $P_{S'} = G(\mathcal{O})$. The group P_t is the t-th congruence subgroup of $G(\mathcal{O})$.

For arbitrary S' , it is well known that $P_{S'}/P_{S',1}$ is the $\mathbb{F}_q$-rational points of the reductive group whose Dynkin diagram is the subdiagram of (10) determined by S' . For $t \geq 1$, set

$$\ell_t = \{g-1 \mid g \in P_t\} = < g_\beta \mid ht_{S'}(\beta) \geq t > . \tag{16}$$

The set ℓ_t is an $\mathcal{O}$-lattice in g and if $2t \geq t'$, then $P_t/P_{t'}$ is isomorphic to $\ell_t/\ell_{t'}$ via the map $i(g) = g-1$. It is easy to check that

$$\ell_{t+ht(0,1)} = \pi \ell_t . \tag{17}$$

This periodicity allows us to define ℓ_t for all integers t . If ℓ is an $\mathcal{O}$-lattice in g , let $\ell' = \{y \in g \mid tr(y\ell) \subseteq P\}$. In our setting, $\ell_{t'} = \ell_{-t+1}$. Let ψ be a additive character of F with conductor P . For $t \geq 1$, we can identify the characters of P_t/P_{t+1} with the cosets ℓ_{-t}/ℓ_{-t+1} by viewing the coset $x + \ell_{-t+1}$ as the character ψ_x of P_t/P_{t+1} given by

$$\psi_x(g) = \psi(tr(x'(g-1))) , \qquad (x' \in x + \ell_{-t+1}) . \tag{18}$$

Definition. For a fixed S' and $t \geq 1$, a coset $x + \ell_{-t+1}$ in ℓ_{-t} is said to be nondegenerate if $x + \ell_{-t+1}$ does not contain any nilpotent elements.

Consider now the following classes of representations which we define to be the smallest nondegenerate representations:

i) $(P_{S'}, \Omega)$: $P_{S'}$ a compact parahoric subgroup and Ω an irreducible cuspidal representation of $P_{S'}/P_{S',1}$ **(19)**

ii) $(P_{S',t}\Omega)$: $t \geq 1$ and $\Omega = \psi_x$, where $x + \ell_{-t+1}$ is nondegenerate.

A pair (J,Σ) consisting of an open compact subgroup J of G and an irreducible representation Σ of J is said to be a nondegenerate representation if there is a pair (L,Ω) as in (19) such that $J \supset L$ and the restriction $\Sigma_{|L}$ contains Ω.

Conjecture. Given any irreducible admissible representation (Π,V) of G, there is a smallest nondegenerate representation (L,Ω) such that the restriction of Π to L contains Ω.

The evidence for the conjecture is that it is true for GL_2, GL_3, GL_4, and GL_5, see [50,67,72,73]. Moreover, recent work of Bushnell [10] has shown that for supercuspidal representations, the conjecture is true provided the minimal conductoral exponent of Π is greater than 3. It should be noted that the nondegenerate representations are very closely related to Howe's notion of an essential K-type, see [48], and it is not implausible that nondegenerate representations can be used to extend the results of §4 of [48] in a manner very much as envisioned there. As additional evidence for the plausibility of such a conjecture, it is not hard to generalize the conjecture to encompass other split reductive groups and with more work to some nonsplit reductive groups. Here the evidence for an "analogous" conjecture is [85,86,87,88]

Split Evidence: B_2, B_3, C_3, and D_4 when $p \neq 2$ and G_2 for $p \neq 2, 3$.
Nonsplit Evidence: $U(2,1)$ when $p \neq 2$.

The truth of the "analogous" conjecture for the above groups $G = \mathcal{G}(F)$, can be used as an anchor for the classification of all irreducible admissible representations of G. Here, for each smallest nondegenerate representation (L,Ω), we classify those irreducible representations of G which contain (L,Ω). This problem is well known to be equivalent to determining the irreducible finite dimensional representations of the Hecke algebra $\mathcal{H}(G/L,\Omega)$ of compact supported functions f on G with values in $\operatorname{End} V_\Omega$ such that $f(kgk') = \Omega(k)f(g)\Omega(k')$ for $g \in G$ and $k,k' \in L$. In the case of $U(2,1)$ and $GSp(4)$, this is done by exhibiting Hecke algebra isomorphisms in exact analogy with those proved for GL_n in [51].

PS. The conjecture formulated in this article has been proved. Details will appear elsewhere.

References

[10] C.J. Bushnell: Oral communication to A. Moy.

[13] H. Carayol: Representation supercuspidal de GL_n ; C.R. Acad. Sci. Paris, Serie A. 288 (1979), 17 - 20.

[45] H. Hijikata: Some supercuspidal representations induced from parabolic subgroups, in: Automorphic Forms of Several Variables, Taniguchi Symposium, Katata 1983, edited by I. Satake and Y. Morita; Birkhäuser - Boston-Basel-Stuttgart, 1984.

[48] R. Howe: Some results on the representation theory of GL_n over a p-adic field; Pac. Jour. of Math. 73 (1977), 479 - 538.

[50] R. Howe: Classification of irreducible representations of GL_2 ; IHES notes (1978).

[51] R. Howe, A. Moy: Harish-Chandra Homomorphisms for p-adic Groups, CBMS Regional Conference, Series in Mathematics 59, Amer. Math. Soc. Providence, Rh. I., 1985.

[67] P. Kutzko: On the supercuspidal representations of GL_2 ; Amer. J. Math. 100 (1978), 43 - 60.

[72] P. Kutzko: Oral Communication.

[73] P. Kutzko, O. Manderscheid: On the supercuspidal representations of GL_4 , I; Duke Math. J. 52 (1985), 841 - 867.

[82] F. Mautner: Spherical functions over p-adic fields II; Amer. Jour. of Math. 86 (1964), 171 - 200.

[85] A. Moy: Representations of U(2,1) over a p-adic field, preprint.

[86] A. Moy: Representations of GSp(4) over a p-adic field, preprint.

[87] A. Moy: Minimal K-types for GSp(6) over a p-adic field, preprint.

[88] A. Moy: Minimal K-types for G_2 over a p-adic field, preprint.

[92] G. Prasad, M. Raghunathan: Topological central extensions of semi-simple groups over local fields; Annals of Math. 119 (1984), 143 - 201.

[101] J. Shalika: Representations of the two by two unimodular group over local fields, IAS notes (1966).

Contemporary Mathematics
Volume **86**, 1989

PREUVE DE LA CONJECTURE DE LANGLANDS LOCALE NUMERIQUE POUR GL(n)

par G. Henniart, Orsay

Ceci est une version à peine modifiée, mais à laquelle j'ai ajouté quelques références, d'une lettre à G. Laumon datée du 27 mars 1986. J'y explique comment l'on prouve la conjecture de Langlands numérique locale pour GL(n) : il y a "autant" de représentations du groupe de Weil-Deligne d'un corps localement compact non archimédien F, de degré $n \geq 1$ donné, que de représentations admissibles irréductibles de $GL(n,F)$.

Cher Laumon,

Voici quelques détails sur ma preuve de la conjecture locale numérique de Langlands, que je déduis de ta théorie des transformées de Fourier locales [76]. Nous en avions discuté à l'automne dernier, et je les ai exposés à Augsbourg en décembre et à Göttingen la semaine dernière.

Soit donc F un corps local à corps résiduel fini de cardinal q. On choisit une clôture séparable algébrique $\bar{F}$ de F et on note W_F le groupe de Weil de $\bar{F}$ sur F. Soit n un entier, $n \geq 1$. Langlands a conjecturé l'existence d'une bijection canonique entre l'ensemble $G_F^0(n)$ des classes d'équivalence de représentations complexes, continues et irréductibles, de dimension n, de W_F et l'ensemble $A_F^0(n)$ des classes d'équivalence de représentations admissibles irréductibles cuspidales de $GL(n,F)$. Cette bijection doit en particulier être compatible à la torsion par les (quasi-)caractères de $F^\times$ et préserver les (exposants des) conducteurs.

Grâce aux résultats de Clozel et Arthur [1, 16, 17] sur la théorie du changement de base pour $GL(n)$, je peux prouver, <u>quand</u> F <u>est de caractéristique nulle,</u> l'existence d'une application <u>injective</u> de $G_F^0(n)$ dans $A_F^0(n)$ possédant ces deux propriétés : on utilise essentiellement que W_F est prorésoluble; cependant je n'ai pas de construction unique et canonique et je ne sais pas, par exemple, prouver dans tous les cas la préservation des facteurs ε de paires. Pour la suite, j'ai besoin d'un peu plus que ce qui précède. Je fixe une $\mathbb{Z}_p$-extension $\tilde{F}$ de F, totalement ramifiée, et pour tout n et toute extension finie K de F dans $\tilde{F}$, je construis une injection de $G_K^0(n)$ dans $A_K^0(n)$, compatible à la torsion par les caractères de $K^\times$ et préservant les exposants, ces injections étant compatibles aux changements de base K'/K, où K' est une extension finie de K dans $\bar{F}$. Quand F est de caractéristique non nulle, j'utilise une technique de relèvement à la caractéristique nulle, à la Kazhdan [58] pour, là aussi, construire une injection $G_F^0(n) \to A_F^0(n)$ possédant les propriétés demandées. Cependant cette construction n'est pas utile pour les démonstrations de ce qui suit.

Si l'on veut, on peut énoncer la conjecture de Langlands numérique locale comme disant qu'une application injective comme plus haut est aussi bijective. C'est ce que je démontre, par des arguments de comptage inaugurés par Tunnell et Koch. Une conséquence, qui semble difficile à démontrer directement, est que si $\Pi \in A_F^0(n)$ alors on peut trouver une suite d'extensions $F = F_0 \subset F_1 \subset \ldots \subset F_k$, F_i cyclique sur F_{i-1}, telle que le changement de base de Π à F_k ne soit plus cuspidal.

Les arguments de comptage reposent sur la remarque que les éléments de $G_F^0(n)$ (resp. $A_F^0(n)$) de conducteur (de Swan, disons) donné forment un nombre fini d'orbites sous l'action des caractères non ramifiés de $F^\times$. C'est assez facile à voir pour $G_F^0(n)$ et pour $A_F^0(n)$ cela découle de la correspondance entre représentations essentiellement de carré intégrable (admissibles et irréductibles, bien sûr) de $GL(n,F)$ et représentations (toujours admissibles et irréductibles) de $D^\times$, où D est un corps gauche central sur F et de degré réduit n. Cette correspondance est prouvée par Deligne, Kazhdan et Vigneras, quand F est de caractéristique nulle, dans [3]. Si F est de caractéristique positive, on utilise l'argument, suggéré par Kazhdan, de réduction à la caractéristique nulle : c'est assez facile en ce cas parce que les algèbres de Hecke pour $GL(n,F)$ ont été étudiées en détail par Howe et Moy [51] et qu'on dispose pour elles de présentations agréables par générateurs et relations. Ici encore, les propriétés essentielles de la correspondance entre représentations de $GL(n,F)$ et représentations de $D^\times$ sont qu'elle est compatible à la torsion par les caractères de $F^\times$ et qu'elle préserve les conducteurs.

Maintenant, les représentations admissibles irréductibles de $D^\times$ sont, à peu de choses près, des représentations de groupes finis et l'on peut, sans trop de difficultés, dénombrer celles qui ont telle ou telle propriété. Ce fut l'objet d'un travail de H. Koch (Bemerkungen zur numerischen lokalen Langlands-Vermutung I). Dans ce papier, il réduit la conjecture de Langlands numérique à l'énoncé suivant, qui est celui que je démontre. On fixe un entier $n \geq 1$, un entier $j \geq 0$ et on note $C_{n,j}$ le nombre d'orbites, sous l'action des caractères non ramifiés de $F^\times$, de (classes de) représentations Σ, complexes, continues, irréductibles, de W_F, de dimension $\dim \Sigma$ divisant n, d'exposant de Swan $sw(\Sigma)$ vérifiant $\frac{n}{\dim \Sigma}\, sw(\Sigma) \leq j$, et dont la restriction au groupe d'inertie I_F est irréductible (ce qui revient à dire que Σ n'est pas équivalente à une de ses tordues par un caractère non ramifié de $F^\times$).

Théorème: On a $C_{n,j} = q^j(q-1)$.

C'est clair pour $n = 1$. Si n est une puissance de p , m premier à p , et que $C_{n,j} = q^j(q-1)$ quel que soit le corps F , alors (*) $C_{nm,j} = C_{n,j}$ quel que soit le corps F . Cela donne le théorème si n est premier à q ; mais en ce cas le théorème était déjà connu grâce à la paramétrisation de Howe de $G_F^0(m)$ et $A_F^0(m)$ pour m premier à q . Koch et Zink démontrent aussi le théorème pour n égal à la caractéristique résiduelle p de F , donc pour n divisible par p mais pas par p^2 . Cependant dans la suite je n'utiliserai pas ces résultats mais seulement la relation (*) pour m premier à p . Cela évite tout argument de comptage pénible.

Je démontre le théorème en distinguant deux cas : celui où F est de caractéristique strictement positive, qui découle de tes résultats sur les transformées de Fourier locales et celui où F est de caractéristique nulle, que je traite par réduction à l'autre cas : ces derniers arguments sont plus difficiles que ceux de Kazhdan car il s'agit d'une réduction dans le "mauvais" sens.

Supposons donc d'abord F de caractéristique strictement positive. On remarque d'abord que dans la définition de $C_{n,j}$ on peut remplacer les représentations complexes par des représentations $\bar{\mathbb{Q}}_\ell$-adiques, sans changer l'énoncé. Ta théorie des transformées de Fourier locales montre qu'on a une bijection naturelle, pour tout entier $n \geq 1$ et tout entier $j \geq 1$, entre les éléments de $G_F^0(n)$ d'exposant de Swan j , et les éléments de $G_F^0(n+j)$ d'exposant de Swan j , en outre cette bijection est compatible à l'action des caractères non ramifiés de $F^\times$. L'existence d'une telle bijection et la relation (*) forment l'essentiel des arguments ci-après.

Pour $j = 0$ les représentations à dénombrer sont des caractères modérément ramifiés d'où aisément $C_{n,j} = (q-1)$. Il s'agit donc de prouver, pour $j \geq 1$, que la différence $D_{n,j} = C_{n,j} - C_{n,j-1}$ vaut $(q-1)^2q^{j-1}$. Pour cela on raisonne par récurrence sur $r = v_p(n)$. Grâce à la relation (*), on peut supposer $n = p^r$. Alors le résultat est vrai pour $r = 0$ par la théorie du corps de classes, comme déjà remarqué. Supposons donc $r \geq 1$ et notons p^s le pgcd de n et j , posant en outre $j = p^s k$. Alors $D_{n,j}$ est la somme, pour α variant de 0 à s , du nombre d'orbites sous l'action des caractères non ramifiés de $F^\times$, d'éléments de $G_F^0(p^{r-\alpha})$, d'exposant de Swan $p^{s-\alpha}k$, dont la restriction à I_F est irréductible. Par ta transformée de Fourier, c'est aussi la somme, pour α variant de 0 à s , du nombre d'orbites d'éléments de $G_F^0(p^{r-\alpha} + p^{s-\alpha}k)$, d'exposant de Swan $p^{s-\alpha}k$, dont la restriction à I_F est irréductible. Si $s < r$, k est premier à p , et cette dernière somme n'est autre que $D_{n+j,j}$, qui par hypothèse de récurrence vaut $(q-1)^2q^{j-1}$. Il reste donc à traiter les cas $s = r$. Pour cela on utilise le résultat suivant:

Lemme. Soit β un entier ≥ 1 et $\Sigma \in G_F^0(p^\beta)$, dont la restriction à I_F est irréductible. Si p^β divise $sw(\Sigma)$ alors Σ n'est pas minimale : il existe un caractère χ de $F^\times$ tel que $sw(\chi\Sigma) < sw(\Sigma)$.

Grâce à ce lemme, on voit aisément que si $n = p^r$ divise j, alors $D_{n,j} = (q-1)C_{n,j-1}$, car on tord les éléments que compte $C_{n,j-1}$ par des caractères de $F^\times$ d'exposant de Swan j/n pour obtenir les éléments que compte $D_{n,j}$. On a donc, modulo le lemme, le théorème.

Pour prouver le lemme, on remarque d'abord que si Σ vérifie les conditions du lemme, alors la restriction de Σ au groupe d'inertie sauvage P_F est irréductible. On peut donc, comme dans mon article sur le comportement des facteurs ε sous-torsion, attacher à Σ un élément c_Σ de $(F^\times/U_F^1) \otimes \mathbb{Z}[1/p]$ [42]. L'hypothèse $p^\beta | sw(\Sigma)$ implique que c_Σ est en fait un élément de $F^\times/U_F^1$, d'où l'existence de χ. Plus précisément, on choisit un caractère additif non trivial ψ de F. On peut supposer que Σ se factorise par un quotient fini G de W_F. Le dernier sous-groupe de ramification non trivial de G fixe alors une extension finie galoisienne E de F et la restriction de Σ à W_E vaut n-fois un caractère de W_E, correspondant à un caractère de $E^\times$. La restriction de ce dernier au dernier sous-groupe de congruence de $E^\times$ sur lequel il est non trivial s'écrit $1 + x \longmapsto \psi \circ Tr_{E/F}(cx)$ où $c \in E^\times$ est bien déterminé modulo U_E^1. Comme c est invariant sous l'action de $Gal(E/F)$, il correspond à un élément (c'est c_Σ) de $F^\times/U_F^1 \otimes \mathbb{Z}[1/p]$. Sous l'hypothèse du lemme c_Σ appartient en fait à $F^\times/U_F^1$ et choisissant un caractère χ de $F^\times$ prolongeant le caractère $1 + x \longmapsto \psi(c_\Sigma x)$ du sous-groupe de congruence convenable de $F^\times$, on voit que la restriction à W_E de $\chi^{-1}\Sigma$ est triviale d'où $sw(\chi\Sigma) < sw(\Sigma)$.

Cela termine la preuve du théorème dans le cas où F est de caractéristique non nulle.

Supposons désormais F de caractéristique nulle. On va ramener ce cas au cas précédent. Pour cela on choisit une $\mathbb{Z}_p$-extension totalement ramifiée $\tilde{F}$ de F dans $\bar{F}$. Cette extension est APF au sens de Fontaine et Wintenberger [27,113] et on peut introduire le corps des normes $X = X(\tilde{F}/F)$ qui est un corps local de caractéristique non nulle, de même corps résiduel que F. Le groupe de Galois $Gal(\bar{F}/\tilde{F})$ est naturellement muni d'une filtration par des sous-groupes de ramification (indexée par des nombres réels) et on dispose d'un isomorphisme préservant la ramification entre $Gal(\bar{F}/\tilde{F})$ et $Gal(\bar{X}/X)$ où $\bar{X}$ est une clôture séparable algébrique de X. On a une notion de conducteur de Swan pour les représentations de

$\text{Gal}(\bar{F}/\tilde{F})$ et on transporte à $\text{Gal}(\bar{F}/\tilde{F})$ tous les résultats qu'on a obtenus précédemment sur $\text{Gal}(\bar{X}/X)$. Il ne reste plus qu'à relier, par la restriction, les représentations de $\text{Gal}(\bar{F}/F)$ à celles de $\text{Gal}(\bar{F}/\tilde{F})$ et à utiliser la théorie du changement de base pour obtenir ce qu'on veut.

Pour cela on construit, pour toute extension finie K de F dans $\tilde{F}$ et tout entier $n \geq 1$, une injection de $G_K^0(n)$ dans $A_K^0(n)$, qu'on prolonge de la manière habituelle en une injection de $G_K(n)$ (classes d'équivalence de représentations de dimension n de groupe de Weil-Deligne) dans $A_K(n)$ (classes d'équivalence de représentations admissibles irréductibles de $GL(n,F)$). On effectue ces constructions de façon à obtenir la compatibilité aux changements de base; c'est possible, comme je l'ai signalé au début.

Soit $\Pi \in A_F^0(n)$. Ce qu'il s'agit de prouver est que Π est l'image d'un élément de $G_F^0(n)$ par l'injection qu'on vient de construire. A cause des propriétés du changement de base, il suffit de prouver que $\Pi_K \in A_K(n)$ est image d'un élément de $G_K(n)$, pour K assez grand. Une réduction facile montre qu'on peut même supposer que Π_K appartient à $A_K^0(n)$ pour tout K. On constate que l'exposant de Swan de Π_K se stabilise quand K augmente. Notons s la valeur limite. Utilisant les fonctions de Herbrand pour les extensions APF, on voit qu'il existe une extension assez grosse K de F dans $\tilde{F}$ telle que la restriction à $W_{\tilde{F}}$ donne une bijection de l'ensemble des éléments de $G_K^0(n)$ d'exposant de Swan s sur l'ensemble des éléments de $G_{\tilde{F}}^0(n)$ d'exposant de Swan de s (j'ai utilisé ici des notions et notations évidentes pour $\tilde{F}$). Grâce aux arguments de comptage en caractéristique non nulle, on voit que pour cette extension assez grosse K de F dans $\tilde{F}$, l'injection de $G_K^0(n)$ dans $A_K^0(n)$ induit une bijection pour les éléments d'exposant de Swan s. Par suite $\Pi_K \in A_K^0(n)$ est image d'un élément de $G_K(n)$ ce qui prouve le résultat.

References

[1] J. Arthur, L. Clozel: Exposés au Séminaire sur la formule des traces; I.A.S., Princeton, 1984.

[3] J. Bernstein, P. Deligne, D. Kazhdan, M.-F. Vigneras: Représentations des algèbres centrales simples p-adiques, in: Représentations des Groupes Réductifs sur un Corps Local; Hermann, Paris, 1984, 33 - 117.

[16] L. Clozel: Exposés au cours Peccot; Collège de France, Paris 1984.

[17] L. Clozel, G. Henniart and J. Arthur: Base change for GL_n, preprint 1986

[27] J -M. Fontaine, J.-P. Wintenberger: Le corps des normes de certaines extensions algébriques de corps locaux; C.R.A.S. Paris t. 288. pp. 367 - 370. Extensions algébriques et corps de normes de certaines extensions APF des corps locaux; C.R.A.S., t. 288. pp. 441 - 444.

[42] G. Henniart: Les conjectures de Langlands locales pour GL(n) ; Seminar of Theory of Numbers, Paris, 1982 - 83, Editor M.J. Bertin and C. Goldstein, Birkhäuser, Boston-Basel-Stuttgart, 1985.

[51] R. Howe, A. Moy: Harish-Chandra Homomorphisms for p-adic Groups, CBMS Regional Conference, Series in Mathematics 59, Amer. Math. Soc. Providence, Rh. I., 1985.

[58] D. Kazhdan: Exposé au colloque en la mémoire de Harish-Chandra, I.A.S. Princeton, Mai 1984.

[76] G. Laumon: Transformation de Fourier, constantes d'equations fonctionelles et conjecture de Weil, Publ. Math. I.H.E.S. no. 65 (1987), pp. 131 - 210.

[113] J.-P. Wintenberger: Le corps des normes de certaines extensions infinies de corps locaux, applications; Ann. Scient. Es. Norm. Sup., eème série, t. 16 (1983) pp. 58 - 89.

REFERENCES

[1] J. Arthur, L. Clozel: Exposés au Séminaire sur la formule des traces; I.A.S., Princeton, 1984.

[2] E. Artin, J. Tate: Class field theory; Benjamin, Reading, Mass., 1974.

[3] J. Bernstein, P. Deligne, D. Kazhdan, M.-F. Vigneras: Représentations des algèbres centrales simples p-adiques, in: Représentations des Groupes Réductifs sur un Corps Local; Hermann, Paris, 1984, 33 - 117.

[4] I. N. Bernštein, A. V. Zelevinskii: Representations of the group GL(n,F) where F is a local non-archimedean field; Russian Math. Surveys 31 (1976), 1 - 68 (transl. from Uspekhi Mat. Nauk 31, no. 3 (1976), 5 - 70).

[5] I. N. Bernštein, A. V. Zelevinskii: Induced representations of reductive p-adic groups I; Ann. Scient. ENS 10 (1977), 441 - 472.

[6] A. Borel: Admissible representations of a semisimple group over a local field with vectors fixed under an Iwahori subgroup; Inv. Math. 35 (1976), 233 - 259.

[7] R. Brauer: On primitive projective groups, in: Contributions to Algebra, ed. by H. Bass, P. J. Cassidy, J. Kovacic; Ac. Press, 1977, 63 - 82.

[8] J. Buhler: Icosahedral Galois representations; Lecture Notes in Mathematics 654, Springer Verlag, Berlin-Heidelberg-New York 1978.

[9] C. J. Bushnell: Hereditary orders, Gauss sums and supercuspidal representations of GL_N ; J. für Reine und Angew. Math.

[10] C. J. Bushnell: Oral communication to A. Moy.

[11] C. J. Bushnell, A. Fröhlich: Gauss sums and p-adic division algebras; Lecture Notes in Mathematics 987 Springer Verlag, Berlin-Heidelberg-New York 1983.

[12] C. J. Bushnell, A. Fröhlich: Non-abelian congruence Gauss sums and p-adic simple algebras; Proc. London Math. Soc. (3), 50 (1985), 207 - 264.

[13] H. Carayol: Representation supercuspidal de GL_n ; C. R. Acad. Sci. Paris, Serie A, 288 (1979), 17 - 20.

[14] H. Carayol: Representations cuspidales du groupe lineaires; Ann. Sci. ENS 17 (1984), 191 - 225.

[15] P. Cartier: Representations of p-adic groups: A survey; Proceedings of Symposia in Pure Math. Vol. 33 (1979), part 1, 111 - 155.

[16] L. Clozel: Exposés au cours Peccot; Collège de France, Paris, 1984.

[17] L. Clozel: Manuscrit, 1985.

[18] L. Corwin: Representations of division algebras over local fields; Advances in Math. 13 (1974), 259 - 267.

[19] L. Corwin: Representations of division algebras over local fields II; Pacific Journ. Math. 101, (1982), 49 - 70.

[20] L. Corwin, R. Howe: Computing Characters of Tamely Ramified Division Algebras; Pac. J. Math. 73 (1977), 461 - 477.

[21] L. Corwin, A. Moy, P. J. Sally Jr.: Degrees and formal degrees for division algebras and GL_n over a local field. To appear in Pacific J. Math.

[22] L. Corwin, P. Sally: Paper in preparation.

[23] L. Corwin, P. J. Sally Jr.: Discrete series characters on the elliptic set in GL_n To appear.

[24] C. W. Curtis, I. Reiner: Methods of representation theory I; Wiley, New York 1981.

[25] P. Deligne: Formes modulaires et réprésentations de GL(2), in: Modular Functions of one variable II(ed. P. Deligne, W. Kutzko), Lecture Notes in Mathematics No. 349, Springer Verlag, Berlin-Heidelberg-New-York 1973, 55 - 105.

[26] P. Deligne: Les constantes des équations fonctionnelles des fonctions L , in: Modular functions of one variable II (ed. P. Deligne, W. Kutzko), Lecture Notes in Mathematics 349, Springer Verlag, Berlin-Heidelberg-New York 1979, 501 - 597.

[27] J.-M. Fontaine, J.-P. Wintenberger: Le corps des normes de certaines extensions algébriques de corps locaux; C.R.A.S. Paris t. 288, pp. 367 - 370. Extensions algébriques et corps de normes de certaines exensions APF des corps locaux; C.R.A.S., t. 288, pp. 441 - 444.

[28] A. Fröhlich: Central extensions, Galois groups, and ideal class groups of number fields; Amer. Math. Soc., 24, Providence, Rh. I., 1983.

[29] A. Fröhlich: Galois module structure of algebraic integers; Springer Verlag, Berlin-Heidelberg-New York 1983.

[30] A. Fröhlich: Principal orders and embedding of local fields in algebras; Proc. London Math. Soc. (3), 54 (1987), 247 - 266.

[31] A. Fröhlich: Tame representations of local Weil groups and of chain groups of local principal orders; Heidelberger Akad. d. Wiss. (1986), Springer-Verlag, Berlin-Heidelberg-New York, 1 - 100.

[32] I. M. Gelfand, D. A. Kazhdan: Representations of the group GL(n,K) where K is a local field, in: Lie groups and their representations; J. Wiley and Sons, London 1975.

[33] P. Gérardin: Représentations du groupe SL(2) d'un corps local (d'après Gel'fand, Graev et Tanaka); Séminaire Bourbaki no. 332, nov. 1967, Benjamin, 1969.

[34] P. Gérardin: Cuspidal unramified series for central simple algebras; Proc. Symp. Pure Math. 33, Amer. Math. Soc., Providence, R. I., 1979.

[35] P. Gérardin, W.-C. W. Li: Fourier transforms of representations of quaternions; J. reine angew. Math. 359 (1985), 121 - 173.

[36] P. Gérardin, W.-C. W. Li: A functional equation for degree two local factors; Canadian Math. Soc. Bull. 28 (3) (1985), 355 - 371.

[37] P. Gérardin, W.-C. W. Li: Identities on quadratic Gauss sums, preprint.

[38] P. Gérardin, W.-C. W. Li: Degree two monomial representations of local Weil groups and correspondences (preprint).

[39] R. Godement, H. Jacquet: Zeta functions of simple algebras; Lecture Notes in Mathematics 260, Springer Verlag, Berlin-Heidelberg-New York 1972.

[40] Harish-Chandra: Admissible distributions on reductive p-adic groups, in: Lie Theories and Their Applications, Queens Papers (Quenn's University, 1978), 281 - 347.

[41] G. Henniart: Représentations du groupe de Weil d'un corps local; L'Ens. Math., IIe série, 26 (1980), 155 - 172.

[42] G. Henniart: Les conjectures de Langlands locales pour GL(n); Seminar of Theory of Numbers, Paris, 1982 - 83, Editor M. J. Bertin and C. Goldstein, Birkhäuser, Boston - Basel - Stuttgart, 1985.

[43] G. Henniart: Galois ε-factors modulo roots of unity; Inv. Math. 78 (1984), 117 - 126.

[44] G. Henniart: On the local Langlands conjecture for GL(n) : the cyclic case; Ann. Math. 123 (1986), 143 - 203.

[45] H. Hijikata: Some supercuspidal representations induced from parahoric subgroups, in: Automorphic Forms of Several Variables, Taniguchi Symposium, Katata 1983, edited by I. Satake and Y. Morita; Birkhäuser - Boston - Basel, Stuttgart, 1984.

[46] R. Howe: Representation theory for division algebras over local fields (tamely ramified case); Bull. AMS 77 (1971), 1063 - 1066.

[47] R. Howe: Kirillov theory for compact p-adic groups; Pac. J. Math. 73 (1977), 365 - 382.

[48] R. Howe: Some qualitive results on the representation theory of GL_n over a p-adic field; Pac. Jour. of Math. 73 (1977), 479 - 538.

[49] R. Howe: Tamely ramified supercuspidal representations of GL_n ; Pac. J. Math. 73 (1977), 437 - 460.

[50] R. Howe: Classification of irreducible representations of GL_2 ; IHES notes (1978).

[51] R. Howe, A. Moy : Harish-Chandra Homomorphisms for p-adic Groups, CBMS Regional Conference, Series in Mathematics 59, Amer. Math. Soc. Providence, Rh. I., 1985.

[52] B. Huppert: Endliche Gruppen. Springer Verlag, Berlin-Heidelberg-New York 1967.

[53] I. M. Isaacs: Character Theory of Finite Groups. Academic Presss, New York 1976.

[54] H. Jacquet: Principal L-functions of the linear group, in: Automorphic forms, representations and L-functions (A. Borel & W. Casselman ed.), Proc. Symp. P. Math. 33 vol 2 (American Math. Soc. Providence, 1979), 63 - 86.

[55] H. Jacquet, R. P. Langlands: Automorphic Forms on GL(2); Lecture Notes in Mathematics no. 114, Springer Verlag, Berlin-Heidelberg-New York, 1970.

[56] H. Jacquet, I. I. Piatetski-Shapiro, J. Shalika: Conducteur des représentations du groupe linéaire; Math. Ann. 256 (1981), 199 - 214.

[57] D. Kazhdan: On lifting, in: Lie Groups Representations II, Lecture Notes in Math. 1041 (Springer, 1983).

[58] D. Kazhdan: Exposé au colloque en la mémoire de Harish-Chandra; I. A. S. Princeton, mai 1984.

[59] H. Koch. Classification of the primitive representations of the Galois group of a local fields; Invent. math. 40 (1977), 195 - 216.

[60] H. Koch: Bemerkungen zur numerischen lokalen Langlands-Vermutung; Prepublication AdW der DDR, P-Math 28/81, Berlin, 1981.

[61] H. Koch: Eisensteinsche Polynomfolgen und Arithmetik in Divisionsalgebren über lokalen Körpern; Math. Nachr. 104 (1981), 239 - 251.

[62] H. Koch: Zur Arithmetik von Divisionsalgebren über lokalen Körpern; Math. Nachr. 100 (1981), 9 - 19.

[63] H. Koch: Bemerkungen zur numerischen lokalen Langlands-Vermutung; Trudy Math. Inst. Steklov 163 (1984), 108 - 114.

[64] H. Koch, E.-W- Zink: Zur Korrespondenz von Darstellungen der Galoisgruppen und der zentralen Divisionsalgebren über lokalen Körpern (der zahme Fall); Math. Nachr. 98 (1980), 83 - 119.

[65] H. Koch, E.-W. Zink: Bemerkungen zur numerischen lokalen Langlands-Vermutung II; Trudy Math. Inst. Steklov 163 (1984), 115 - 117.

[66] T. Kondo: On Gaussian sums attached to general linear groups over finite fields; J. Math. Soc. Japan, 15 (1963), 244 - 255.

[67] P. Kutzko: On the supercuspidal representations of GL_2 ; Amer. J. Math. 100 (1978), 43 - 60.

[68] P. Kutzko: The Langlands conjecture for GL(2) of a local field; Ann. Math. 112 (1980), 381 - 412.

[69] P. Kutzko: The exceptional representations of GL_2 ; Composito Math. 51 (1984), 3 - 14.

[70] P. Kutzko: On the restriction of supercuspidal representations to compact open subgroups; Duke Math. J. 52 (1985), 753 - 764.

[71] P. Kutzko: Towards a classification of the supercuspidal representations of GL_N ; London J. Math., to appear.

[72] P. Kutzko: Oral Communication.

[73] P. Kutzko, O. Manderscheid: On the supercuspidal representations of GL_4, I; Duke Math. J. 52 (1985), 841 - 867.

[74] P. Kutzko, D. Manderscheid: On intertwining operators for $GL_N(F)$, F a non-archimedean local field; pre-print.

[75] R. P. Langlands: Base Change for GL(2); Ann. of Math. Studies 96, Princeton University Press, 1980.

[76] G. Laumon: Manuscrit, automne 1985.

[77] W.-C. W. Li: On the representations of GL(2) I, ε-factors and n-closeness, J. reine angew. Math. 313 (1980), 27 - 42; II, ε-factors of the representations of GL(2)×GL(2), J. reine angew. Math. 314 (1980), 3 - 20.

[78] W.-C. W. Li: Barnes identities and representations of GL(2) II, Non-archimedean local case; J. reine angew. Math. 345 (1983), 69 - 92.

[79] I. G. Macdonald: Symmetric functions and Hall polynomials; Oxford University Press, 1979.

[80] I. G. Macdonald: Zeta functions attached to finite general linear groups; Math. Annalen 249 (1980), 1 - 15.

[81] G. W. Mackey: Unitary representations of group extensions I, Acta Math. 99 (1958), 265 - 311.

[82] F. Mautner: Spherical functions over p-adic fields II; Amer. Jour. of Math. 86 (1964), 171 - 200.

[83] C. Moreno: Matching theorems for division algebras and GL_n ; Not published.

[84] A.Moy: Hecke algebra isomorphisms for GL_n over a p-adic field; pre-print.

[85] A. Moy: Representations of U(2,1) over a p-adic field, preprint.

[86] A. Moy: Representations of GSp(4) over a p-adic field, preprint.

[87] A. Moy: Minimal K-types for GSp(6) over a p-adic field, preprint.

[88] A. Moy: Minimal K-types for G_2 over a p-adic field, preprint.

[89] A. Moy: Local constants and the tame Langlands correspondence; Amer. J. Math., 108 (1986), 863 - 930.

[90] O. Ore: On a special class of polynomials, Trans. AMS 35 (1933), 559 - 584.

[91] H. Prado: Représentations de $GL(2,\mathbb{R})$ et identités de type Barnes pour la fonction; C. R. Ac. Sc. Paris t. 300, Série I, no. 4 (1985), 97 - 100.

[92] G. Prasad, M. Raghunathan: Topological central extensions of semi-simple groups over local fields; Annals of Math. 119 (1984), 143 - 201.

[93] H. G. Quebbemann, W. Scharlau, M. Schulte: Quadratic and hermitian forms in additive and abelian categories; J. Alg. 59 (1979), 264 - 289.

[94] I. Reiner: Maximal Orders; Academic Press, 1975.

[95] J. F. Rigby: Primitive linear groups containing a normal nilpotent subgroup larger than the center of the group; J. London Math. Soc. 35 (1960), 389 - 400.

[96] J. Ritter: An explicit Brauer formula for local Galois characters.

[97] F. Rodier: Représentations de GL(n,k) où k un corps p-adique; Séminaire Bourbaki (1981/1982)587.

[98] J. Rogawski: Representations of GL(n) and division algebras over a p-adic field; Duke Math. J. 50 (1983), 161 - 196.

[99] J. P. Serre: Corps locaux, 2 ième éd., Hermann, Paris, 1968.

[100] J. P. Serre: Une "formule de masse" pour les extensions totalement ramifieeś de degreé donné d'un corps local; C. R. Acad. Sci. Paris, Série A, 286 (1978), 1031 - 1036.

[101] J. Shalika: Representations of the two by two unimodular group over local fields; IAS notes (1966).

[102] S. Shirai: On the central class field mod m of an algebraic number field; Nagoya-Math. J. 71 (1978), 61 - 85.

[103] A. J. Silberger: The Langlands Quotient Theorem for *p*-adic group; Math. Annalen 236 (1978), 95 - 104.

[104] M. Tadić: Solution of the unitarizability problem for general linear groups (non-archimedean case); preprint Max-Planck-Institut für Mathematik Bonn, 1985.

[105] J. T. Tate: Fourier analysis in number fields and Hecke's zeta functions; Thesis, Princeton University 1950. Also: Algebraic Number Theory (J. W. S. Cassels & A. Fröhlich edd.) (Academic Press, London 1967) 305 - 347.

[106] J. T. Tate: Local constants, in: Algebraic Number Fields (A. Fröhlich ed.), Academic Press, London 1977, 89 - 132.

[107] J. T. Tate: Number theoretic background, in: Automorphic Forms, Representations, and L-Functions (Corvallis), Proc. Symp. Pure Math. 33, AMS, Providence 1979.

[108] M. J. Taylor: On Fröhlich's conjecture for rings of integers of tame extensions; Invent. Math. 63 (1981), 41 - 79.

[109] J.-L. Waldspurger: Algèbres de Hecke et induites de représentations cuspidales pour GL(N) ; Crelle Journal 370 (1986), 127 - 191.

[110] A. Weil: Sur certain groupes d'operateurs unitaires, Acta Math. 111 (1964), 143 - 211.

[111] A. Weil: Basic Number Theory; Springer Verlag, Heidelberg-New York, 1973.

[112] A. Weil: Exercises dyadiques; Inv. Math. 27 (1974), 1 - 22.

[113] J.-P. Wintenberger: Le corps des normes de certaines extensions infinies de corps locaux, applications; Ann. Scient. Ec. Norm. Sup., 4ème série, t. 16 (1983), pp. 58 - 89.

[114] K. Yamazaki: On projective representations and ring extensions of finite groups; J. Fac. Sc. Univ. Tokyo, Sect. I, 10, 1963/64, 147 - 195.

[115] A. Zelevinsky: Induced representations of reductive p-adic groups II: On irreducible representations of GL(n) ; Ann. Sci. ENS 13 (1980), 165 - 210.

[116] E. W. Zink: Weil-Darstellungen und lokale Galoistheorie; Math. Nachr. 92 (1979), 265 - 288.

[117] E. W. Zink: Lokale projective Klassenkörpertheorie - Grundbegriffe und erste Resultate; Preprint R-Math-1/82, Akad. d. Wiss. d. DDR, Berlin, 1982.

[118] E. W. Zink: Remarks on a local Langlands-conjecture; Preprint R-Math-14/80, Akad. d. Wiss. d. DDR, Berlin, 1982.

[119] E. W. Zink: Lokale projektive Klassenkörpertheorie II; Math. Nachr. 114 (1983), 123 - 150.

[120] E. W. Zink: Über das Kirillov-Dual endlicher nilpotenter Ringe; Preprint R-Math-03/85, Akad. d. Wiss. d. DDR, Berlin, 1985.

ABCDEFGHIJ -- 89